Water in the Middle East and in North Africa

Springer

Berlin
Heidelberg
New York
Hong Kong
London
Milan
Paris
Tokyo

Fathi Zereini • Wolfgang Jaeschke (Eds.)

Water in the Middle East and in North Africa

Resources, Protection and Management

With 136 Figures and 40 Tables

Springer

Editors:

PD Dr. habil. Fathi Zereini
University of Frankfurt
Institute of Mineralogy
Department of Environmental Analytics
Georg-Voigt-Str. 16
60054 Frankfurt, Germany
E-mail: *zereini@kristall.uni-frankfurt.de*

Professor Dr. Wolfgang Jaeschke
University of Frankfurt
Center for Environmental Research (ZUF)
Georg-Voigt-Straße 14
60325 Frankfurt, Germany
E-mail: *jaeschke@zuf.uni-frankfurt.de*

ISBN 3-540-20771-6 Springer-Verlag Berlin Heidelberg New York

Library of Congress Cataloging-in-Publication Data Applied For

A catalog record for this book is available from the Library of Congress.
Bibliographic information published by Die Deutsche Bibliothek
Die Deutsche Bibliothek lists this publication in die Deutsche Nationalbibliographie; detailed bibliographic data is available in the Internet at <http://dnb.ddb.de>.

Springer-Verlag Berlin Heidelberg New York
Springer-Verlag is a part of Springer Science+Business Media

springeronline.com

Printed in Germany

Cover Design: Erich Kirchner, Heidelberg
Typesetting: Camera-ready by the editors

Printed on acid free paper 30/3141 – 5 4 3 2 1 0

Foreword

Freshwater is the most precious and valuable natural resource essential for all kind of life on the land surface. Especially man depend on the availability and quality of the water, it is vital for the socio-economic growth of societies and nations as well as for the sustainability of the environment. Freshwater resources are not distributed in an equal way over the continents. Regions with surplus facing areas with shortage or even lack of water. From their nature arid and semi-arid areas are characterized by scarcity of water with all the affiliated unfavourable attributes. In the Middle East this lack of sufficient water coupled with the high growth rates of population constitutes serious water problems which has become e real political issue.

The German-Arabic Society for Environmental Studies, Frankfurt, in co-operation with the Centre for Environmental Studies at the University Frankfurt has raised up this problems for a scientific discussion. With the Symposium on "Water of the Middle East and North Africa: Resources, Protection and Management" in Frankfurt on November 25 and 26, 2002 a platform was created, which enabled participants - scientists and politicians - from twelve countries to discuss the water problems facing the region. It was the intension of the organisers to deepen the knowledge about the hydrologic regimes, to increase our understanding of the complex integrated processes and to draw the necessary conclusions with regard to an integrated sustainable water management. Jointly scientists and politicians could talk in a friendly and frankly atmosphere and try to find answers for acceptable solutions. It has proved that still we need additional comprehensive studies, which has not only to consider the hydrologic aspects, but should include also the socio-economic and political issues. On behalf of the participants I would like to thank the organisers for bringing about this important Symposium. We especially appreciate the big and successful endeavours of the two organising institutions as well as the personal commitment of the main coordinators Dr. habil. Fathi Zereini and Prof. Dr. W. Jaeschke.

It is a great merit of the organisers that the contributions presented at the Symposium are now available in the printed version. With this book a wide readership will have access to the questions, research priorities, methods, knowledge and results acquired by different groups. It will activate an interdisciplinary exchange and stimulate new research work, which hopefully will help to develop new strategies to solve the challenging problem of water scarcity in the Middle East. The discussion and exchange with decision-makers and politician, opened so successfully at the Symposium, have to be continued, in order to work out practical concepts for an integrated water management. With that the scientific results can be directly transposed to the benefit of the peoples affected.

Karlsruhe, October 2003

Prof. Dr. Heinz Hötzl
University of Karlsruhe, Germany

Preface of the Editors

Due to scarce water resources and the resulting acute lack of water, which exists already today, the problem of fresh water is an explosive issue in the Middle East. Besides approaches at the level of politics, also science can be a driving force in the search for strategies to ease the problem. A more efficient use of existing resources, more effective protection measures for water quality monitoring, and improvement of water management with the objective of fair distribution are requirements for a solution of the water shortage in the Middle East. Another requirement is the use of the latest environmental technologies for treatment and reuse of waste water.

„Water of the Middle East and North Africa: Resources, Protection, and Management“ was the topic of the first environmental symposium of the German-Arabic Society for Environmental Studies (DaG-Umweltstudien), which took place in cooperation with the Center for Environmental Research (Zentrum für Umweltforschung, ZUF) on November 25 and 26, 2002 at Johann Wolfgang Goethe-University, Frankfurt am Main. More than 120 scientists from different fields of study from universities and research institutes and representatives from different development organizations, the economy, and politics from the Arab countries and Germany participated in this international symposium. The main topics were presented in 25 talks and 15 posters. At the conclusion of the conference, many participants expressed the desire to present the differenct aspects of the discussed problem in a book.

This book presents a comprehensive overview of the current subject "Water of the Middle East and North Africa". The water problem is described and discussed in four main chapters and different case studies. The four main chapters are "Water resources“, "Water protection“, "Water management“, and "Water as source of conflict“. 39 scientists from Algeria, Egypt, Germany, Israel, Jordan, Lebanon, Palestine, Switzerland, Syria, Turkey, UK, and West Africa have contributed to this book. Their interdisciplinary articles comprise the most recent research results on the present water situation in the Middle East and North Africa and their importance for a peaceful coexistence of the population. With this topic, the book addresses experts as well as the interested public.

The editors would like to thank the authors and reviewers for the successful cooperation during the professional and on-schedule compilation of this book. We thank Prof. E. Salameh, Jordan University; Dr. W. Ali, Karlsruhe University; Dr. Gh. Abu Ju'ub, RWTH-Aachen; Dr. K. Tielboerger, University of Potsdam; and

Dr. K. Schelkes, BGR, Hannover, by whose help the book could be expanded with topics that were not discussed during the symposium. We are also especially indebted to Dr. Ch. Witschel, Springer publishing house, who has made publication of this book possible. We would also like to sincerely thank Ms. L. Tonarelli, Springer publishing house, and Mr. H. H. Dülfer, Center for Environmental Research, Frankfurt am Main University, for their editorial work.

Frankfurt am Main, October 2003

PD Dr. habil. Fathi Zereini
Institute of Mineralogy / Environmental Analytic
and
Prof. Dr. Wolfgang Jaeschke
Center for Environmental Research, Frankfurt, Germany

Contents

1 Water Resources 1

1.1 Natural Scarcity of Water Resources in the Semi-arid and Arid Middle East and its Economical Implications
Heinz Hötzl 3

1.2 Exploitation of Fossil Aquifers and Future Water Supplies in the Middle East
Elias Salameh 15

1.3 Hydrogeological Investigations in the North-Eastern Dead Sea area, Suweimeh, Jordan
W. Ali, J. Glaser, M. Thiel, H. Hötzl, H. Werz 21

1.4 The Jordan River: Natural Flow and Current Consumption by the Riparian Countries
Ghassan Abu Ju'ub & Kurt Schetelig 31

1.5 Hydrology and management of Lake Kinneret aimed at water quality protection
Moshe Gophen 41

1.6 The Water Crisis in the E. Mediterranean – and Relation to Global Warming?
Pinhas Alpert 55

1.7 Surface Water in Jenin Governorate-Palestine
Ghassan Abu Ju'ub & Kurt Schetelig 63

1.8 Herodion-Beit Fajjar well field of high quality and resource-productivity groundwater endangered by severe over-pumpage
Qannam Z, Merkel B and Lange T 75

1.9 Water Resources, Protection and Management in Palestine
Abu Safieh, Yousef 87

1.10 The role of transient flow and drainage on regional groundwater flow pattern in semi-arid and arid climates
Klaus-Peter Seiler 101

1.11 Why no Middle East water wars: global solutions to local deficits?
Tony Allan 111

2 Water Protection 127

2.1 Management, protection and sustainable use of groundwater – results of long-term technical co-operation in the Middle East
Schelkes K, Hobler M, Schmidt G & Steinbach V 129

2.2 Groundwater Vulnerability Mapping in the Arab Region
Margane A, Hobler M, Droubi A, Rajab R, Subah A & Khater AR 145

2.3 Intermittent Water Supply and Domestic Water Quality in the Middle East
Fuad Hashwa & Sima Tokajian 157

2.4 Sewage water treatments and reuse in Israel
M. Ben-Hur 167

2.5 Integrated Wastewater Concepts for Olive Mills in the West Bank, Palestine
Silke Drescher 181

2.6 Distribution Pattern of some inorganic pollutants in groundwater and soil of a sewage farm at NE-Cairo/Egypt
Abo el Abas, Yosry M 191

3 Water Management 205

3.1 Water Utilization in Semi-Arid Zone, The Hula Valley (Israel): Pollutant removal, agriculture and Ecotourism Management.
Moshe Gophen 207

3.2 Regionalization of hydrogeological Data for use in groundwater modelling: Application to "Insel Hengsen" and at northern Algeria
Blel Azouzi and Jürgen Schröter 227

3.3 Water Resources Management in Palestine: Political, technical and financial obstacles
Fawzy Naji 239

3.4 Conceptual and numerical flow models of the Western Aquifer - Palestine / Israel
Clemens Messerschmid 251

3.5 Ecosan – Introduction of Closed-Loop Approaches in Wastewater Management and Sanitation – A Supra-Regional GTZ - Project
Christine Werner, Heinz-Peter Mang, Jana Schlick, Papa Abdoulaye Fall 263

3.6 IMPETUSWest Africa An integrated approach to the efficient management of scarce water resources in West Africa – Case studies for selected river catchments in different climatic zones –
P. Speth and M. Christoph 275

3.7 Capacity Building in Water Management in Palestine – Experience of InWEnt Palestinian Water Sector Training programme (1995 – 2000)
Ismail Al Baz 287

3.8 Efficient groundwater-management for organizations with a small financial budget – the continuous improvement method
Kurt von Storch 295

4 Water as Source of Conflict 307

4.1 Water conflict and water management in the Middle East
Jamill Sabbagh 309
4.2 Syria and Turkey in Water Diplomacy (1962-2003)
Marwa Daoudy 319
4.3 Water Issue Among the Riparian States of Euphrates and Tigris Transboundary Rivers
Gurer, Ibrahim 333
4.4 Control, Allocation and Consumption of Water Resources in the Jordan Basin: The Struggle over a Scarce and Vital Resource in the Context of the Middle East Conflict
Margret Johannsen 345
4.5 Water Conflicts in the Middle East: The Twin Rivers Orontes and Litani
Ghassan Abu Ju'ub & Rafig Azzam 353

Index..363

List of Contributors

Abdoulaye Fall, P., Dr.
Deutsche Gesellschaft für Technische Zusammenarbeit (GTZ) GmbH,
Dag - Hammarskjoeld – Weg 1-5, Postfach 5180,
65726 Eschborn, Germany, ecosan
e-mail: ecosan@gtz.de

Abo el Abas, Yosry, Dr.
Freisinger Str. 4
10781 Berlin
e-mail: yosry Aboelabas [yosryaboelabas@yahoo.com]

Abu-Safieh, Yousef, Dr.
Minster of Environmental Quality Authority
Palestine National Authority, Palestine
e-mail: pena@rannet.com

Abu Ju'ub, Ghassan, Dr. Ing.
Department of Engineering Geology and Hydrogeology(LIH),
Aachen Univeristy of Technology (RWTH).
Aachen.Germany.
e-mail: gh_ab_jb@yahoo.com
or: abu@lih.rwth-aachen.de

Al Baz, Ismail, Dr.
InWent, Capacity Building International Germany
Lützowufer 6-9
D-10785 Berlin, Germany
e-mail:ismailalbaz@nets.com.jo

Ali, Wasim, Dr.
Department of Applied Geology
Kaiserstr. 12
76128 Karlsruhe,Germany
e-mail: wasim.ali@agk.uka.de

Allan, Tony, Prof. Dr.
King's/SOAS London Water Research Group
e-mail: [ta1@mistral.soas.ac.uk]

Alpert, Pinhas, Prof. Dr.
Department of Geophysics and Planetary Sciences
Tel-Aviv University, Tel-Aviv 69978, Israel
e-mail: pinhas@cyclone.tau.ac.il

Azouzi, Blel, Dr.
Universität Bremen –FB
Klagenfurther Strass, 28359 Bremen, Germany
e-mail: azouzi@uni-bremen.de

Azzam, Rafig, Prof. Dr.
Department of Engineering Geology and Hydrogeology (LIH),
Aachen University of Technology (RWTH),
Aachen,Germany.
e-mail: azzam@lih.rwth-aachen.de

Ben-Hur, Memi
Institute of Soil, Water and Environmental Science,
The Volcani Center, POB 6
Bet Dagan, 50250 Israel
e-mail: [meni@volcani.agri.gov.il]

Christoph, Michael, Dr.
Institut für Geophysik & Meteorologie,
Universität zu Köln,
Kerpener Str. 13, D-50923 Köln, Germany
e-mail: christoph@meteo.uni-koeln.de

Daoudy, Marwa, Dr.
Research Fellow, Development Studies, School of Oriental and African Studies (SOAS) London.
Current e-mail: md37@soas.ac.uk
Permanent e-mail: daoudy@hei.unige.ch
Current address: c/o Prof. John Anthony Allan, SOAS,
Thornhaugh Street, London WC1H 0XG
Permanent Address: Department of Political Science,
Graduate Institute for International Studies,
Rue de Lausanne 132, Case Postale 36,
1211 Geneva 21, Switzerland

Drescher, Silke, Dipl.Wirt.-Ing.
Project Officer Solid Waste Management
EAWAG/ SANDEC
P.O. Box 611
8600 Duebendorf, Switzerland
e-mail:Drescher Silke
[Silke.Drescher@eawag.ch]

Droubi, Abdallah, Dr.
The Arab Centre for the Study of Arid Zones and Dry Land (ACSAD)
P.O. Box 2440
Damascus, Syria
e-mail: droubi@scs-net.org

Glaser, Jens, Dipl. Geol.
Department of Applied Geology
Kaiserstr. 12
76128 Karlsruhe,Germany

Gophen, Moshe, Prof. Dr.
Senior Scientist - MIGAL,
Galilee Technological Center, and Kinneret Limnological Laboratory
POB 831 Kiryat Shmone (11016) Israel
e-mail: Moshe Gophen
[gophen@migal.org.il]

Gurer, Ibrahim, Prof. Dr.
Gazi University
Faculty of Engineering & Architecture
Department of Civil Engineering
06570 Maltepe, Ankara, Turkey
e-mail : gurer@gazi.edu.tr

Hashwa, Fuad, Prof. Dr.
Dean, School of Arts and Sciences
Chair, Division of Natural Sciences
Microbiology & Biotechnology
Lebanese American University
Byblos Campus
P. O. Box 36 - F-80,
Byblos, Lebanon
e-mail: fhashwa@lau.edu.lb

Hobler, Manfred, Dr.
Federal Institute for Geosciences and Natural Resources (BGR)
Stilleweg 2
D-30655 Hannover, Germany
e-mails: m.hobler@bgr.de

Hötzl, Heinz, Prof. Dr.
Department of Applied Geology
Kaiserstr. 12
76128 Karlsruhe,Germany
e-mail: heinz.hoetzl@agk.uka.de

Jaeschke, Wolfgang, Prof. Dr.
Zentrum für Umweltforschung
J. W. Goethe-University
Georg-Voigt-Str. 14
D-60054 Frankfurt am Main, Germany
e-mail: jaeschke@zuf.uni-frankfurt.de

Johannsen, Margret, Dr.
Institute for Peace Research and Security Policy at the University of Hamburg (IFSH)
Falkenstein 1
D-22587 Hamburg, Germany
e-mail: johannsen@rrz.uni-hamburg.de

Khater, Ahmed R., Dr.
Research Institute for Groundwater,
National Water Research Center
Ministry of Water Resources and Irrigation
NWRC Building
El Kanater El Khairiya, Egypt 13621
e-mail: drkhater@hotmail.com

Lange, Torsten
UFZ Leipzig-Halle GmbH
Theodor-Lieser-Str. 4
06120 Halle, Germany
e-mail: tlange@halle.ufz.de

Mang, Heinz-Peter
Deutsche Gesellschaft für Technische Zusammenarbeit (GTZ) GmbH,
Dag - Hammarskjoeld – Weg 1-5, Postfach 5180,
65726 Eschborn, Germany, ecosan
e-mail: Heinz-Peter.Mang@gtz.de

Margane, Armin, Dr.
Federal Institute for Geosciences and Natural Resources (BGR)
Stilleweg 2
D-30655 Hannover, Germany
e-mail: armin.margane@bgr.de

Merkel, Broder, Prof. Dr.
Technische Universität Bergakademie Freiberg
Institute of Geology
Gustav-Zeuner-Str. 12
09599 Freiberg/Sachsen, Germany
e-mail: merkel@geo.tu-freiberg.de

Messerschmid, Clemens
Research and coordination advisor
P.O.Box 38383, Jerusalem 91383
e-mail: Clemens@susmaq.org

Naji, Fawzy, Dr.
DFN-GEOCONSULTING
Eulenstr. 13
D-38114 Braunschweig, Germany
E-mail: dfn-naji@t-online.de
fawzynaji@hotmail.com

Qannam, Ziad, Dr.
Technische Universität Bergakademie Freiberg
Institute of Geology
Gustav-Zeuner-Str. 12
09599 Freiberg/Sachsen, Germany
e-mail: qannam@web.de

Rajab, Refaat, Dr.
The Arab Centre for the Study of Arid Zones and Dry Land (ACSAD)
P.O. Box 2440
Damascus, Syria
e-mail: refaat@scs-net.org

Sabbagh, Jamill, Dr.
RegioComun - Institut für integrierte Raumentwicklung am Geographischen Institut
Johannes Gutenberg-Universität
Becherweg 21
55099 Mainz, Germany
e-mail: Jamill Sabbagh [J.Sabbagh@geo.uni-mainz.de]

Salameh, Elias, Prof. Dr
Dept. of Env. And Applied Geology
University of Jordan,
POB 9999 Webdeh, 1191
Amman, Jordan
e-mail: [Salameli@ju.edu.jo]

Schelkes, Klaus, Dr.
Federal Institute for Geosciences and Natural Resources (BGR)
Stilleweg 2
D-30655 Hannover, Germany
e-mails: k.schelkes@bgr.de

Schetelig, Kurt, Prof. Dr.
Department of Engineering Geology and Hydrogeology(LIH),
Aachen University of Technology (RWTH),
Aachen, Germany.
e-mail: sche@lih.rwth-aachen.de

Schlick, Jana
Deutsche Gesellschaft für Technische Zusammenarbeit (GTZ) GmbH,
Dag - Hammarskjoeld – Weg 1-5, Postfach 5180,
65726 Eschborn, Germany, ecosan
e-mail: ecosan@gtz.de

Schmidt, Gerhard
Federal Institute for Geosciences and Natural Resources (BGR)
Stilleweg 2
D-30655 Hannover, Germany
e-mails: g.schmidt@bgr.de

Schröter, Jürgen
Universität Bremen –FB
Klagenfurther Strass, 28359 Bremen, Germany
e-mail: schroe@uni-bremen.de

Seiler, Klaus-Peter, Prof. Dr. Dr. habil.
GSF National Research Centre
D-85758 Neuherberg, Gremany
e-mail: seiler@gsf.de

Speth, Peter, Prof. Dr.
Institut für Geophysik & Meteorologie, Universität zu Köln,
Kerpener Str. 13, D-50923 Köln, Germany
e-mail: speth@meteo.uni-koeln.de

Steinbach, Volker, Dr.
Federal Institute for Geosciences and Natural Resources (BGR)
Stilleweg 2
D-30655 Hannover, Germany
e-mails: v.steinbach@bgr.de

Subah, Ali
Ministry of Water and Irrigation (MWI)
P.O. Box 2412
Amman 11183, Hashemite Kingdom of Jordan
e-mail: Ali_Subah@mwi.gov.jo

Thiel, Martin, Dipl. Geol.
Department of Applied Geology
Kaiserstr. 12
76128 Karlsruhe,Germany

Tokajian, Sima, Dr.
Post Doctoral Research Fellow
Microbiology & Biotechnology
Lebanese American University
Byblos Campus
P. O. Box 36 - F-80,
Byblos, Lebanon

von Storch, Kurt, Dr.
Jenaer Str. 16
D-65205 Wiesbaden, Germany
e-mail: klik.vonstorch@t-online.de

Werner, Christine
Deutsche Gesellschaft für Technische Zusammenarbeit (GTZ) GmbH,
Dag-Hammarskjoeld-Weg 1-5, Postfach 5180,
65726 Eschborn, Germany, ecosan
e-mail: Werner Christine 4412 [Christine.Werner@gtz.DE]

Werz, Heike, Dipl. Geol.
Department of Applied Geology
Kaiserstr. 12
76128 Karlsruhe,Germany
e-mail: heike.werz@agk.uka.de

Zereini, Fathi, PD Dr. habil.
Institute of Mineralogy / Environmental Analytic
J. W. Goethe-University
Georg-Voigt-Str. 16
D-60054 Frankfurt am Main, Germany
e-mail: zereini@kristall.uni-frankfurt.de

List of Referees

Ali, Wasim, Dr.
Department of Applied Geology
Kaiserstr. 12
76128 Karlsruhe,Germany
e-mail: wasim.ali@agk.uka.de

Alkazaza, A. Dr.
Deutsches Orient-Institut
Neuer Jungfernstieg 21
D-20354 Hamburg, Germany

Azzam, Rafig, Prof. Dr.
Department of Engineering Geology and Hydrogeology (LIH),
Aachen University of Technology (RWTH),
Aachen,Germany.
e-mail: azzam@lih.rwth-aachen.de

Abu Ju'ub, Ghassan, Dr. Ing.
Department of Engineering Geology and Hydrogeology(LIH),
Aachen Univeristy of Technology (RWTH).
Aachen.Germany.
e-mail: gh_ab_jb@yahoo.com
or: abu@lih.rwth-aachen.de

Hashwa, Fuad, Prof. Dr.
Dean, School of Arts and Sciences
Chair, Division of Natural Sciences
Microbiology & Biotechnology
Lebanese American University
Byblos Campus
P. O. Box 36 - F-80,
Byblos, Lebanon
e-mail: fhashwa@lau.edu.lb

Helmers Eckard, Prof. Dr.
Umweltcampus Birkenfeld der FH Trier
-University of Applied Sciences-
P.O. Box 13 80
D-55761 Birkenfeld, Germany
e-mail : helmers@umwelt-campus.de

Jaeschke, Wolfgang, Prof. Dr.
Zentrum für Umweltforschung
J. W. Goethe-University
Georg-Voigt-Str. 14
D-60054 Frankfurt am Main, Germany
e-mail: jaeschke@zuf.uni-frankfurt.de

Naji, Fawzy, Dr.
DFN-GEOCONSULTING
Eulenstr. 13
D-38114 Braunschweig, Germany
E-mail: dfn-naji@t-online.de
fawzynaji@hotmail.com

Salameh, Elias, Prof. Dr
Dept. of Env. And Applied Geology
University of Jordan,
POB 9999 Webdeh, 1191
Amman, Jordan
e-mail: [Salameli@ju.edu.jo]

Schetelig, Kurt, Prof. Dr.
Department of Engineering Geology and Hydrogeology(LIH),
Aachen University of Technology (RWTH),
Aachen, Germany.
e-mail: sche@lih.rwth-aachen.de

von Storch, Kurt, Dr.
Jenaer Str. 16
D-65205 Wiesbaden, Germany
e-mail: klik.vonstorch@t-online.de

Zereini, Fathi, PD Dr. habil.
Institute of Mineralogy,
J. W. Goethe-University
Georg-Voigt-Str. 16
D-60054 Frankfurt am Main, Germany
e-mail: zereini@kristall.uni-frankfurt.de

1 Water Resources

1.1 Natural Scarcity of Water Resources in the Semi-arid and Arid Middle East and its Economical Implications

Heinz Hötzl
University of Karlsruhe 76128 Karlsruhe, Germany

1.1.1 Introduction

The glances of the astronauts back from the universe to our earth and impressive satellite images showing the whole or large part of the globe have just made aware to many people the great variety and differences of our earth surface. It wouldn't have taken all these new pictures. Detailed maps on the physical, geological, vegetation or climatic conditions have demonstrated since long time the different shapes, landforms and environments occurring on our earth. The differences are the result of energetic and physical variations according the position to the sun as well as of the long history of the geologic evolution. These all together contribute to the richness and variety of life, which makes our earth so unique.

However the differences in the environment also cause differences in the availability of basic life conditions. Whether this regards to the distribution of oceans and continents, mountains and lowlands, deserts and tropical forests, or the distribution of basic resources like minerals, ore, oil, natural nutrients or finally the freshwater, which is vital for most of the life on the continents. Therefore preferential habitats occur. This is valid in general for all form of life, but especially it hits our modern human society with its resource-consuming lifestyle and therefore influences the respective socio-economic development.

1.1.2 Scarcity of water in arid and semiarid regions

Per definition arid and semiarid areas are characterized by shortage of water, whereby in arid areas the evaporation surpasses the precipitation in the predominant part of the year. In general the total amount of the precipitation is significantly less than 250 mm per year. The mostly complete lack of surface water due to the high evaporation rate causes hostile life conditions for plants and animals, so that sand and rock deserts prevails. Along the episodic flooded wadis or outcropping aquifers salt soils are developed and in depressions even salt ponds may occur.

Semiarid areas are defined by arid conditions over at least six month and the rest of the year with rather rare or episodic rainfall amounting between 250 and 500 mm per year. Intermittent river may occur during the wet season. Sparse grass with few bushes is the characteristic vegetation. In order to emphasize the scarcity of available water in such region it should be mentioned that the average global precipitation rate amounts around 1100 mm per year and that in the moist mid-latitudes values between 1000 and 2500 mm and in the moist tropics values >3000 mm are common.

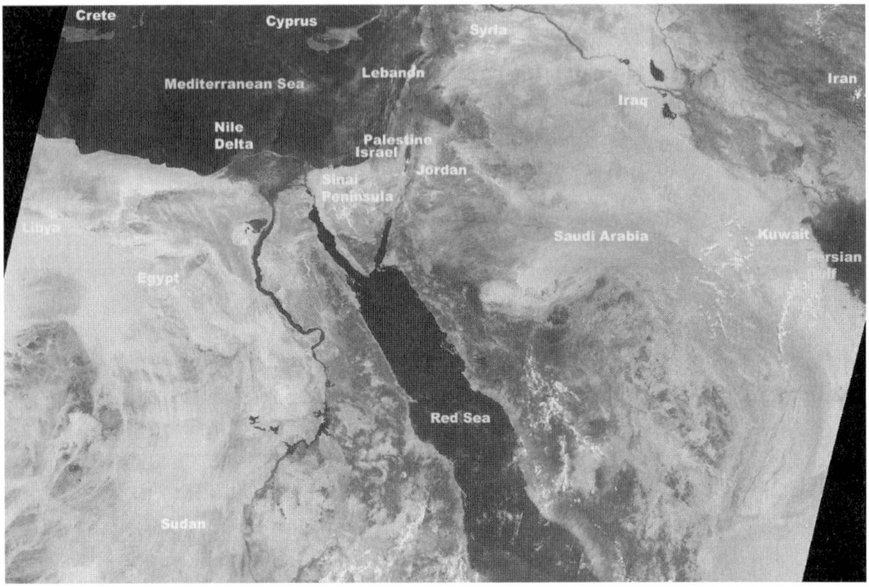

Fig. 1.1.1. Satellite image of Northeast Africa and the Middle East showing in the light grey colours the sand deserts of the Sahara and the Arabian Shelf Platform, in dark grey the dominant rock desert of the Nubian and Arabian crystalline shield, and in black caused by the vegetation cover the Nile Valley and Nile Delta (Terra Sensor, MISR Data, NASA 2000).

Arid and semiarid zones are distributed mainly along the tropics of Cancer and Capricorn (Horse latitudes). The reason behind is the atmospheric circulation in form of convection cells. Air masses from the equator ascending, cooling down and spreading in the higher atmosphere to the north and south, forming by the way the two Hadley cells of air circulation, dive down over the Horse latitudes. This causes very continual high-pressure conditions in these regions, whereby the now once again warming air takes away the available air moisture with the easterlies towards the equatorial region. Because of these effects clouds disperse, no precipitation occurs and under a clear sky evaporation reaches extreme intensity. Especially on the northern hemisphere the arid belt is very well developed in North Africa and adjacent part of Asia. The wide arid core zone forms the Sahara, the Arabian Peninsula, the Middle East as well as parts of Iran (Fig. 1.1.1). Compared with that the semiarid zones represent only narrow transition strips between the arid and the following subtropical zone in the north and the marginal tropical zone in the south.

1.1.3 Available water resources in the Middle East

The region of the Middle East shares both the arid zone, which passes over from Sinai to the Negev and continuing in the southern and central part of Jordan, as well as the semiarid zone including wide part of the coastal plain in Israel and Lebanon, the Palestinian and Jordan mountain areas along the Jordan Rift Valley and the north-western part of Syria. In the subsequent part of this contribution attention will be drawn on a general water-balance of the watershed along the Jordan River and the Israeli coastal plain.

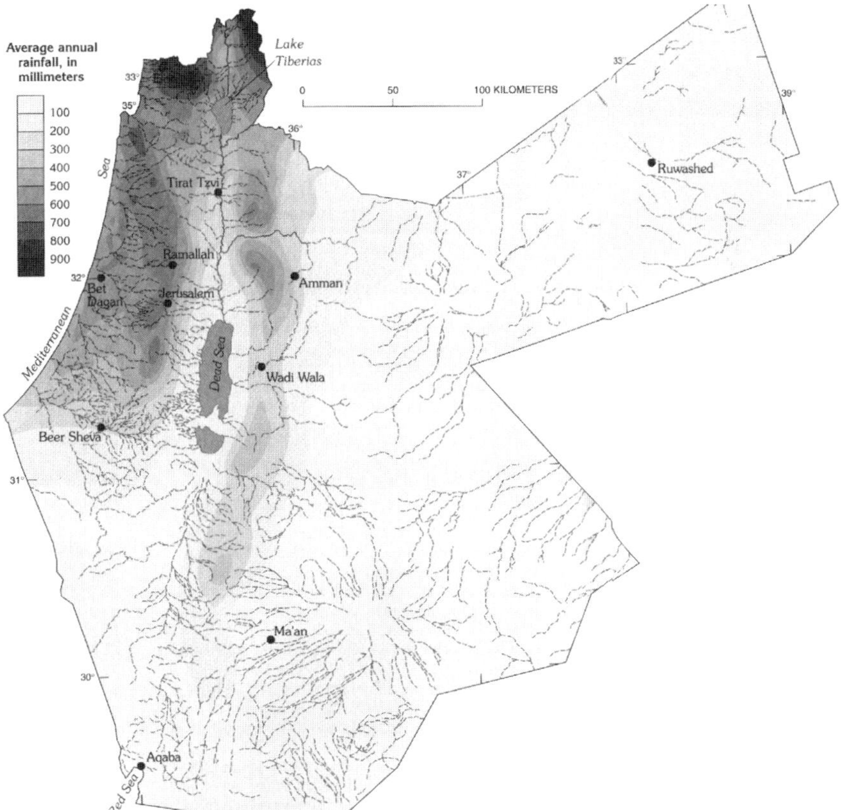

Fig. 1.1.2. Distribution of the average yearly rainfall amount in parts of the Middle East (Exact 1998).

In Fig. 1.1.2 the average annual rainfall in millimetres (after EXACT 1998) is presented. Precipitation rates of more than 800 mm can be found in the high mountains of the Lebanon and Anti-Lebanon, where the main springs of the Jordan River are located. Rainfall of 500 to 800 occur only in the Northern part of Israel and along the escarpment west of the Jordan Rift Valley down to Jerusalem, while most of the coastal plain and the escarpment east of the Rift Valley belongs to the semiarid region with precipitation of 250 to 500 mm per year. Arid conditions occur south of Beer Sheba, in the southern part of the Rift Valley including the near surrounding of the Dead Sea and of course the Jordan desert in the southeast.

Surface water

Resulting from the precipitation in the semiarid areas in winter and early springtime mainly intermittent, partly also permanent surface runoff occurs. The most important dewatering system is the Jordan River, which originates in the Lebanon Mountains and flowing in a first step into the Lake Tiberias. In former times there was a permanent outflow from the Lake Tiberias and the Jordan River reached the Dead Sea. Additional rivers drain the western escarpment outside the Jordan river watershed towards the coastal plain, where so far the water isn't used for irrigation it is infiltrating into the underground and alimenting the aquifers. Within the Jordan watershed there are mainly intermittent rivers in the wadis. In total the available surface runoff amounts to about 1400 Mio m^3 per year, of that more than 600 Mio m^3 per year belong to the inflow of the Jordan River into the lake Tiberias (DOMBROWSKI 1995, HOTTINGER 1992, SHUVAL 1992).

The available surface runoff is now more or less completely used for water supplies and irrigation purposes. In dry years we even can observe an overexploitation of the existing resources stored in the Lake Tiberias, so that the lowering of the sea water table reaches a critical value with regard to the complex balance with saltwater intrusion. That happens in spite of additional replenishment by deviating water from the Yarmuk River to the lake. Fortunately intercalated wet years can still raise the water table above the critical level. Water from the Lake Tiberias is the main freshwater resource for whole Israel. It is pumped from the lake to the coastal plain and distributed there by the national water carrier from the north to the Negev in the south.

Similar intensive use of surface water prevail downstream the Lake Tiberias. Runoff from the escarpments through numerous wadis towards the Jordan Rift Valley are collected in channels like the Ghor channel on the Jordan side and distributed for irrigation purposes. In spite of sparing use and partly reuse of waste water (Fig. 1.1.3) the available natural surface water in the Jordan Rift Valley is nearly emptied, so that only small amounts of the remnant salty water reaches the Dead Sea nowadays. This causes another damage by lowering the Dead Sea level (ARKIN & GILAT 1999, SALAMEH 2001), which has fallen down more than 20 m in the last twenty years.

Fig. 1.1.3. Storage pond for treated waste water, which is used for irrigation, Wadi Shuayb downstream the City of Salt, Jordan (Photo: HÖTZL 2001).

Groundwater

In the semiarid areas the seasonal rainfall also contributes to the recharge of the aquifers (LLOYD 1995), so that these natural resources still renewed within the hydrologic cycle. The recharge rate varies and depends from the soil, relief and specific climatic conditions. In general 10 to 30 percent of the rainfall contributes to the groundwater recharge.

The most important aquifers of the region are from west to the east:

Coastal aquifers: mainly sands and fine gravels of Quaternary and Tertiary age in the coastal plain between the Mediterranean Sea and the rise of the mountains.

Mountain aquifers: mainly limestone and dolomites of Cretaceous, partly Jurassic and Lower Tertiary age forming a large anticline structure, which finds its repetition in the morphology between the coastal plain and the Jordan Rift Valley. Due to the water divide along the crest of the mountain ridge the aquifer is divided in the „Western Mountain Aquifer", draining towards the coastal plain, and the „Eastern Mountain Aquifer", discharging towards the Jordan Rift Valley.

Shallow aquifers in the Jordan Rift Valley: Quaternary sandy and silty layers of alluvial, aeolian and lacrustine origin, partly with high mineralised or brackish water,

Aquifers of the eastern rift escarpment: Triassic, Jurassic and Cretaceous sandstones, limestone and dolomites dipping to the west towards the Jordan Rift Valley.

Table 1.1.1. Available freshwater resources in the aquifers (extended after Dombrowsky 1995).

Aquifer	Safe Yield in Mio m^3/year	Used in Mio m^3/year
Coastal Aquifer	280	280
Western Mountain Aquifer	335	335
Eastern Mountain Aquifer	125	100
Shallow Aquifers	80	50
Aquifers of the Eastern Escarpment	75	50

Fig. 1.1.4. Ein el Sultan, Jericho, capture of the spring for municipal water supply and irrigation purposes (Photo: Hötzl 1999).

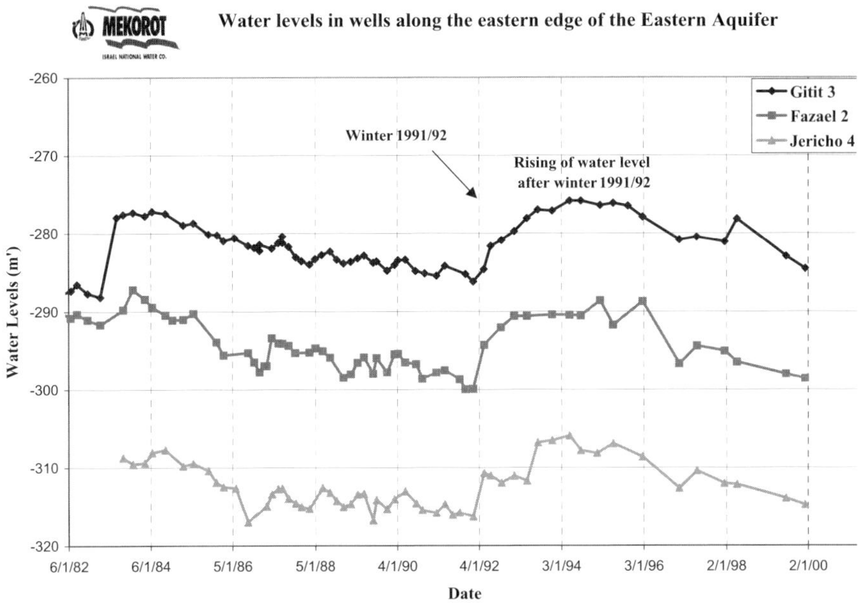

Fig. 1.1.5. Hydrographs from groundwater observation wells in the Jericho area with declining levels due to overexploitation. Recovery takes place in extreme years only (GUTTMAN 2000).

Table 1.1.1 shows the estimated safe yield of these aquifers and the used share. The figures emphasize the dramatic situation regarding the exploitation. It seems that only in the eastern part along the Jordan Rift Valley additional capacities are available, however especially in these regions the real remaining resources are not very clear due to the interference with salt and brackish water, partly resulting from mixing with ascending brines. In the past the groundwater was mainly used by capturing their natural outflow in the springs (Fig. 1.1.4). In the last thirty years a lot of additional wells were drilled in order to satisfy the increasing demand. The overexploitation with regard to the recharge of average hydrologic years can be seen from the declining water levels in Fig. 1.1.5. Only few wet years still enable to recreate to former conditions. Beside the quantitative problems qualitative aspects due to human activities become more important. The main problem is still the salinization due to irrigation and evaporation, the discharge of waste water, the adding of salt fertilizers or the draw by of salt-water and brines from the deeper aquifers because of changes of hydraulic conditions due to overpumping (Melloul et al. 1994, 1995, Muskat et al. 1989, Rosenthal et al. 1999, SALAMEH 2001). Contamination by organic pollutant may become a problem in the near future as a consequence of increasing urbanisation in the mountainous area. Detailed long-term observation and hydraulic studies are requested to safe yard the available resources.

In some arid areas like in the Libyan desert or in the Arabian Peninsula fossil groundwater, resulting from former wet climatic periods, serve as important additional resources (ISAAR 1972, HÖTZL & ZÖTL1984, HÖTZL ET AL. 1984). Though there are indications of stronger climatic changes in the young geologic history also in the Near East, fossil water resources are of less importance. The main aquifers are still connected with recent recharge area, so that the balance of the used aquifer systems is still replenished by the seasonal infiltrations of rain water. The only important fossil freshwater resources are located in the southern Jordanian desert. Due to the mainly nomadic life there these resources were not used in the past, however large agricultural enterprises have started now. In connection with the urgent drinking water demand of urbanised areas like the city of Amman priorities should be considered for these not renewable resources.

1.1.4 Socio-economic implications of the water shortage

Arid and semi-arid areas as have been shown are under high water stress. The lack of sufficient water coupled with high population density and/or high population growth rates results in serious water crisis, which can lead to political tension or even armed conflicts, if different parties or states involved. In some of the emerging conflict situation (DOMBROWSKI 1995, HOTTINGER 1992, SHUVAL 1992), we are looking how this was handled or solved in the past. A reasonable, but nowadays not always practicable solution is demonstrated by the development of life in the nature. It is primarily determined by adaptation on the existing conditions of the natural environment. This includes qualitative aspects, where for instance plants adapt to seasonal availability of water, and quantitative aspects, where for instance the amount of water regulates and restricts the number of species and individuals.

Also the human beings in their early history subjugated themselves to the natural conditions. For arid and semi-arid areas nomadism has proved as a practical way of life. The harsh conditions kept the population small and on the other hand the nomads were flexible enough to migrate over longer distance if life was endangered by longer drought periods. Inevitable conflict situation rising up, when nomads moving into more wet areas where already other tribes were settling. Always recurring fights characterize the juxtaposition of peoples in the marginal agricultural areas and the nomads, when the scarcity of water was extending.

The situation in the Middle East has aggravated with the increasing population and their extension towards dry and more hostile regions. A sufficient availability of water has proved to be the key for a prosper economical development of the individuals as wells as of societies and nations. This is valid in general and is especially the case in arid and semiarid areas. There are three main issues of modern human life which are impaired by water scarcity: the personal private demand, the demand for food production and the demand for trade and industry.

The personal private demand has increased with the habits of modern civilisation. The smallest part is the need of potable water, much more is required for the personal hygiene, the cleaning of household and other application in the private sphere. The World Health Organisation recommends a daily amount per capita from 65 litres to guarantee the necessary sanitary conditions for a healthy habitat. Though this amount includes also water beyond the personal demand, it forms a good basis for comparison. The standards in the more humid industrialised countries varies between 120 and 180 litres per day and capita. In arid areas the values exceeds 250 litre. From the total water consumption in Israel for instance a value of 330 litres per day and capita is calculated, while the respective values for the Palestinian West Banks amounts around 60 litres. It emphasis the existing water crisis in the region and raise up the question for additional resources.

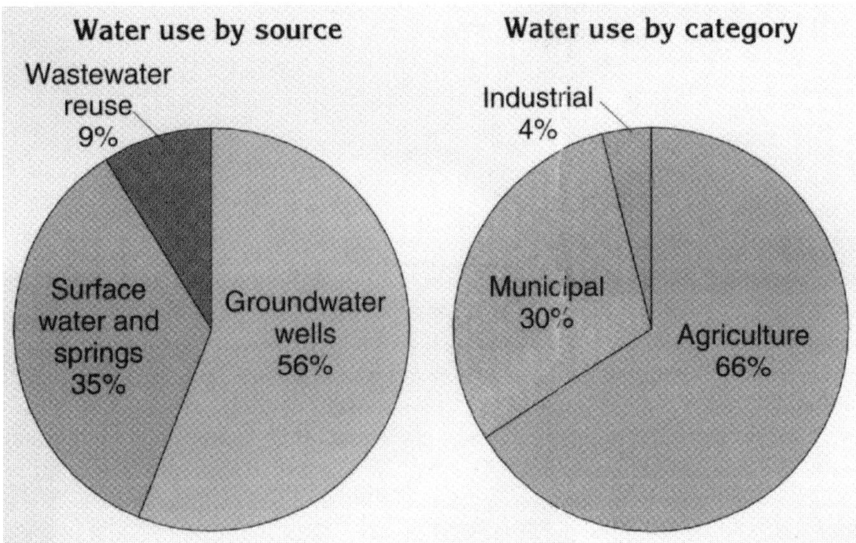

Fig. 1.1.6. Water use by source and by practical application in the Middle East (EXACT 1998)

More than 60 percent of the water consumption in the Middle East region is used for agricultural purpose (Fig. 1.1.6). Though specific fruits are exported, there is still additional demand to increase the agricultural and farming production rate to reach sufficient self-reliant conditions. Still there are wasteful use of the valuable water by inadequate irrigation techniques or the growing of plants with high water consumption (Fig. 1.1.7). But in spite of increasing application of modern irrigation techniques the lack of freshwater is evident in large parts of the region. This fact obstruct the necessary basic economic development and influences the whole economic and social situation in the countries of this region.

Fig. 1.1.7. Agricultural production by a lavish irrigation technique using the unrechargeable resources of fossil groundwater in the southern Jordan desert (Photo: HÖTZL 2002).

The lack of water shows also strong effects on industrial development. In areas where the resources are not enough to cover the necessary food production only limited amount of water will be available for industrial production. Frequently it is overlooked that even those products which seems to have nothing to do with water need high amounts of water for the production process. Thereby industrial infrastructure and further development are missing which could be another pillar of the economic system. With that are connected the job situation. Especially in densely populated areas this can be the source of high unemployment rates showing up the whole extent of the socio-economic interconnections.

1.1.5 Conclusions

The natural scarcity of water in the semi-arid to arid areas of the Middle East causes serious water problems which are connected with strong socio-economic consequences and hinders a sustainable economic development. The natural water resources are already exhausted to a great extent, so that a careful water management is requested, which has to consider the hydrologic and hydrogeologic conditions of water recharge. Because of the fact that no substantial additional resources are available in the region, priorities for water supply according the socio-economic aspects and multinational agreements on a balanced and consistent water distribution have to be established. Further on strategies how to run the future water demand have to be developed.

Water management have to concentrate on

- more efficient water use by optimisation of irrigation techniques, by cultivation of selected crops with low water demand and by restructuring land use and economic systems,
- increase of the available water resources by floodwater storage, by exploration of the last groundwater reserves, by additional reuse of waste water and by desalination of brackish water or sea water.

References

ARKIN Y & GILAT A (1999) Dead Sea sinkholes, an ever-developing hazard. – Environmental Geology, 39, pp 711-722

DOMBROWSKY I (1995) Wasserprobleme im Jordanbecken. – WZB Berlin, FS II 95-403, 1-215

EXACT – Executive Action Team (1998) Overview of Middle East water resources. Water resources of Palestinian, Jordanian and Israeli Interest. – Printed Poster of the U.S. Geological Survey

GUTTMAN J (2000) Multi-lateral Project. - Mekorot, unpub. report

HOTTINGER A (1992) Wasser als Konfliktstoff. Eine Existenzfrage für Staaten des Nahen Ostens. – Europa-Archiv, 6, 153-163

HÖTZL H & ZÖTL J (1984) Hydrogeology of central and southern Arabian Shelf. – In JADO AR & ZÖTL J (eds.) Quaternary Period in Saudi Arabia, vol. 2, pp 246-266

HÖTZL H, JADO AR, MOSER H, RAUERT W & ZÖTL J (1984) Climatic Fluctuation in the Holocene and in the youngest Pleistocene. - In JADO AR & ZÖTL J (eds.) Quaternary Period in Saudi Arabia, vol. 2, pp 301-324.

ISSAR A, BEIN A & MICHAELI A (1972) On the ancient water of the upper Nubian sandstone aquifer in the central Sinai and southern Israel. – J. Hydrology, 17, pp 355-374

LLOYD JW (1995) The reality of recharge in semi-arid and arid areas.- Proc. Intern.Conf. on Water resources management in arid countries, Oman, 2, pp 466-479

MELLOUL AJ & GOLDENBERG LC (1994) Groundwater pollution by airborne toxic contaminants: a factor to be considered in groundwater management. – IAHS, Publ., 220, pp 95-105

MUSKAT L, ROSENTHAL E, RONEN D, & MARGARITZ M (1989) Organic contaminants in Israel coastal aquifer. – Environ. Qual. Ecosyst. Stab. IV-A, pp 360-366

MELLOUL AJ & AZMON B (1995) A graphic expression of salinazation and pollution of groundwater. The case of Israel's groundwater. – Environmental Geology, 30, pp 126-136

ROSENTHAL E, WEINBERGER G & KRONFELD J (1999) Groundwater salinization caused by residual Neogene and Pliocene sea water, Judea group aquifer, southern Israel. – Groundwater, 37, pp 261-270

SALAMEH E (2001) Sources of water salinities in the Jordan Valley Area/Jordan. – Acta hydrochim.hydrobiol, 29 (2001), pp 329-362

SHUVAL HI (1992) approaches to resolving the water conflicts between Israel and her neighbours. – Water Intern., 17, pp 133-143

1.2 Exploitation of Fossil Aquifers and Future Water Supplies in the Middle East

Elias Salameh
University of Jordan, Amman, Jordan

Abstract

The Middle East and North Africa region is overusing its fossil, non-renewable groundwater resources in a way, which can be described as ruinous exploitation to the stored quantities of water and their qualities.

Irrigated agriculture developed on these resources is feeding on the resources base of the MENA nations and their future generations, by exhausting the non-renewable resources for private gains.

It seems that, the concept of intergeneration equity in the field of fossil groundwater resources and the concept of environmental services of water have not yet entered the vocabulary of water planers and decision-makers in the MENA region.

If the over exploitation policies of fossil and non-renewable groundwater resources in the MENA region continue along the same line, that region will soon be suffering of one of its most severe socio-economic, environmental and intergeneration equity problems.

1.2.1 Introduction

Although water is the most abundant among the natural resources worldwide, the Middle East and North Africa (MENA) are among the least blessed areas of the world with respect to the availability of water resources (Salameh and Bakhit 1990). Vast areas in the MENA region are even bedeviled by hyper-aridity.

The natural scarcity of water resources in the area is also greatly aggravated by human activities such as:

1. Alarming growth rates of population resulting in doubling the population of the different countries every 18 to 30 years.
2. Growing degradation of the available resources thus reducing their utility at their original quality.
3. Increasing demand due to higher standards of living, industrialization and irrigation.

Historically, aridity can be considered as a major natural characteristic of the MENA region, with its short rainy seasons and high evaporation rates.

Water with only few exceptions, is scarce and precious. Therefore, it is not surprising that water and irrigation engineering had their birth and beginning in the Middle East (Sousa 1983). In Mesopotamia, water of both the Euphrates and the Tigris Rivers was utilized for land irrigation initiating techniques and construction of irrigation canals; in Egypt, the Nile water was harnessed for irrigation, thus initiating and promoting Egyptian engineering and land surveying. In Jordan, Syria Lebanon and Palestine where 90% of the land is semi arid and with the exception of the very small rivers, such as the Jordan River, Litani, Orontos and their tributaries devoid of perennial water sources, man was obliged to develop special techniques to survive during most of the seasons. Several examples of constructions related to water utilization are well preserved in the area. The oldest known dam in the world stands in Jawa in the northeastern part of Jordan as a witness of the need to collect water; it is 5.5m in height and around 5200 years old (Helms 1981). Another impressive example is the water supply of Petra, 2000 years old, where a spring in Wadi-Musa village, in Jordan was diverted to the city through a system consisting of an inter-connected network of ceramic pipes hanging 4-6 meters above ground on the side rocks serving drinking purposes. Additionally, an open channel was constructed to serve the other uses (Pade 1999, Huneidi 1999). In Yemen, Ma'arib Dam is a historic legend of achievements in technological development.

The abovementioned examples, among many others illustrate the ingenious measures inhabitants of the MENA region have developed to assure their survival in this harsh area throughout the course of history.

Forced by necessity, technicians of all civilizations struggled to develop the hydro-technical infrastructure without which none of the ancient empires could have developed or survived, for that long.

Historic evidence allows the conclusion that the history of man in the MENA region, throughout the last three to four millennia has been determined and shaped largely by one major infra-structural element, namely water. This essential resource has great influence over human life when it is scarce. In the MENA region lifestyles of people, their socio-economics and their conflicts have, since historic times, all been determined and shaped by this basic factor (Naff and Matson 1984).

Rain-fed agriculture developed when the amount of rain proved sufficient to support plant life. Irrigated agriculture was practiced along watercourses such as the Nile, the Euphrates, the Tigris the Jordan and around springs, as well as around oases in the desert areas.

1.2.2 Recent Developments

In the past, the availability of water and the technologies used for its exploitation not only determined lifestyles and socio-economics but also limited the population to the number which, could be supported by the amount of produced food.

In the last few decades, the population growth rate has been very high, not only due to natural growth but as a result of the waves of refugees moving from one place to another especially those coming from Palestine to Jordan.

During the last few decades, the whole development in the MENA region has been concentrated in agriculture, mainly irrigated agriculture, which entails developing water resources. Irrigated agriculture created increasing job opportunities through less expensive investments born by both, private and public sectors. This averted the potential catastrophe of poverty and hunger, and fostered domestic peace in the MENA area. However, with the sharp increase in population and acceleration in agricultural development as well as the establishment of many small, medium-sized and even heavy industries, the available water resources proved to be insufficient to meet aspirations for additional development (Gleick 1999).

Despite the implementation of intensive water projects and conservation measures, water shortage became the major obstacle to development in the MENA region. This put and is still putting specialists and politicians under severe stresses concerning the future of the MENA economic growth (Schiffler 1998).

The development of water resources then expanded to include the groundwater resources, which in the MENA region were, until the sixties of the last century, only utilized by primitive ways such as digging canals, or galleries, or shallow hand-dug wells, compared to today's technologies of deep drilling, strong pumps, and availability of electric power, which have led, in the last few decades, to the abstraction of huge amounts of groundwater, in most places exceeding the safe yields of aquifers. Most aquifers in the area, especially the huge ones are now suffering of over-exploitation.

The resulting imbalances between renewable amounts of water and extractions have, in the last four decades, been gradually leading to depletion and Stalinization of MENA groundwater resources. In other words, the over exploited amounts of water (amounts exceeding the safe yield) are now being taken from limited reserves, which have been partly exhausted during the last decades. Mining processes with foreseen lifetimes are afflicting other aquifers.

That situation is presently aggravated by degradation processes affecting both the surface and the groundwater resources in the different parts of the MENA region, thus limiting their use to purposes, which can do with lower water qualities.

Depletion and degradation of the MENA water reserves are affecting the present generation due to water pollution, lower water quality in household, interruptions in supply, higher pumping cost due to lowered water levels… etc.

It is also greatly affecting the biodiversity in the MENA region by curtailing the environmental services of the water, because most water development projects do not have any provisions for environmental services, which means leaving some water for the environment to conserve nature and maintain biodiversity.

In the near future, coming years, next decade and for next generations it will not only be an interruption in supply, lower water quality and higher cost, but a very limited supply, very low quality and very high cost. Governments will then have to put most of their efforts to reach at an adequate water supply.

Most aquifers are depleting with time and their qualities are deteriorating to an extent where they will not satisfy even irrigation quality requirements. The first cases of such depletion and quality deterioration are already recorded in different parts of the MENA region (Salameh and Bakhit 1990).

Thus, unless immediate actions by aggressive management procedures are applied to preserve ground water quantities and qualities the MENA region will soon be suffering one of its most severe socio-economic, environmental and inter-generation equity problems.

1.2.3 Overexploitation

The groundwater resources of the MENA can be categorized according to their recharge time in two types:

1. Renewable groundwater resources, receiving recharge every year.
2. Non-renewable groundwater resources, which in foreseen times-tens to hundreds of years- are not renewable. Extractions from these resources is quasi a mining process because, they will not be replaced by infiltration water or by other mechanisms. This type of water can, according to its connections to the renewable groundwater resources, be divided into two categories:
   - Fossil groundwater bodies, which receive no recent recharge.
   - Fossil groundwater bodies underlying renewable groundwater bodies.

Threatened by depletion, exploitation of fossil water must be considered as a strategic, economic, and environmental issue, which involves an intergeneration equity dimension. Therefore, any decision to extract fossil water should undergo comprehensive studies aiming at conserving coming generations' rights to these resources, either by keeping them untouched for these generations' disposal or by investing the revenues of their exploitation and use in projects which have to yield enough to generate the same amounts of water from other sources, such as, desalination of sea water or import of water from water rich areas. This implies that projects using fossil water should yield enough revenues to cover next generations' expenditures to obtain quantities of water, equivalent to the extracted amounts.

Failure in the planning of projects for fossil water exploitation results in exhaustion and damage of sources and may prove detrimental to coming generations' rights to them.

Since there are no common rules concerning the exploitation of fossil groundwater and because water is a common good and when it is fossil, has, in addition an **intergeneration dimension**, fossil water should be dealt with, only in a framework of **national planning**. Therefore, ad hoc decisions in that context are in water resources planning never in place and decision should underlie **strict** and **comprehensive economic, strategic, environmental and intergeneration equity considerations**.

This leads us to the question of how much can a nation extract from its fossil or non-renewable groundwater resources.

The answer, in my opinion is that, a nation can extract only that amount of fossil water from which use it can generate enough revenues (to be invested) to make available the same amount of water for next generations' uses.

1.2.4 The Bleak Picture

Within the MENA region there are no aquifers to be developed in addition to those already under use. Therefore, no substitutes for any of the groundwater resources serving municipal supplies are to be found or developed in the area.

This implies that the MENA region from within its area is not able to continue supplying water within the presently supplied amounts, if the groundwater aquifers now in use suffer any further depletion or quality deterioration, such as salinization, which prohibits the use for household purposes. Also, any increase in extraction means further declines in water levels and further quality degradation, or generally, a faster damage of aquifers.

Therefore, to maintain and increase municipal water supply, which are **pressing issues**, at present, and are only expected to sharpen in the future, four options are open:

1. Curtail irrigated agriculture to stop groundwater depletion.
2. Improve irrigation efficiency and conveyance systems.
3. Treat and reuse wastewater in irrigation to substitute fresh water uses in irrigation.
4. Desalinate seawater at a cost of $US 0.6 –1.0 per cubic **meter** .

1.2.5 Summary and Conclusion

- The MENA region is overusing its water resources by **over-extracting** groundwater on the account of its reserves or savings.
- The MENA region is using its groundwater resources in a way, which can be described as **ruinous exploitation** to both the stored water quantity and quality.
- Irrigated agriculture depending on abstracting nonrenewable groundwater is feeding on the **resources base** of the MENA nation by exhausting the nonrenewable groundwater resources for private gains.
- The overexploitation of aquifers during the last decades rendered **desalination of seawater** for municipal water supply (if radical unfavorable solutions are to be avoided), the only practical option for the MENA countries in the coming decades. Desalination costs $US 0.6 to 1.0 per cubic meter at sea shores (for an oil price of around $US 22 per Barrel).
- The extraction of nonrenewable groundwater for irrigation is gradually proving **detrimental to the groundwater resources of the MENA region and to its economy**. The population of the area has to carry the cost of that damage of $US 1.0 per cubic meter of groundwater extracted for use in irrigation.

- The losses **"de-savings"** of the MENA countries water assets due to irrigation with fossil groundwater, estimated in monetary terms amount to **$US 5-7 billions** a year.
- It is emphasized here, that the concept of "**Environmental Services**" of water resources does not seem to be of any concern for water planners and water managers in the MENA region. It is nowhere on their **agenda**. There seems to be a big gap in the **environmental education** in the MENA countries, especially among that group **of policy makers**.

References

Gleick PH (1999) The World's Water 1999-1999. Island Press Washington DC p 307

Helms SW (1981) Lost City of the Black Desert.Methuen London p 263

Huneidi M (1999) Hydraulic System in the Siq and the areas flanking it. Occident and Orient Vol *4* No 1&2 German Institute of Archaeology Amman

Naff T, Matson RC (1984) Water in the Middle East – Conflict or Cooperation.Westview Press London p 235

Pade M (1999) The Nabataean hydraulic System at Petra. Occident and Orient Vol *4* No 1&2 German Institute for Archaeology Amman

Salameh E, Bakhit MA (eds) (1990) Water resources of the Arab World and their strategic importance. (in Arabic) University of Jordan Amman p 311

Schiffler M (1998) The economics of groundwater management in arid countries – Theory, international experience and case studies of Jordan. Frank Cass Publishers London p 394

Sousa A (1983) History of Mesopotamian Civilization in the light of irrigational agricultural projects. (in Arabic) Vol *I* p 572 Vol *II* Al Huriya Printing House Baghdad p 422

1.3 Hydrogeological Investigations in the North-Eastern Dead Sea Area, Suweimeh, Jordan

W. Ali, J. Glaser, M. Thiel, H. Hötzl, H. Werz

Department of Applied Geology, Karlsruhe University

Abstract

The Sweimeh Area at the north-eastern shoreline of the Dead Sea represents the margin between the Jordan Valley to the west and the escarpment of the Jordan mountains to the east. This study is a part of the activity of the Karlsruhe team within the course of the Multinational BMBF Joint Research Project Water Resources Evaluation for a Sustainable Development in the Jordan Rift Basin as part of the German – Israeli – Jordanian – Palestinian Joint Research Program, 2nd Phase

The objective of this study is the detailed knowledge and the understanding of the hydrogeological conditions along the cross section from Jerusalem to Amman via Jericho.

The outcropping formations in the study area range from the Triassic Zerqa Ma'in Group to the Quaternary Lisan Marl Formation and young basaltic flows. The study area is structurally intensively affected by the development of the Jordan Rift Graben System.

Hydrogeologically, the area comprises the three regional important aquifer complexes of Jordan. The Zerqa Ma'in and the Kurnub Sandstone Group are summarised within the Deep Sandstone Aquifer Complex, the Ajlun Group belongs to the Upper Cretaceous Aquifer Complex and the alluvial deposits together with the basalts build the Shallow Aquifer Complex.

15 springs are hydrochemically investigated in the Sweimeh area. The results show a dramatical increase of salinization from the east towards the Dead Sea in the West of the study area. Additionally, the results show very different chemical compositions. Therefore, two main groups of springs can be distinguished. All spring waters have a high content of TDS which are classified as medium saline to brine. In all springs a fluctuation of cation distribution is observed between two work sample cycles within a time of two weeks. Ionic ratios show different origins and influences of the spring waters. The investigation of stable isotopes confirm the differences in the spring water composition.

The execution of pumping tests close to the study area in Wadi Kafrien permitted the determination of the hydraulic parameters of the Kurnub Sandstone Formation. The evaluation resulted transmissivities in a range from $1{,}34 * 10^{-3}$ m²/s to $7{,}97 * 10^{-3}$ m²/s.

1.3.1 Introduction

The Sweimeh area (Fig. 1.3.1) is located about 40 km south west of the capital Amman in the Jordan Valley at the north eastern shoreline of the Dead Sea. This study is part of the activity of the „German-Israeli-Jordanian-Palestinian Joint Research Project for the Sustainable Utilisation of Aquifer Systems".

Fig. 1.3.1. Satellite image with the position and size of the study area at the Dead Sea, scale 1:143 000, (Landsat TM, 1984)

The main results are part of the Msc.Thesis of Jens Glaser (2002) and Martin Thiel (2002). Prior investigations for more understanding of the aquifers along the hydrogeological section Amman - Jerusalem were undertaken by WOLFER (1998), LENZ (1999), ROTHER (1999) and BECKER (2000).

The main objectives of this study are:

- Detailed geological mapping of the Sweimeh area and Wadi Udeimi including a WNW-ESE cross-section.

- The hydrochemical characteristics of spring waters and the evaluation of wells and pumping tests in the study area.
- Investigation of the different existing aquifers and their relationship to structural systems by hydrochemical parameters.
- Investigation of stable isotopes in the spring waters and the salinization processes.

1.3.2 Geology

The outcropping strata in the study area is at Triassic and Lower Cretaceous age. Alluvial deposits in wadis and the Jordan Valley plain are of quaternary age. The Triassic Zerqa Group is only represented by the Dardur and Ain Musa Formation which are combined into the Sweimeh Sandstone Group. It consists of calcareous and dolomitic sandstone intercalated with dolomite. The Lower Cretaceous Kurnub Sandstone consists of partly calcareous, poorly cemented, varicoloured fine sandstone.

Fig. 1.3.2. Boundary between KURNUB SANDSTONE and AJLUN FORMATION in the study area (Photo J. Glaser)

The transition between Zerqa and Kurnub can not be fixed lithologically and is mapped morphologically (Fig. 1.3.2). The Ajlun Group, only represented by the A1/A2 Naur Formation is of early Cenomanian age and consists of well bedded calcite and dolomite with intercalated clayey siltstones. The Quaternary deposits are represented by the Lisan Marl Formation (JV 3) with well bedded laminated grey-white silt facies intercalated with gypsum (Fig. 1.3.3) and the alluvial deposits consisting of poor to well rounded gravel (JV 1).Two alluvial fans are mapped at the Dead Sea shoreline.

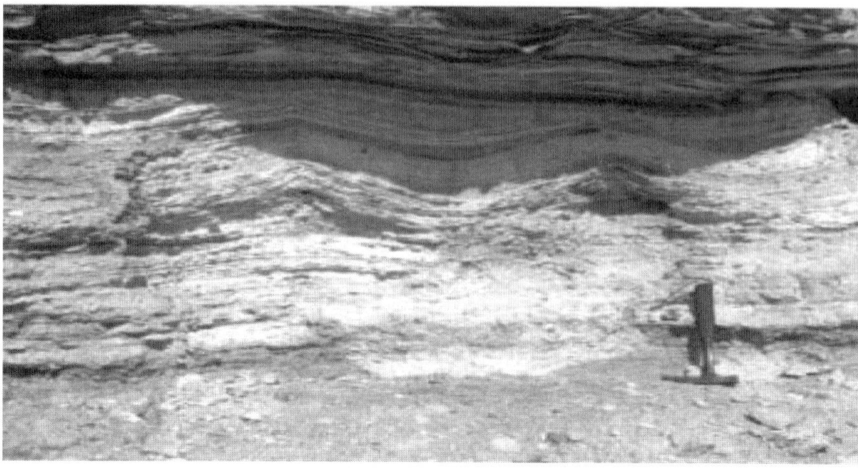

Fig. 1.3.3. Laminated Lisan Marl Formations (Photo J. Glaser).

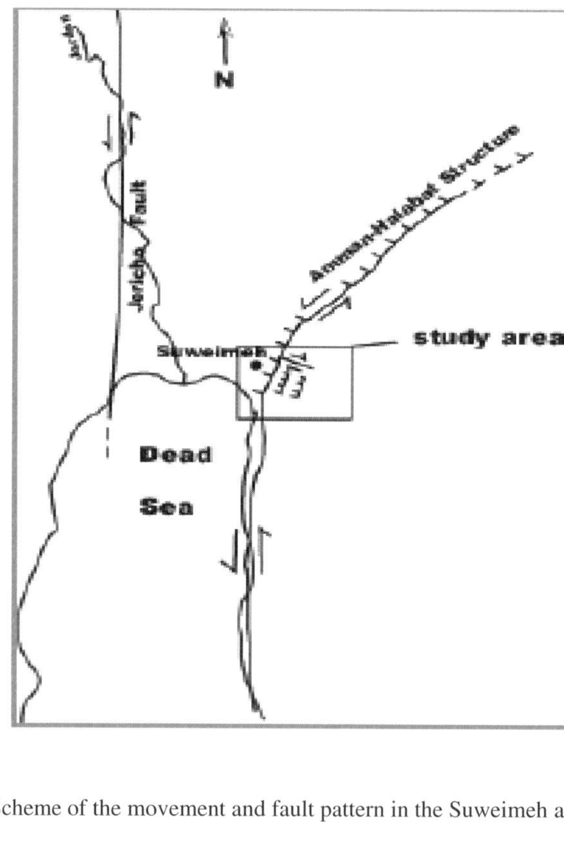

Fig. 1.3.4. Scheme of the movement and fault pattern in the Suweimeh area

1.3.3 Structural elements

The Dead Sea Transform Fault in the study area is not that well exposed as in many other parts of the Dead Sea pull-apart basin, where the dislocation plane is clearly visible (Fig. 1.3.4). However the effects of this dominating fault are recognizable in the north-south trending faults occuring above all in the south-west of the study area.

The Amman-Halabat Structure is not exposed because of the coverage by soil and gravels. It extends from the north-eastern part of the Dead Sea close to the village of Suweimeh passing the north of the village Al Addasiyeh and going further to Amman. This fault zone is 1-5 km wide and consists of many structural elements including faults and folds controlled by the same stress system. The downthrow of this fault is to the north-west with a vertical displacement up to 400 m (SHAWABEKEH 2001).

1.3.4 Hydrochemistry

The main recharge of the aquifers takes place along the highlands and the plateau. During the sampling period, the discharge of the springs has not changed significantly despite several precipitation events. The discharge ranged from drippy springs up to springs with a discharge of 0,75 l/s. Conspicuous are the high electric conductivities of the spring waters which stem from evaporitic layers within the single formations on one hand and on the structure fault relation of some springs close to the Dead Sea on the other hand. The electric conductivities range from 1730 μS/cm up to 58800 μS/cm and the temperatures from 17,6°C to 29,4°C.

15 springs were investigated in the Sweimeh area (Fig. 1.3.5). The hydrochemical investigations show a dramatical increase of salinization towards the Dead Sea and very different chemical compositions. The 15 sampled springs in the Sweimeh area can be divided after the content of TDS into three groups of springs:

- Two springs (S 01, S 04) show an extremely high content of TDS with a maximum value of 40670 mg/l which is about 5 to 10 times higher than in other springs in particular the springs S 02 and S 03 just about 100 m in distance. These two springs may depend on the Amman-Zerqa Fault. The high values of salinity suggest an extended and long flow path of the groundwater. The chemical composition of anions differs drastically to all other springs.
- Three springs (S 02, S 03, S 15) additionally discharging out of the alluvial deposits below an elevation of –300 m MSL. They also show a very high content of TDS at about 4000 to 4500 mg/l. These springs are dominantly influenced by dissolution process of soluble salt minerals in the deposits of the Dead Sea occurring about below –180 m MSL.
- In springs discharging out of the Triassic and Lower Cretaceous Aquifers (S 05, S 06, S 07, S 08, S 09, S 10, S 11, S 12, S 13, S 14), the content of TDS is at about 1200 up to 3400 mg/l. Most of these springs are located in the Kurnub

Sandstone Formation, especially in the transition zone between the Kurnub and Zerqa Sandstone. Only two springs (S 12, S 14) are related to the Zerqa Sandstone Formation. Some of these springs are heavily used for drinking and irrigation (S 07, S 09, S 13). S 07 and S 13 actually have the lowest content of TDS of all sampled spring waters.

In all springs a fluctuation of cation distribution is observed between two work sample cycles within a time of two weeks.Ionic ratios shows different origins and influences of the spring waters.

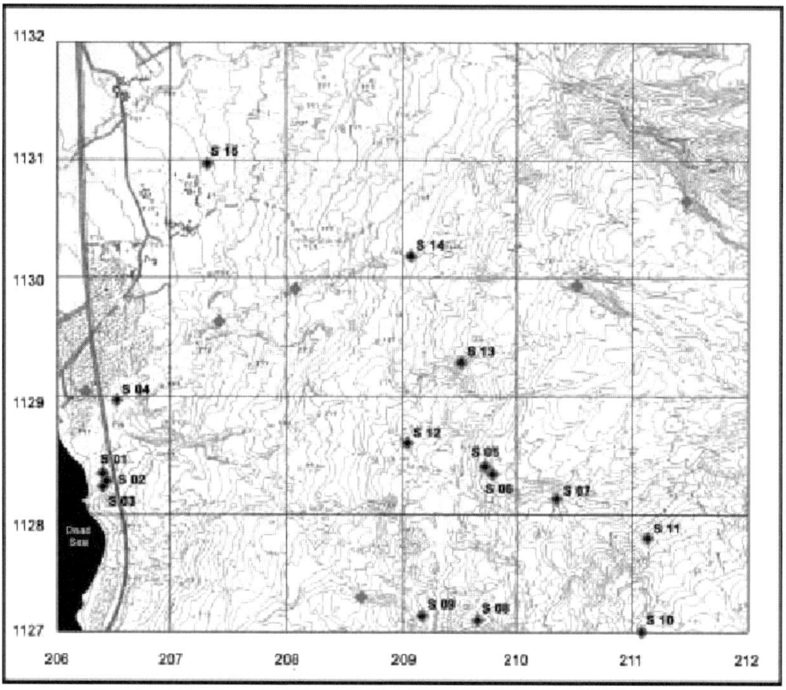

Fig. 1.3.5. Location of the springs in the Sweimeh Study Area.

The investigation of stable isotopes confirm the differences in the spring water composition (Fig. 1.3.7).

Most isotopic distribution of the spring samples are located at or close to the characteristic continental rain relationship of : $\delta D = 8\ \delta\ ^{18}O + 10$ as well as between the meteoric water line $\delta D = 8\ \delta\ ^{18}O + 10$ and east Mediterranean water line $\delta D = 8\ \delta\ ^{18}O + 22$.

Spring water samples of S 08, S 09, S 10 and S 11 (signed with circles) have a trend to the thermal groundwater line $\delta D = 3{,}97\ \delta\ ^{18}O + 16{,}9$. This suggests a thermal character of the groundwater of spring S 08, S 09, S 10 and S 11 and the non-thermal character of the other spring water samples.

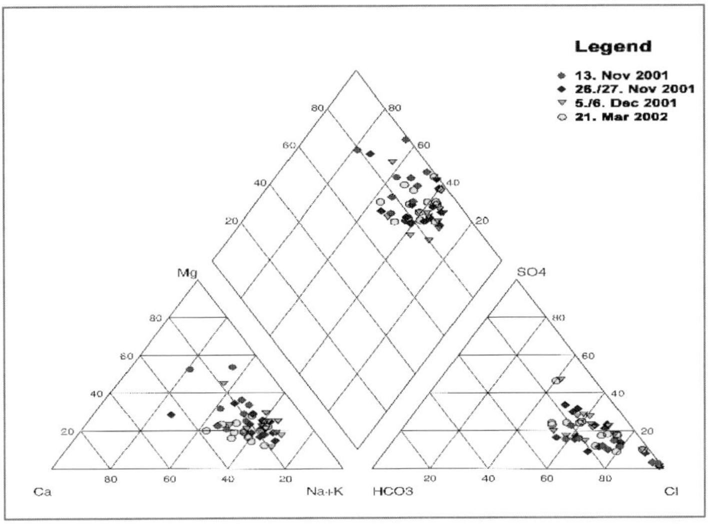

Fig. 1.3.6. Piper Diagram of the Chemical Distribution of the Springs in the Sweimeh Area

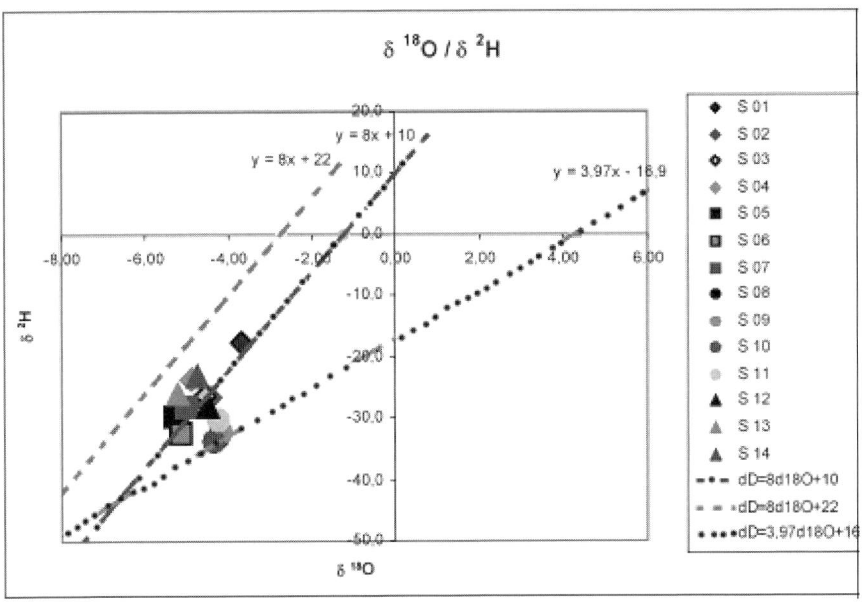

Fig. 1.3.7. Results of stable isotopes analysis

1.3.5 Pumping test in Wadi Kafrin near the Study area:

Due to the wide variations of the hydrological parameters even in the north of Jordan one aim of this study was to get detailed informations on hydraulic properties. During the period of 10.10. – 30.12.2001 three wells were drilled on behalf of the MWI in Wadi Kafrien to provide the Hotels at the Dead Sea with potable water. Wadi Kafrien is situated 32 km south-west of the capital Amman and the three wells are located approximately 14 km north-east of the study area. On account of the proximity of the wells to the study area the analysis of these pumping tests are comprised in this study. These wells are the Kafrien Project Wells 8A, 9A and 10A. The distance between the three wells varies between 500 m to 1000 m.

The aim of the MWI was to drill down in the Kurnub Sandstone Formation, because this formation guarantees a good quantity as well as a good quality of potable water.The final depths were for Well 8A 530 m bgl, for Well 9A 500 m bgl and for Well 10A 500 m bgl.

The analysis of the drawdown-time-courses of the three wells (Fig. 1.3.8) indicate, that the Kurnub Sandstone does not behave like a fractured rock formation. All the wells behave similarly.

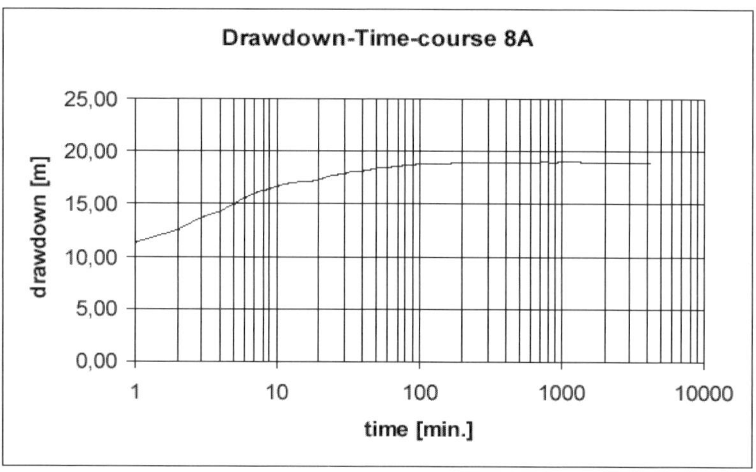

Fig. 1.3.8. Drawdown-Time-course of well Kafrien 8A as exemple for the drilled wells

The different graphs do not show a typical course following the double-porosity concept developed by BARRENBLATT ET AL. (1960). The Tab.1.3.1 shows the transmissivity and the K-values of the three wells. The saturated thicknesses for the Kurnub Formation were estimated with 177 m for Well 8A, 203 m for Well 9A and 275 m for Well 10A respectively.

Table 1.3.1. Transmissivity and the K-values of the three wells in Wadi Kafrien

Hydraulic Parameters	Well 8A	Well 9A	Well 10A
$T_{(Theis)}$ [m²/s]	$6{,}01 \cdot 10^{-3}$	$2{,}21 \cdot 10^{-3}$	$1{,}81 \cdot 10^{-2}$
$K_{(Theis)}$ [m/s]	$3{,}39 \cdot 10^{-5}$	$1{,}09 \cdot 10^{-5}$	$6{,}59 \cdot 10^{-5}$
$T_{(Cooper-Jacob)}$ [m²/s]	$5{,}07 \cdot 10^{-3}$	$1{,}34 \cdot 10^{-3}$	$7{,}97 \cdot 10^{-3}$
$K_{(Cooper-Jacob)}$ [m/s]	$2{,}86 \cdot 10^{-5}$	$6{,}60 \cdot 10^{-5}$	$2{,}90 \cdot 10^{-5}$

The calculated hydraulic parameters for the wells in Wadi Kafrien differ from the ones modelled by the BGR in 1997 for northern Jordan. Therefore two reasons can be given. The first one is that the hydraulic parameters of the BGR-report are modelled parameters what means they are generalised. The other reason is the wide local fluctuation of the hydraulic parameters and rock properties in the Jordan Valley.

1.3.6 Acknowledgement

This paper is a part of the activities of the German - Israeli - Jordanian - Palestinian Joint Research Program Water Resources Evaluation for a Sustainable Development in the Jordan Rift Basin sponsored by the German Federal Ministry of Education, Science, Research and Technology (BMBF). The authors are grateful for administrative and financial support from the Ministry (Dr. J. Heidborn) and the Forschungszentrum Karlsruhe (Dr. W. Robel, Dr. H. J. Metzger, Frau S. Proboscht). Many thanks to the colleagues in the joint research program in the region for their continuous support and special thanks for all who provided necessary data for this study. For their support in supervising the diploma students in the field we thank Prof. E. Salameh from the Jordan University and Dr. H. El-Nasser the Minister of Water and Irrigation in Jordan and their staff. Special thanks to PhD-candidat Mrs. C. Werchau from the Department of Applied Geology Karlsruhe for the reviewing of the paper and useful discussions.

References

Abu Makhlouf K (2001) The Origin of Groundwater Salinity along a Profile extending from Baqa ´a to Dead Sea. Faculty of Graduate Studies University of Jordan Amman

Becker K, Ali W, Hoetzl H (2001) Study on water quality in the area of Wadi Shueib, Jordan Valley, Jordan. WIR 10 International Symposium on Water–Rock Interaction Villasimius Italy Vol 1 pp 457-459 Balkema Netherlands

Bender F (1974) Geology of Jordan. Contributions to the Regional Geology of the Earth Gebrüder Borntraeger Berlin

Burdon DJ (1959) Handbook of the Geology of Jordan;Accompany and explain the 1:250.000 geological map east of the Reift by A. M. Quennell. Govt. Hashemite Kingdom of Jordan 82 p

El-Naser H (1991) Groundwater Resources of the Deep Aquifer Systems in NW-Jordan. Hydrogeological and Hydrochemical Quasi 3-Dimensional Modelling. Lehr- und Forschungsbereich Angewandte Geologie und Hydrogeologie der Universität Würzburg 144 p Würzburg

Japan International Cooperation Agency JICA (1995) Ministry of Water and Irrigation, Hashemite Kingdom of Jordan. The Study on Brackish Groundwater Desalination in Jordan. Draft Final Report Supporting Report Yachiyo Engeneering Co Ltd Mitsui Mineral Development Engeneering Co Ltd Tokyo

Lenz S (1999) Hydrogeological Investigation along the Wadi al Kafrein and the Kafrein Reservoir, Jordan. Master Thesis Department of Applied Geology University of Karlsruhe AGK Karlsruhe

Macdonald M, & Partners (1965) East Bank Water Resources; Hydrogeological survey of the Madaba Ma ´an area. Vol *1-3*. Central Water Authority CWA Hashemite Kingdom of jordan Amman

Masri M (1963) Reports on the Geology of the Amman-Zerqa Area. Central Water Authority CWA unpublished Amman

National Water Master Plan Of Jordan NWMP (1977) Vol 1 Main Report Natural Resources Authority (Amman). German Agency for Technical Cooperation LTD (Frankfurt) Bundesanstalt für Geowissenschaften und Rohstoffe Hannover

Niemi TM, Ben-Avraham Z, Gat JR (1997) The Dead Sea, The Lake and ist Setting. Oxford University Press New York

Parker DH (1970) The Hydrogeology of the Mesozoic, Cainozoic Aquifers of the Western Highlands and Plateau of East Jordan. Technjical Report Vol *2* Hydrogeology United Nations Development Program Food and Agriculture Organisation of the United Nations Rome

Quennell AM (1951) The Geology and Mineral Resources of former Transjordan. Colon Geol Min Resources London

Rother M (1999) Hydrogeological Investigations in the North-Eastern Dead Sea Area, Jordan. The Thermal Springs along Wadi Zerqa Ma ´in & Ain ez Zara. Master Thesis Department of Applied Geology University of Karlsruhe AGK Karlsruhe

Salameh E, Rimawi O (1984) Isotopic Analysis and Hydrochemistry of the Thermal Springs along the Eastern Side of the Jordan Dead Sea-Wadi Araba Rift Valley. Journal of Hydrology *73* 1984 pp129-145 Amsterdam

Salameh E, Udluft P (1985) The Hydrodynamic Pattern of the Central Part of Jordan. Geologisches Jahrbuch pp 39-53 Hannover

Technical Cooperation (1996) Project No.89.2105.8 Groundwater Recources of Northern Jordan Vol 1 Rainfall, Spring Discharge and Baseflow Part 2 Spring Discharge in northern Jordan. Water Authority of Jordan Amman Federal Institute for Geoscience and Natural Resources Hannover Water Authority of Jordan Amman

Wolfart R (1959) Geology and Hydrogeology of the Irbid District (Hashemite Kingdom of Jordan). Bundesanstalt für Bodenforschung Hannover

Wolfer J (1998) Hydrogeological Investigations along the Jerusalem-Jericho Transect (Wadi El Qilt), Israel/West Bank. Master Thesis Department of Applied Geology University of Karlsruhe AGK Karlsruhe

1.4 The Jordan River: Natural Flow and Current Consumption by the Riparian Countries

Ghassan Abu Ju'ub & Kurt Schetelig
Department of Engineering Geology and Hydrogeology (LIH), University of Technology (RWTH), Aachen

Key Words

Jordan River, Upper Jordan, Lower Jordan, water management, water distribution, water conflict, Yarmuk River, Lake Tiberias, Dead Sea, Dan River, Banias River, Hasbani River, Palestine, Israel, Jordan, Syria, Lebanon, Middle East, Israeli National Water Carrier.

Abstract

The Jordan River is located in the Middle East. It is an area characterised by scarcity of its water resources. This resulted in a continuous mismanagement and conflict over the river's water.

The absence of any agreement between the riparian countries over the usage of the river complicated the problem. The result is a serious damage to the environment and more disputes in the area.

This paper identifies the physical characters of the Jordan River System. The distribution of its water between the riparians according to previous plans and the current usage by these riparians are also discussed. In addition, the environmental damages to the river and its surroundings are also dealt with.

1.4.1 Introduction

The Jordan is a small river, but a well-known one, mainly for two reasons:

- Because it is a holy river, where Jesus the Christ was baptised by John the Baptist.
- Because it is located in an area suffering from scarcity in water resources, which makes every drop of its water counts.

The area where the Jordan River is located is known in Arabic as Belad As-Sham (As-Sham Land). This area includes Historical Palestine, Syria, Lebanon and Jordan. At present the river has five riparian countries: Palestine, Israel, Syria, Jordan and Lebanon.

These five countries have a total area of about 312,000 km^2 and a total population of about 35 millions. Syria has more than half of the total area and about half of the total population.

The competition over the usage of the Jordan's water started long time ago, remarkably in the early years of the 19th century. Nowadays, the dispute over this river is very well known, it is mainly between Israel on one side with Palestine, Syria, Jordan and Lebanon on the other.

1.4.2 General Description of the Jordan River

The Jordan originates at the footsteps of Mount Hermon (Jabal Ash-Sheikh). The sources of the river are at about 2200 m amsl [11] (above mean sea level) high. The source of the river is near the point where the borders between historical Palestine-Syria-Lebanon meet. The first three main tributaries of the river are Dan, Banias and Hasbani. They join each other north of Lake Al-Huleh inside Israel.

The river discharges in the Dead Sea, which is the lowest point on earth with a current water level of around 414 m bmsl (below mean sea level), see figure 1.4.1.

The total length of the river is around 350 km [7, 11]. This is quite long compared to the straightforward length, which is about 228 km [8, 10]. This difference is because the river meanders a lot in its way, especially in its lower part (lower Jordan) between Lake Tiberias and the Dead Sea. This meandering, together with the current small flow quantities makes the lower Jordan unsuitable for navigation.

The total catchment area of the Jordan River is about 18,300 km^2 [7]. Most of it is in Syria and Jordan. Its total annual flow is around 1.3 $Mrd.m^3$ [11]. This is a variable quantity and depends on the rainfall. It can be 1.5 $Mrd.m^3$/yr. [7] or more in rainy years.

This flow quantity shows that the Jordan is indeed a small river compared to the large rivers in the region: The Nile with 84 $Mrd.m^3$/yr., Euphrates with 36 $Mrd.m^3$/yr., and Tigris with 49 $Mrd.m^3$/yr. [2].

The Jordan River system includes several elements, the main ones are:

- The Jordan River itself, which is normally sub-divided into two main parts: upper Jordan, and lower (or main) Jordan.
- Lake Tiberias.
- Yarmuk River.
- Dead Sea.

1.4.3 Upper Jordan

The part of the Jordan River between its main source on Mount Hermon and Lake Tiberias is known as Upper Jordan.

As mentioned above, the river has three main tributaries at its origin at the footsteps of Mount Hermon:

A. Dan (Al-Qadi): Originates inside historical Palestine (present day Israel). Its average annual flow is about 250 Mm^3 (Million cubic meters) [7]. It is the largest between the three tributaries in terms of water quantities and contains as much water as the Banias and Hasbani together. Its flow quantity is almost stable and does not fluctuate much between the different years, in contradiction to the other two tributaries. The Dan is the shortest of the three tributaries.

B. Banias: This tributary originates in Syria (Golan Heights). It has an average annual flow of 125 Mm^3 [7], but it fluctuates [11] a lot between one year and another depending on precipitation.

C. Hasbani: This one originates in Lebanon (south Lebanon). Its average annual flow is almost equal to the Banias (around 125 Mm^3) [7] and also fluctuates a lot [11]. It is the longest of the three. Two smaller tributaries form the Hasbani: Hasbayya and Al-Wazzani.

The three tributaries join each other 6 km south of Mount Hermon, in a place north of Lake Al-Huleh to form the main stream of the Upper Jordan Jordan.

Lake Al-Huleh was a small lake located north of Lake Tiberias. It does not exist any more, because it was dried up by the Israeli's in the 1950's.

After forming its main stream, the Upper Jordan receives additional water quantities of about 360 Mm^3/yr. [2, 11 based on Israeli World bank study] from tributaries (mainly wadis) discharging in it and in Lake Tiberias from both sides. Finally this part of the Jordan River discharges in Lake Tiberias.

1.4.4 Lake Tiberias (Sea Genezareth, Sea of Galilee)

This lake has a pear-like shape and it is the main fresh water body inside historical Palestine. It works as a natural storage reservoir for the water of the Upper Jordan.

The lake is located in northern Israel. It has an area of about 166 km^2 [2] and storage capacity of around 4000 Mm^3 [2], that means an average storage depth of around 25 m.

The level of the lake keeps decreasing. Its original elevation was about 210 m bmsl [10], the current is about 214 m bmsl. Lake Tiberias is shown in figure 1.4.1.

The total annual water flow into the lake (after evaporation) is around 660 Mm^3 [2, 11 based on Israeli World Bank study]. This quantity is variable and depends a lot on the precipitation.

Israel uses almost all these water quantities. Only around 70 Mm^3/yr. of water is released from the lake into the Main (Lower) Jordan. This water is saline and of bad quality.

The main reason of salinity are the saline springs and wadis which pass first through the beds of salty ancient seas before reaching Lake Tiberias and Lower Jordan. Other reasons are evaporation and the return water from irrigation (drainage water).

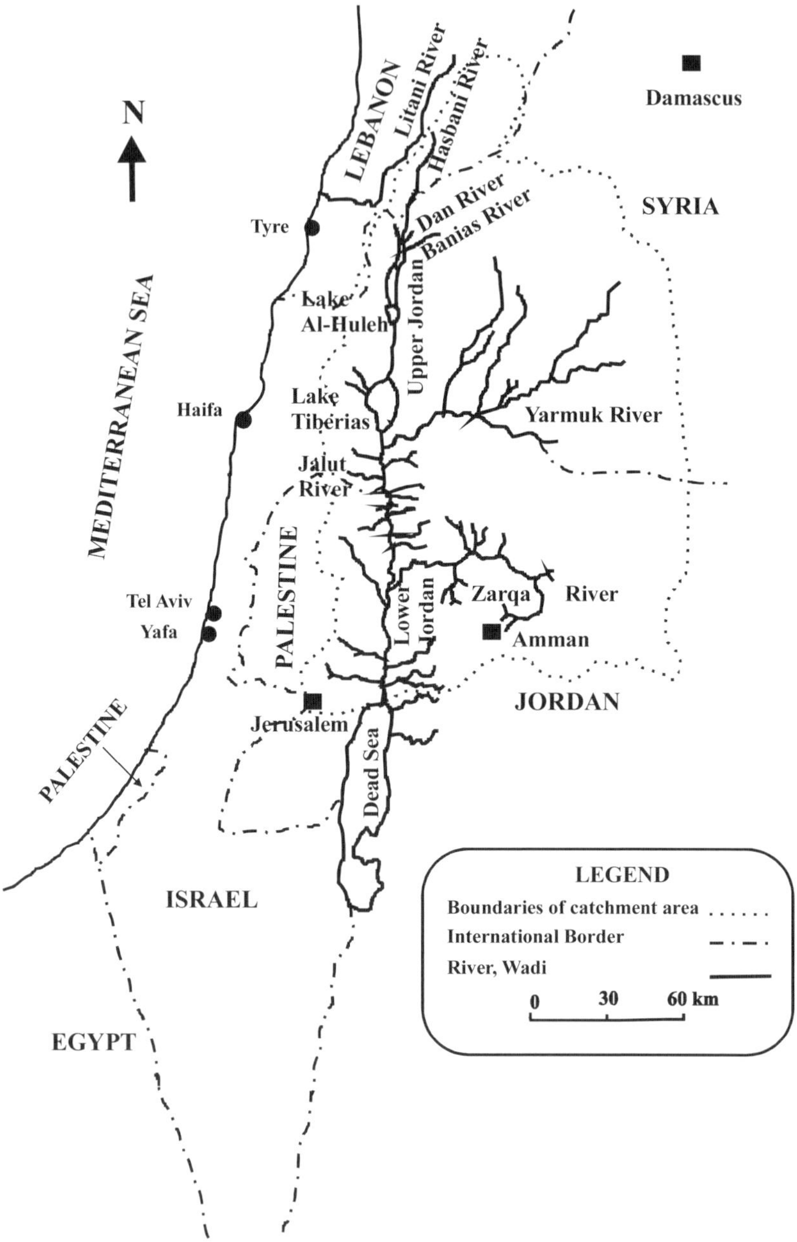

Fig. 1.4.1. The Jordan River System, modified after [4]

Israel diverts most of the lake's water through its National Water Carrier, which was built shortly after the establishment of the state of Israel in 1948. At that time the Israelis wanted to build a water pipeline to transfer the water from the country's water-rich north to the centre and the south. The construction of the Israeli National Water Carrier started in 1953 and was completed in 1964.

The Carrier diverts the water from the northern part of Lake Tiberias. It has a capacity of more than one Mm^3/day. In the 1980's it conveyed 420-450 Mm^3/yr. [7]. The remaining water of the lake, which is not diverted through the carrier, is used in areas nearby the lake for different purposes, mainly agricultural.

The main aim of the carrier is to transport the water from Lake Tiberias to the northern part of An-Naqab (Negev) desert in the south. The water is to be used there for irrigation purposes. The carrier aims also at supplying other parts of Israel with water, mainly the densely populated central coastal plain.

1.4.5 Lower (Main) Jordan

The part of the Jordan River between Lake Tiberias and its final discharge in the Dead Sea is known as the Lower (Main) Jordan.

After the Jordan River leaves Lake Tiberias, it receives its largest and most important tributary, the Yarmuk River, and it joins the Jordan 10 km south of the Lake.

Later on, and through its way towards the final destination in the Dead Sea, the river receives additional water quantities of about 245 Mm^3/yr. [1]. These quantities come from its tributaries (wadis) on both sides, but mainly from the east. The most important tributaries here are:

A. Zarqa River: Comes from the eastern side. It is around 75 km long, and has an average annual flow of 95 Mm^3/yr. [7]. Jordan constructed the King Talal Dam on its stream, with a storage capacity of 78 Mm^3.

B. Jalut River: This one comes from the west and it is smaller than the Zarqa River. It originates from Marj Bin Amer (Esdaralon valley), Bisan Plain and the Mountains of Jalboon. Most of its catchment is in Israel with a small part located in Jenin Governorate in Palestine.

Out of these 245 Mm^3/yr., Jordan uses 80-100 Mm^3/yr. Israel uses the water of the wadis on its side, estimated at 20-30 Mm^3/yr. Palestine uses nothing.

The extensive usage of the water of Upper Jordan and the tributaries of Lower Jordan turned the Lower Jordan into a small stream of unclean water, a little bit better than a ditch of wastewater [7].

Finally the Jordan River discharges in the Dead Sea, carrying with it only 200-300 Mm^3/yr. of very salty, bad quality water.

The sharp reduction in the water quantities reaching the Dead Sea caused very serious ecological damage to this very special phenomenon on earth.

1.4.6 Yarmuk River

The Yarmuk River is the largest and most important tributary of the Jordan. It originates mainly from the Hoaran Plain, mostly in southern Syria but also in northern Jordan. Most of it and its tributaries are located inside Syria. Later on it forms the Syrian-Jordanian border for a distance of around 40 km. Finally it forms the Jordanian-Israeli border for few km before joining the Jordan River.

The total length of the Yarmuk is around 75 km. Its average annual flow is 400-500 Mm^3 [7], this quantity is variable and depends on the annual rainfall.

Jordan and Syria have signed several treaties regarding the usage of the Yarmuk water in the years 1953, 1977, 1987 and 1998. These treaties included several items, the most important of them was; probably; the decision to establish Al-Maqaren or Al-Wehda Dam. This dam was not constructed because of the Israeli objection.

It is believed that Jordan and Syria use around 300 Mm^3/yr. of the Yarmuk water. Israel uses around 85 Mm^3/yr. Palestine uses nothing. Only 60-70 Mm^3/yr. of very salty, bad quality water flow from the Yarmuk into Lower Jordan [2].

1.4.7 Dead Sea

The Dead Sea is located in the centre of the Jordan Valley, which is the lowest and most well-known part of the Syrian-African Rift extending from southern Turkey in the north to the south of the African continent.

South of the Dead Sea Wadi Araba runs towards the Gulf of Aqaba via the watershed between the Dead Sea and the Red Sea.

The Dead Sea is a unique phenomenon. It is the lowest point on earth with a current elevation of around 414 m bmsl. Its original area was around 1010 km^2.

The water of the Dead Sea is very salty. The salt content is around 250,000 ppm (parts per million), i.e. 25%. This is more than 7 times the salinity of the ocean (around 35,000 ppm) and about 70% of saturation.

During the last few decades the sea witnessed very serious negative effects, resulting mainly from the sharp decrease in the water quantities reaching it. Some of the negative effects and dangers facing the Dead Sea are:

A. At this time it receives only about 1/3 of the original water quantities. The other 2/3 used to come from the Jordan River, but not any more. This 1/3 of the original discharge comes either from the wadis discharging directly in the Sea or from the little quantities still flowing in the Lower Jordan. Today the runoff of the Lower Jordan River comes only to about 200-300 Mm^3/yr.[7]. This represents only about 15-25% of the total discharge without diversions by the riparians, mainly by Israel.

B. The reduction of water quantities causes a continuous sharp decline in its water level with an average decrease of about half a meter per year [7]. Its level decreased from 392 m bmsl in the early 1960's to 414 m bmsl in 2001 [2].

C. Another negative effect of the reduction of the water quantities is the sharp reduction in its area, which is shrinking constantly and is now 25% less than the original one [2]. Its southern part (the one below the Tongue "Lisan") does not exist any more. It dried up in the late 1970's [7], now there are only some salt swamps.
D. A third effect of this reduction is the sharp increase in salinity, which has lots of negative effects.

All of that resulted in a serious danger and damage to the unique ecological system in and around the Dead Sea.

The only positive effect of the higher salinity is a certain reduction of costs when exploiting the mineral resources from the Dead Sea, as do Jordan and Israel. The most common minerals there are Magnesium Chloride, Calcium Chloride and Sodium Chloride (common salt). The Dead Sea may represent the biggest Magnesium deposit in the world, considering the technically easy access and processing.

1.4.8 Shares in the Jordan River's Water

The water of the Jordan River is extremely important to the surrounding countries, which always had contradicting ambitions in it. Therefore, the distribution of this river's water and the allocation of a certain quantity for each riparian was always a matter of concern and importance.

Several plans were made throughout the 20th century to distribute the Jordan River's water between its 5 riparian countries: Palestine, Israel, Jordan, Syria and Lebanon. These plans are mentioned in several literatures. The Johnston Plan of 1955 is considered widely as the most important between them.

Johnston is an American, who came to the Middle East as an envoy of the former US president Eisenhower. He stayed in the area for 2 years (1953-1955). Finally he introduced his own plan, which is summarised in table 1.4.1 below.

Table 1.4.1. Distribution of the Jordan River's water between the riparian countries according to Johnston Plan.

Country	Share in Jordan River's water (Mm^3/yr.)
Jordan	720*
Israel	400
Syria	132
Lebanon	35
Total	1287

* The share of Palestine and Jordan was set together in Johnston Plan at 720 Mm^3/yr., because at that time the Palestinian West Bank was ruled by Jordan. The share of Palestine was estimated by some literature to be around 257 Mm^3/yr., some others estimated it to be 215 Mm^3/yr.

1.4.9 Current Usage of the River

The actual usage of the Jordan River is much different from the shares allocated by Johnston or other plans. See table 1.4.2.

Table 1.4.2. Present day usage of the Jordan River's water by its riparian countries.

Country	Usage from the Jordan River's water (Mm^3/yr.)
Israel	700
Palestine	0.0
Other Arabic countries: Jordan, Syria and Lebanon	410
Total	1110

Israel uses at least 300 Mm^3/yr. more than its share. Palestine uses nothing, because the Israeli restrictions do not allow the Palestinians to reach the Jordan River, thus preventing them from using its water or cultivating their land along it.

The Palestinians were using some of the Jordan River's water before 1967, but when the Israelis occupied the West Bank in that year they destroyed or confiscated all the 140 pump stations, which the Palestinians installed on the river to use some of its water.

Syria, Jordan and Lebanon use at present around 410 Mm^3/yr., from which Lebanon uses around 10 Mm^3/yr., all from the Hasbani. The remaining 400 Mm^3/yr. are used by Jordan and Syria, from which around 300 Mm^3/yr. are from the Yarmuk. All the water Syria uses comes from the Yarmuk and much of Jordan's usage come from there too, except around 80-100 Mm^3/yr. from the wadis on the eastern side of the Lower Jordan.

Referenes

[1] Abu Ju'ub Gh (2002) The Jordan River System is in Danger. In: Zwischen impermeabel und permeabel. Heft *83* Lehrstuhl fuer Ingenieurgeologie und Hydrogeologie-Aachen University of Technology (RWTH) Aachen

[2] Abu Ju'ub Gh (2002) Water Resources in Jenin Governorate / North of the West Bank – Palestine, A Part of the Water Problem in the Middle East. Ph D thesis Heft *83* Lehrstuhl fuer Ingenieurgeologie und Hydrogeologie-Aachen University of Technology (RWTH) Aachen

[3] Abu Ju'ub Gh, Nashashibi M, Da'as A, Bjerved L and Hammad T (1998) Regional Plan for the West Bank Governorates-Water and Wastewater Existing Situation. Palestinian Ministry of Planning and International Cooperation (MOPIC) Ramallah

[4] El-Musa Sh (1997) Water Conflict. Institute for Palestine Studies Washington DC

[5] El-Musa Sh (1996) Negotiating Water: Israel and the Palestinians. Institute for Palestine Studies Washington DC

[6] FAO (Food and Agriculture Organization of the United Nations). Country profiles-Information about the water sector in the world's countries, and in the Near East region. Internet sites:
http://www.fao.org/waicent/faoinfo/agricult/agl/aglw/aquastatweb/countries/index.htm, and:
http://www.fao.org/waicent/faoinfo/agricult/agl/aglw/aquastatweb/regions/neast/index.htm

[7] Libiszewski S (1995) Water Disputes in the Jordan Basin Region and Their Role in the Resolution of the Arab-Israeli Conflict. ENCOP Occasional Paper No 13 ETH Zurich

[8] Murakami M (1995) Managing Water for Peace in the Middle East: Alternative Strategies. The United Nations University Hong Kong

[9] Schetelig K. and Abu Ju'ub Gh (2001) Konfliktpotential Wasser in Nahost, insbesondere in Israel und Palaestina (in German). In: Nahost Konfliktpotential Wasser (Seminar). Aachen

[10] Van Edig A (1999) Aspects of Palestinian Water Rights. Ramallah Center for Human Rights Studies Ramallah

[11] Water and Environmental Studies Center (WESC)/An-Najah National University (1995) Middle East Regional Study on Water Supply and Demand Development. German Agency for Technical Cooperation (GTZ) Nablus

1.5 Hydrology and Management of Lake Kinneret Aimed at Water Quality Protection

Moshe Gophen

MIGAL-Galilee Technological Center
POB 831 Kiryat Shmone (11016) Israel

Abstract

Long term (1969-2001) data record of hydrological parameters, nutrient and plankton in Lake Kinneret, were statistically analyzed. Trophic relations between food web compartments were quantitatively considered to evaluate ecological forces. Monthly data of inflow discharges and lake volume were used to calculate residence time values in relation to rainfall and water level. The seasonal fluctuations of the hydrological parameters, nutrients, and plankton inventories represent typical subtropical pattern: high level in winter and low in summer months. It was found that nitrogen inventories in the lake declined and the biomass of grazable phytoplankton was enhanced since early 1980's. Dissolved phosphorus was decreased mostly in summer months when the lake is nutrient limited, as a result of phytoplankton uptake. Zooplankton was declined until 1993 and increased later. The most abundant zooplanktivorous fish, Lavnun (Bleak, *Acanthobrama spp.*) densities increased during 1993-95 and subsidized fishery caused zooplankton enhancement and consequently suppression of primary production. Lavnun removal might be efficient to enhance zooplankton grazing capacity and algal suppression if external phosphorus loads from the drainage basin are reduced. The major effect on nutrient inventories in the lake is due to inputs from the drainage basin.Consequently, management implications aimed at lake water quality protection should be mostly directed towards nutrient removal in the drainage basin and loads reduction by pumping water for supply. The water storage capacity in the lake is significant to ensure maximum utilization.

1.5.1 Introduction

Kinneret, the only natural freshwater lake in Israel, supplies ~30% of the national water demands and ~55% of drinking water requirements. The drainage basin area of Kinneret is 2730 km^2 and is located mostly northern to the lake (fig. 1.5.1). The total surface area of the geographically defined as „Hula Valley" is about 200 km^2. The Hula valley's altitude in its northern part is between 150-170 m and in the southern part of the valley 61-65 above sea level. From Hula valley, River Jordan flow southward 15 km, from +60 m to 214 mbsl into Lake Kinneret.

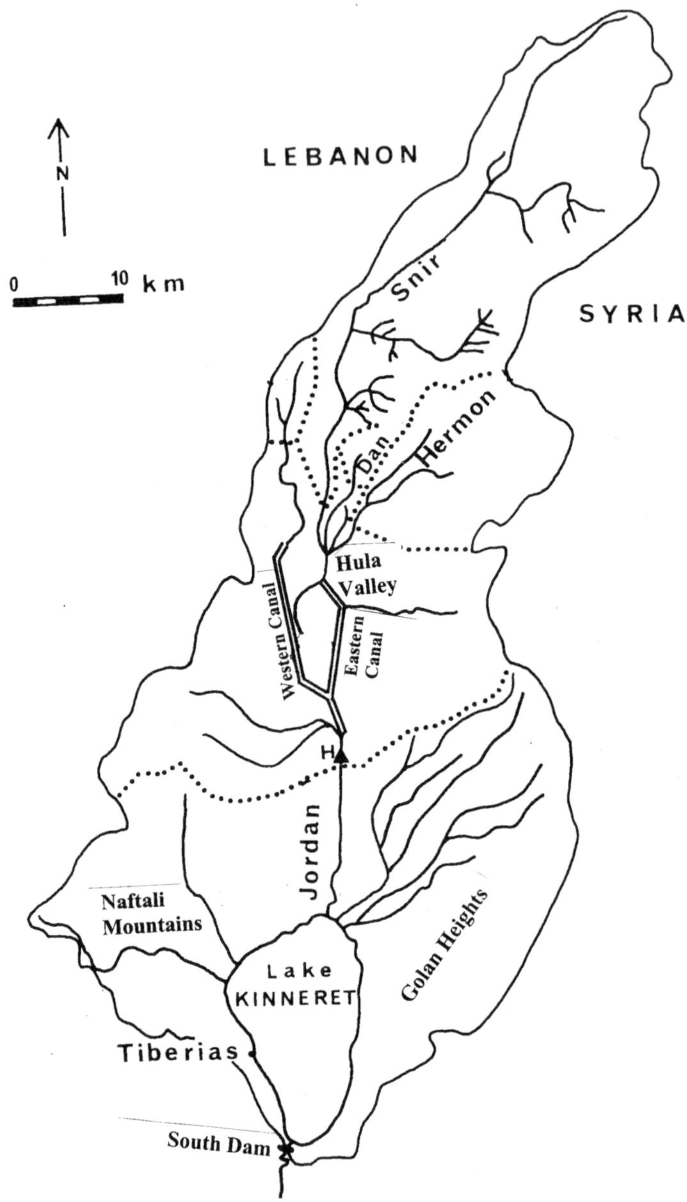

Fig. 1.5.1. Map of the Lake Kinneret drainage basin with indication of the hydrological subunits (doted lines), northern headwaters and other rivers, geographical regions, and Huri Bridge Station on Jordan River (H).Modified from Mero (1978).

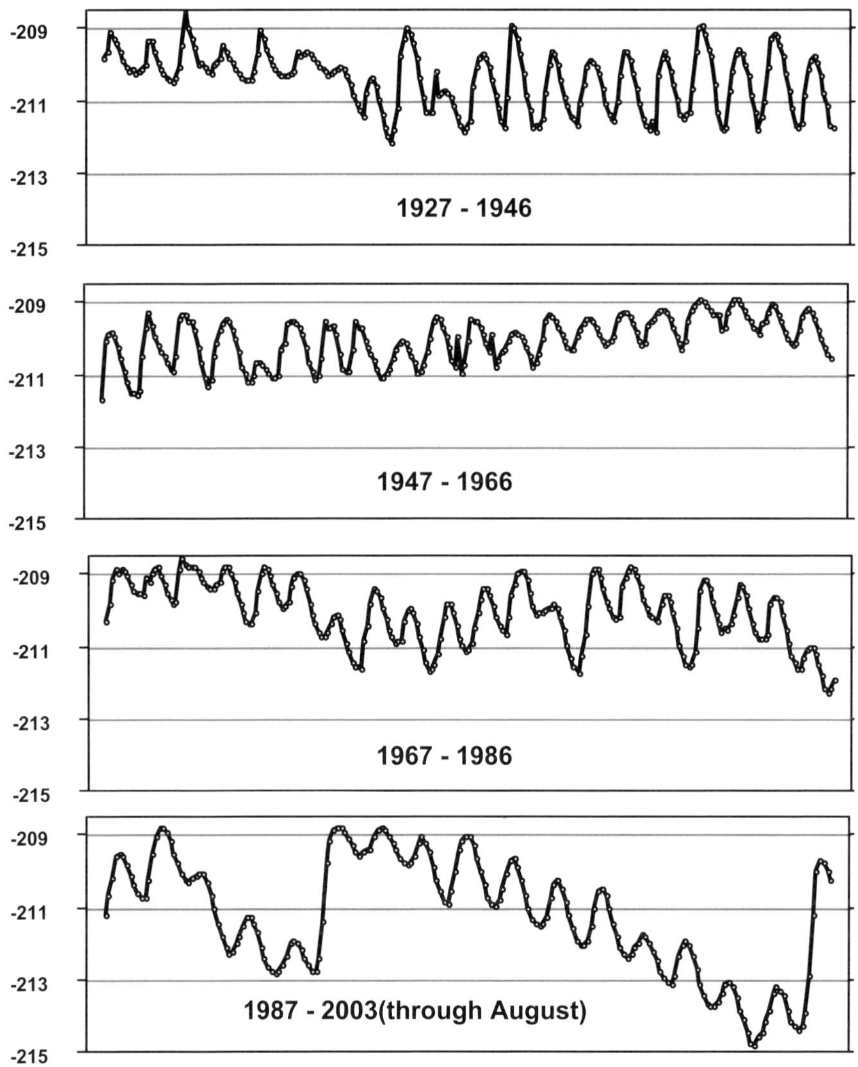

Fig. 1.5.2. Monthly averages of Kinneret Water Level during 1927-2003 (through August). Data source-Hydrological service of the state of Israel.

Lake Kinneret is stratified from May to mid December (anoxic hypolimnion) and totally mixed during Mid December through April. During the last 60 years the Kinneret ecosystem has undergone natural and man-made modifications and high fluctuations of hydrological parameters: construction of the south Dam (1932/3); drying of Hula Lake and swamps (1950's); salty springs diversion (1964); construction of the National Water Carrier (NWC) (operation -1964); implementation of the Hula Project (1994-1998); water level fluctuations between 208.57 and

214.87 mbsl Figs. 1.5.2 & 1.5.3); high inflow discharges ($>10^9 m^3$ per annum) and droughts ($<260 \times 10^6$ m^3 per annum); low (333 mm/y) and high (1060 mm/y) rainfal in the drainage basin (fig. 1.5.4); low (10-50 g/m^2) and high (170 – 412 g/m^2) monthly averages of the biomass *Peridinium spp* during its blooming season; decline of zooplankton biomass during 1969 – 1993 and increase afterwards; low (1983, 1992) and high (1970, 1979, 1999) epilimnetic temperature. Nevertheless, the pattern of seasonal distribution of hydrological, chemical and biological parameters cosistently represented subtropical climate conditions of the Kinneret region: high levels in winter and low in summer months (Fig. 1.5.5) but high hypolimnetic inventories of dissolved phosphorus, ammonium, sulfides and CO_2 in summer – fall period as a result of the thermal and chemical stratification. The lake ecosystem maintain a high level of multiannual stability and did not deviate from resilient fluctuations (Gophen 2000). During 1969-2003, hydrological parameters indicates a high amplitude of changes but consistent seasonal pattern (Fig. 1.5.5). Monthly means of the hydrological parameters are: lake volume- 4117 mcm (= $10^6 m^3$), inflow discharges- 62.6 mcm/month, and rainfall- 70 mm/month. Monthly value of residence time ranged between 0.8 – 51.6 years (fig. 1.5.6).

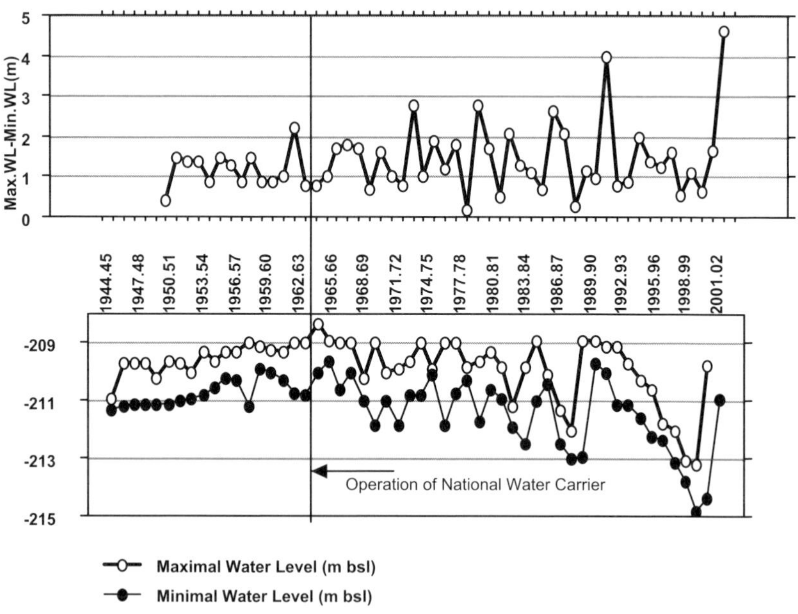

Fig. 1.5.3. Maxima minus minima of water level (upper panel) during 1950/51-2001/02 maxima and minima of water level in Lake Kinneret during 1944/45-2001/02 (lower panel). Operation of the National Water Carrier is indicated. Data source-Hydrological service of the state of Israel.

Table 1.5.1. Lake Kinneret Water Balance (10^6 m^3 /year = mcm/year): averages (1969 – 2002) for droughts (1972/73, 1978/79, 1985/86, 1988 – 1991, 1998 – 2002) and all other years. Data source Hydrological Service, State of Israel (1969-2002).

Source	Droughts	Other Years
Northern to Huri Bridge		
Dan River	193	271
Snir (Hatzbani) River	42	128
Hermon (Banias) River	59	123
All other rivers	96	132
Total	390	654
Consumption Northern to Huri Bridge	110	130
unmeasured losses Northern to Huri Bridge	38	14
Huri Bridge Discharges	242	510
Southern to Huri Bridge		
Golan Heghts	77	145
Yarmuch Diversion	17	25
Eastern and Lower Galilee	56	75
Sublacustrine and direct Rain	56	75
Total	206	320
Total Lake Kinneret Inflows		
Winter	260	530
Summer	188	300
Total	448	830
Total Lake Kinneret Losses		
Dam Overflow	19	143
Evaporation	280	280
NWC*	244[#]	345
Local consumption	71[#]	62
Total	614[#]	830

* National Water Carrier

[#] Punping was 166 mcm/y above natural recharging with consequent water level decline.

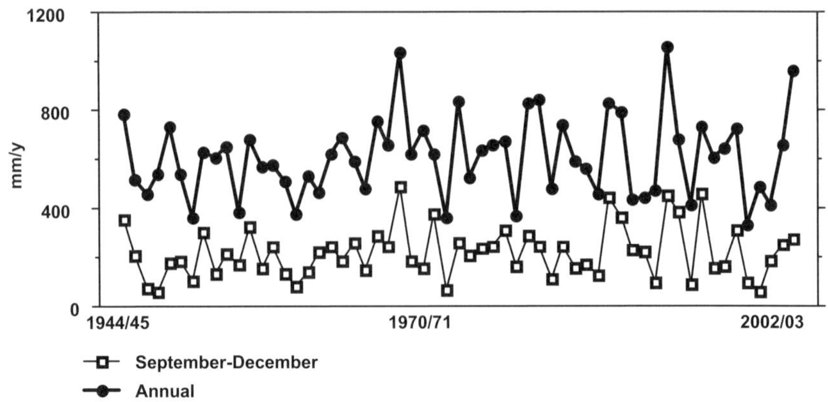

Fig. 1.5.4. Annual (Septaember-June)(upper line, full cycles) and periodical (September-December, empty squares)rainfall (mm) in Dafna(middle part of Kinneret drainage basin) during 1944/45-2002/03. Data source-Moti Peres.

The lake is exploited for its fishing by ca 200 licensed fishermen which remove commercially an average of 1832 ton of fish (108 kg/ha) per annum (Table 1.5.1). The zooplanktivorous Lavnun (Bleak, *Acanthobrama spp.*) comprised 40-60% by weight of total catches and >50% of the stock biomass (Walline et al. 1993). Only 8 species out of 24 recorded in the lake (Ben-Tuvia 1978) are commercially fished.

Fishing policy is aimed at both water quality protection and fish production (Gophen 2003). As a result of optimal reproduction condition during the winter of 1991/92 the lake was occupied by a dense population of sub-commercial size Lavnun resulted in elimination of commercial fishing. Consequently, zooplankton biomass and community grazing capacity were declined to the lowest level ever recorded .A recommendation to subsidize removal of sub-commercial sized bleaks was implemented during 8 consecutive years (1995-2002) and about 4000 tons of small bodied Lavnun were removed (ca 400-500 t/y) and buried (Gophen 2002). Three years later, proportion of marketable sized (>13 cm TL) individuals in the population was enhanced and the commercial fishery of lavnun was renewed. The fishery management program is aimed at reduction of fish predation pressure on zooplankton to enhance nanophytoplankton grazing capacity and to intensify grazing of *Peridinium* by fish together with ensuring fair income to fishermen.Therefore the management program include enhancement of the native and valuable species, *Sarotherodon galilaeus* by stocking of small (>5g/ind) fingerlings, and protection of spawners in breeding grounds during reproduction season and elimination of Silver Carp and *Oreochromis aureus* from stocking program (Gophen 2003).

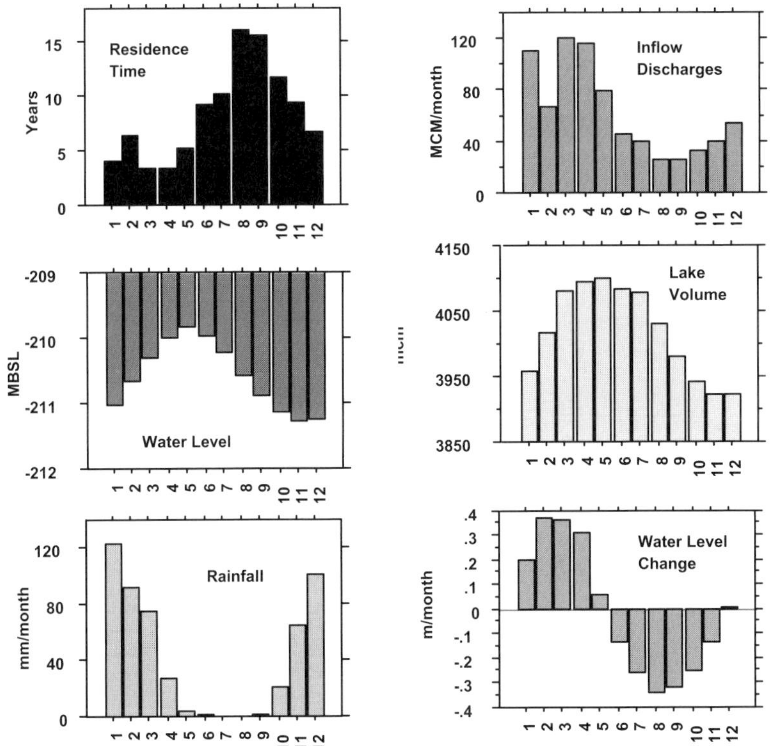

Fig. 1.5.5. Monthly (1969 – 2001) means of Residence Time (in years), inflow discharges, (mcm/month), (10^6 m^3=mcm), lake water level (mbsl altitude), and water level change (m/month) in Lake Kinneret and rainfall (Hula Valley) (mm/month).
Data source-Hydrological service of the state of Israel.

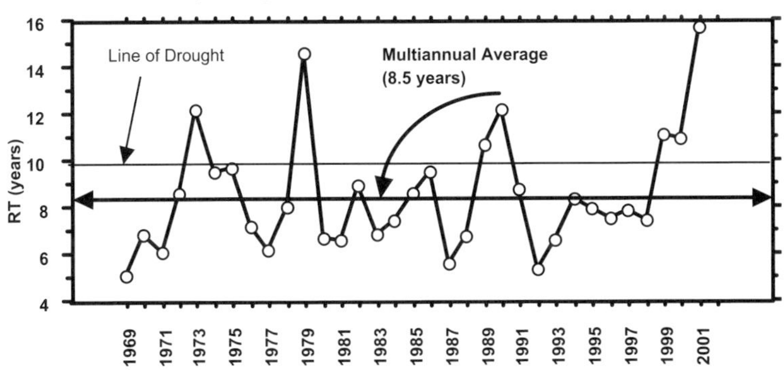

Fig. 1.5.6. Annual (1969 – 2001) averages of monthly values of Residence Time (in years) in Lake Kinneret. Drought and mean levels are indicated.Data source-Hydrological service of the State of Israel.

1.5.2 Methods

Sampling and analysis of zooplankton, phytoplankton, chlorophyll, primary production, and nutrients were carried out routinely (1969 – 2001), and methods are given in previous reports and Gophen et al. (1999). Limnological data on Lake Kinneret was implied from Lake Kinneret Data Base(LKDB 1969-2002) (Schlichter manager): phytoplankton- Pollingher and Zohary; zooplankton - Gophen; primary production - Yacobi and Berman; nutrients - Nishri. Data sources of river discharges, rainfall, pumping rate from Kinneret, water consumption in the upper Galilee, and evaporation values, were taken from annual reports of the Hydrological and Meteorological Service, state of Israel, M. Meron, Y. Tsipris and M. Peres (unpublished data).

1.5.3 Results and Discussion

Water Balance (Table 1.5.1, Figure 1.5.2)

The drainage basin of Lake Kinneret comprised of 4 hydrological subunits (Fig. 1.5.1).The northern headwaters of Lake Kinneret are three major rivers, Hatzbani (Snir), Dan, and Banias (Hermon River) and several smaller rivers (among others, Orvim, Yardenon, Ayun). The three major headwaters combine into one Jordan River which also collect discharges from small rivers on both sides of the Hula Valley and split into two man made canals (western and eastern canals) which cross the Hula Valley. At the south end of the Hula Valley these two canals join together into one Jordan River (Fig. 1.5.1). The relative (%) contribution of these headwaters within the northern sources are: Dan-41%, Hatzbani-20% (consumption by Lebanon is excluded), Banias-19%, and all others-20%. The northern headwaters contribute 61% and 39% of the total Kinneret budget in normal and dry year respectively. During droughts those headwaters contribution to the northern sources are: Dan-49%, Hatzbani-11%, Banias-15% and all others-25%. Results in Table 1.5.1 indicate a significant difference between droughts and normal years. During droughts, about 10% of measured discharges from northern sources, for unknown reason, are naturally deleted from the measured discharge at the Huri bridge station which is located on the south end of the Hula Valley on Jordan River. The other sources of the Kinneret total budget are the followings: Runoffs through three major rivers from the Golan Heights (17%), diversion from Yarmuch River (located southern to Kinneret) (3%), runoffs through two major rivers from the eastern slops of lower Galilee (Naftali Mountains (9%) and from sub-lacustrine fluxes and direct rainfall (9%). Inflows during winter months comprised 64% and during summer 36% in normal years. It should be noted that Israel supply 50 mcm Kinneret water per annum to the State of Jordan through the south dam.

Data in Table 1.5.1 indicate 35-40% reduction of river discharges during droughts. Evaporation rate is unchanged in dry and normal seasons of which 33% evaporate in winter months and 67% in the summer period.

The rate of pumping through the NWC is limited by two factors: 1) Water availability (as dictated by WL altitude)and 2) energy price. The second factor constrain limit the pumping to night time when electricity is cheape. The consideration of WL and water availability factor is more complicated. If water budget is balanced (inputs similar to outputs) the level is not changed. If pumping exceed inputs, the WL decline. The water demands comprised of agricultural, drinking and coastal aquifer recharging. Of the three channels of requirements, the drinking water supply is fixed and supply to agriculture and recharging under drought regime is reduced. Nevertheless if droughts continue more than one or two years the Kinneret budget is negative and water level decline.

The main operations to control WL are an open south dam to prevent WL increase above the upper formulated redline and close dam together with pumping (NWC and local consumers) reduction to prevent WL lowering below the redline when the total balance is negative. The uppermost and lowermost redlines are formulated by the water commission of Israel and it is a formal instruction in charge. The upper line to was instructed at 208.9 mbsl in early seventies to prevent damage to constructions located close to the shoreline at the time of instruction and was accompanied by restriction of constructing closer that 50 m to the shoreline when WL altitude is 208.90 mbsl. The lower limit was changed four times during the last 25 years from 212 to (presently) 215.5 mbsl. These changes were the outcome of two major factors: 1) droughts and water shortage, and 2) significant improvement of our limnological knowledge of the Kinneret ecosystem. Nonetheless, no deterioration of Kinneret water quality as a result of lowering WL below 213 and later at below 214.5 mbsl was indicated. Results in Fig. 1.5.3 represent the impact of NCW operational regime on WL fluctuations in Lake Kinneret.

Residence Time (RT) effects (fig. 1.5.6)

The hydrological parameters (RT values, lake volume, lake water level, inflow discharge, and rainfall), represent a seasonal pattern which is typical to the subtropical region (Fig. 1.5.5). Rainy season is from September through May: 36% during September-December and 64% during January- May (figs. 1.5.4 & 1.5.5). Major nutrient inventories (such as TP, TN) in the lake are mostly affected by internal and external inputs. As a result of the close relations that were indicated between inflow discharges and nutrient inventories in the lake (Gophen 2002; 2003) it is suggested that the external inputs are dominant factor. Inflow discharges predominantly affect the nutrient inventories and consequently on algal production and water quality: higher inflow is accompanied by shorter RT and greater nutrient inventory. In winter, high inflow discharges were measured, lake volume is high and RT values are short (fig. 1.5.5). In summer, discharges are low, the lake volume is low, and RT values are therefore longer. The shorter RT in winter coincide with maximal water and nutrient inputs and the opposite in summer months. The long term record (1969-2002) clearly show increase of concentrations of *Peridinium*, and zooplankton biomass, TP, TN, TIN, and epilimnetic DO concentrations when RT become shorter in winter months (Gophen 2003). On the other hand cyanophytes biomass, % of nanoplankton, and % of zooplankton predation relative to herbivory declined when RT become shorter . The decline of nutrient and algal inventories when RT became longer was documented. RT values did not indicate a clear multiannual (1969-2001) trend in spite of consistent lowering of WL during 1992-2002 and consequently lake volume (figs. 1.5.2, 1.5.6). Lowest water levels (WL) coincide with shortest RT and decline of epilimnetic nutrients. WL was low during droughts and high in years with heavy rainfall (figs. 1.5.2 & 1.5.3). Rainfall, inflow discharges and nutrient (TN, TP) inputs in winter 2002/3 (December-May) were exceptionally high: 875 mm (average 517 mm), 900 mcm (average 550 mcm), 2678 t (annual average 1584 t), 225t (annual average 120 t), respectively and *Peridinium* reappeared in high biomass. During 5 months in winter 2003 water level increased from 214.42 to 209.72 mbsl (fig. 1.5.2) (4.7 m = 790 mcm storage) and RT values were exceptionally low. Because of the major effect of external inputs and to a lesser extend the internal processes on nutrient loads in Lake Kinneret the recommended management, aimed at lake water quality protection, should be mostly directed towards nutrient removal in the drainage basin and/or loads reduction by pumping water, with nutrient, for supply with no apprehensiveness of water quality deterioration as a result of low WL. No negative effect on water quality was indicated when WL was exceptionally low during 1927-2002 (fig. 1.5.2).

Jordan By-pass Diversion (JBPD)

The program known as "JBPD" was discussed in two terms: in early 1980's and in early 1990's . In both cases the project was rejected as a result of prediction of unacceptable low water level and shortning of residense time. It was assumed that these two factors might have a negative impact on water quality. Presently we have much more information on the lake ecosystem response to low water level and shorter RT. The data set of biological, chemical and physical parameters indicates no prominent negative effect on water quality under low (214.87 mbsl was the lowest experienced) WL regime. The long term RT data set analysis prominently show the seasonal pattern which is mostly related to climatological conditions in this subtropical region: heavy rainfall, high inflow discharges, high nutrient levels and plankton densities in winter months and the opposite in summer. It was documented that decline of inflow discharges was followed by improvement of water quality. Therefore it is suggested that partial removal of Jordan River water (with nutrient included) northern to the lake might reduce nutrient inputs to the lake and improve water quality. The diversion of Jordan waters northern to the lake would have a beneficial effect on hydrological management of the lake in case of floody seasons . Under such circumstances water level is increasing and storage capacity become smaller. The by-pass system divert part of the inflow discharges and storage capacity increase with no interference to to the routine operation of the NWC. JBPD's waters are proposed to be accumulated in an operational reservoir and fluxed into the NWC 's network during day time when NWC is not operated due to eapensive energy price. The low salinity of the Jordan waters (20 ppm Cl^-) are preferred for reclamation of the coastal aquifers by recharging. The space between the upper and the lower WL redlines (215.50 - 208.90 mbsl) is 6.6 m which is a volume of $1.1 x 10^9$ m^3. Such a volume is giving the water managers significant security and reliability of drinking water supply.

Results of linear computations of storage capacities and WL in Lake Kinneret under three simulations are presented in table 1.5.2. Results in table. 1.5.2 indicate the followings: Under inflow regime of 1200 or 1000 mcm/y a quantity of 386 or 186 mcm/y can not be consumed and must be released through the south dam in order not to increase WL above 208.9 mbsl and the operation of JBPD ensure full consumption. Under normal inflow discharges (600, 700, 800 mcm/y) and consumption (400 mcm/y) WL will be lowered to 211.40 mbsl and with JBPD operation to 212.6 mbsl. Under drought regime (inflows of 400 or 500 mcm/y) and normal consumption (400 mcm/y) , restricted pumping (200 mcm/y) or JBPD operation the WL will be declined to 212.6, 211.4, and 213.8 mbsl respectively.

Table 1.5.2. Commercial fisheries in Lake Kinneret (averaged for 1988-1999) (SD): annual landings(t) and market value in 10^3 \$USD (%) Data source-Sandovski and Shapira (1999)

Species(common name)	Annual landings		Market Value	
	ton	%	10^3 US-\$	(%)
Sarotherodon galilaeus (Galilee St. Peter's Fish)	377 (110)	21	1750	48
Oreochromis aureus (Jordan St. Peter's Fish)	124 (63)	7	345	9
Tristramella.spp (Tritram's St. Peter's Fish)	58 (26)	3	87	2
Mugil spp. (Gray Mullet)	162 (79)	9	825	23
Barbus spp. (Barbel)	39 (23)	2	82	2
Hypophthalmichthys molitrix (Silver Carp)	64 (50)	3	32	1
Cyprinus carpio (Common carp)	83 (54)	5	242	7
Acanthobrama spp. (Bleak)	924 (314)	50	277	8
Total:	1832 (362)	100	3640	100

All those WL altitudes were documented where no negative changes due to the low WL were indicated. Nature gave us two cases of heavy rainfall seasons after consecutive three years of drought: 1991/2 after 1989-1991 and 2002/03 fter 1998-2002. In those two cases the WL came to the lowest altitude ever recorded at that time: 213 mbsl in 1991 and 214.87 mbsl in 2001. In the first case 521.3 mcm of water were not consumed and released through the south dam during 1991/92 and 1992/93 but in the second case about 800 mcm were stored in the lake during 5 months (2003) with no release through the south dam and the WL increased up to 0.82 mbsl underneath the upper redline. The two cases differ from each other by the information we had on the lake response to low WL.

Table 1.5.3. Annual storage capacities (mcm/y) and WL change after subtraction of water consumption (pumping by NWC and local consumers), evaporation (280 mc/y). JBP diversion considered as 200 mcm/y. WL starting point is the level predicted for the lowest altitude in the present season (210.9 mbsl). Simulations of three water utilization are presented: 1) 200mcm/y-pumping; 2) 400 mcm/y-pumping, and 3) 400 pumping plus 200 JBPD. Seven inflow discharges between 400 – 1200 mcm/y are simulated. The volume of 100 cm WL is equal to 168 mcm.

Inflow(mcm/y)	Lake storage (+= gain, -= loss) under three simulated consumptions: 200mcm/y pumping,, 400 mcm/y pumping and 600 mcm/y (=400mcm/y+200mcmcm/y by JBPD). Water level changes are given (m) in parenthesis		
	200mcm/y	400mc/y	600 mcm/y
1200	+720(4.3)	+520(3.1)	+320(1.9)
1000	+520(3.1)	+320(1.9)	+120(0.7)
800	+320(1.9)	+120(0.7)	-80(0.5)
700	+220(1.3)	-80(0.56)	-280(1.7)
600	+120(0.1)	-80(0.5)	-280(1.7)
500	+20(0.1)	-180(1.1)	-380(2.3)
400	-80(0.5)	-280(1.7)	-480(2.9)

The limnological information that was collected during 1991-2001 backed the decision to lower WL in 2001 (214.87 mbsl) below the 1991 record (213 mbsl). Consequently, storage capacity increased from 690 mcm to 1100 mcm in which about 800 mcm of extra waters were stored during winter 2003. Nevertheless, the present WL is high (0.82 m below the upper redline) and there is an urgent need to increase storage capacity in the lake. It is a critical requirement of the governmental authorities to formulate an appropriate management design aimed at maximum utilization of the Kinneret water resources (to increase storage capacity) in order to eliminate release of wasted waters.

Food web manipulation

The nutrient impacts on algal growth in Lake Kinneret was found to be effective (Berman et al. 1995; Pollingher et al. 1988) as well as fish predation on zooplankton communities (Gophen 2003; Gophen et al. 1999). P-limitation for algal production in Lake Kinneret was documented among others by Serruya et al. (1980), Pollingher et al. (1988), and Berman et al. (1995).

The biomass of herbivorous zooplankton declined since 1970-1993 and significant increase occurred later (Gophen 2003). On the other hand, nanophytoplankton enhancement started in early 1980's. During the 1970's when zooplankton grazing capacity was high - nanophytoplankton biomass was low. Nanophytoplankton biomass was enhanced as a result of increasing levels of Part-P i.e. algal cells and their photosynthetic activity (PP) were accelerated. In spite of prosperity of edible algae during the 1980's, the biomass of hebivorous zooplankton declined until early 1990's with increasing values of cladoceran small/large ratios (Gophen 2002; 2003). It is suggested that those two parameters reflect an intensive predation pressure of fish on zooplankton. The increase of the ratio between small and large - bodied cladocerans was leveled off during the 1990's when Lavnun removal program was carried out (Gophen 2003). Lavnun removal enhanced zooplankton grazing capacity but did not reduce grazable algal stock biomass. Nevertheless, it is suggested that PP was suppressed and improvement of water quality took place. The efficiency of water quality improvement by Lavnun removal might be even higher if phosphorus inventory would be lower.

The period of 1998–2001 was a sequence of droughts: low discharges, decline of water level and lake volume, and high RT values. Nutrient inputs were approximatelly 50% lower then the multiannual averages and lake inventories of

P and N declined. On the other hand, zooplankton biomass chlorophytes, diatoms and cyanophytes were enhanced. It is suggested that the zooplankton enhancement was initiated by reduction of fish predation pressure (Lavnun removal) and recycled P by zooplankters enhanced phytoplankton which partly improved zooplankton production as a feed- back effect.

Lake Kinneret is a mesotrophic lake with eutrophic level during the *Peridinium* bloom (February-June) (Serruya et al. 1980). Lavnun removal was successfully operated resulted in an increase of fish body size and partial renewal of commercial fishery with enhancement of zooplankton grazing capacity.

1.5.4 Conclusions

Food web manipulation in Lake Kinneret included normal commercial removal of 950-1200 tons Lavnun per annum and additional subsidized landings of sub-commercial sized specimen as needed (400-500 t/annum during 1995-2003) aimed at enhancement of zooplankton grazing capacity. On the other hand zooplankton enhancement increases levels of P recycling. Therefore it is required to reduce external P loading from the drainage basin together with Lavnun removal.

The major effect on nutrient loads in Lake Kinneret is due to inputs from the drainage basin and to a lesser extend to internal processes. Consequently, management implications aimed at lake water quality protection should be mostly directed towards nutrient (mostly P) removal in the drainage basin and lake loads reduction by pumping water for supply and/or Jordan by-pass operation. If removal of external P loads is efficiently achieved, Lavnun removal might be beneficial.

If Lavnun removal will be sufficient, zooplankton enhancement might have positive impact on water quality by algal suppression. Zooplankton enhancement during 1994-2001 and not 10 years earlier is indicating two conclusions: 1) lake Kinneret is an ecosystem where biomanipulation can be successfully implemented; 2) zooplankton in Lake Kinneret is not food limited but top-down controlled by predators (predation limited).WL can be lowered below 214 mbsl to increase storage capacity in the lake which is significant to ensure maximum utilization.

Acknowledgements

I am grateful to colleges and technicians in the Kinneret Limnological Laboratory for assistance and collaboration during 34 years of Kinneret studies.

References

Ben-Tuvia, A., 1978. Fishes. In: Lake Kinneret Monographiae Biologicae (C. Serruya ed.) Junk publishers, The Hague. Vol.32; pp. 407-430.

Berman, T., L. Stone, Y.Z. Yacobi, B. Kaplan, M. Schlichter, A. Nishri & U. Pollingher. 1995. Primary production and phytoplankton in Lake Kinneret: A long-term record (1972-1993). Limnology and Oceanography, 40:1054-1076.

Gophen, M. 2000. Lake Kinneret (Israel) ecosystem:long-term instability or resiliency? Water, air and soil pollution, special section: Environmental challenges, (S. (Belkin Guest editor), Kluwer Academic Publisher.123/1-4. 323-345.

Gophen, M. 2002. The management of Lake Kinneret (Israel): Water supply, water quality and food web structure perspectives. P.roceedings (Volume 2) of The Water and Environmental Research ICWRER 2002 (G. H. Schmitz, ed.), Dresden, Germany. Vol. 2 pp.275 – 285.

Gophen, M. 2002. Hydrology and Management of Water Quality: the Case of Lake Kinneret (Israel). Proceedings of the Conference Residense Time in Lakes: Science, Management, Education P. Bruni and R. DeBernardi eds.). Bolsena, Italy, October, 2002. (in press).

Gophen, M., Walline, P., Ostrovsky, I., Azoulay, B. & Easton, J.. 1999 a. Water quality and Fisheries management in Lake Kinneret, (Israel). In: Theoretical Reservoir Ecology and its applications, (J.G. Tundisi,and M. Straskaba eds.). International Institute of Ecology,Brazilian Academy.

Hydrological Service of the State of Israel, 1969-2003 Annual Reports.

Pollingher, U., T. Berman, B. Kaplan, & D. Scharf, 1988. Lake Kinneret phytoplankton: response to N and P enrichments in experiments and in nature. Hydrobiologia, 166: 65-75.

Sandovsky, Z., & J Shapiro (eds.) 1999. The fisheries and aquaculture of Israel, in figures. The State of Israel, Minisrtry of Agriculture-Department of Fisheries. 54 p.

Serruya, C., M. Gophen, & U. Pollingher. 1980. Lake Kinneret: Carbon Flow Patterns and Ecosystem Management. Arch. Hydrobiol. 88(3):265-302.

Walline, P., S. Pizanty, M. Gophen and T. Berman. 1993. The Ecosystem of Lake Kinneret, Israel. ICES.C.M. 1990/L:39 p.1-8.

1.6 The Water Crisis in the E. Mediterranean – and Relation to Global Warming?

Pinhas Alpert

Dept. Geophysics & Planetary Sciences, Tel-Aviv University, ISRAEL

Abstract

There is an on-going debate in Israel on the current political background of the water crisis with Lebanon whether the water shortage in recent years in this region has to do with global warming. On the face of it the analyses of temperature as well as rainfall trends are *not* in agreement with global warming trends. It is the purpose of this Note to show that the observed changes can indeed be linked directly to global warming predictions.

1.6.1 The Temperature Paradox

Winter temperatures in Israel have decreased significantly in 1964-1994 [Ben-Gai et al., 1999] in contradiction to the dominant increasing trends in most world regions as predicted by global warming.

The summer temperatures significant increases are nearly canceled on an annual basis by the winter decreases. The summer results based on 40 stations with daily maximum and minimum temperatures in Israel, show in accordance with global warming prediction [IPCC, 2001] that the summer minimum and maximum temperatures have increased by 0.26 and 0.21 K/dec. The corresponding winter trends are –0.12 and –0.38 K/dec. Hence, in both seasons the diurnal temperature range decreases.

Also, the annual temperature amplitude increases yielding warmer summers and colder winters [Ben-Gai et al., 1999].

It should be noted that summer temperature increases fit the general trends over the Mediterranean (Xoplaki et al., 2003; Price et al., 1999). Summer lower-tropospheric temperatures fit the aforementioned findings. For instance, summer 850 mb temperature show warming trends of 0.13 K/dec (Saaroni et al, 2003, the NCEP reanalysis) and 0.12 K/dec (Ben-Gai et al. 2001a, for Israeli radiosonde). In addition extreme high summer 850 mb temperature become more frequent in recent years (Saaroni et al, 2003).

1.6.2 The Rainfall Paradoxes

As to rainfall, again mixed trends are found over the E. Mediterranean (EM, in brief) but the mixed trends are now on a spatial/geographic basis not temporal; central-south Israel show rainfall increases while in the neighboring regions such as Jordan, Turkey, Syria decreases clearly dominate [Steinberger and Gazit-Yaari, 1996].

A third feature which seems paradoxical is the increase in extreme daily rainfall over the Mediterranean particularly over the west and central in spite of significant decreases in the totals. This was explained by the change in the rainfall distributions [Alpert et al. 2002].

1.6.3 Discussion

In this note we wish to explain the seemingly contradicting messages deduced from the temperature and rainfall trends in this highly conflicted region. And to show that these results can indeed be traced to global warming changes and particularly to the North Atlantic Oscillation (NAO) and other tele-connections and the associated synoptic changes. Figure 1.6.1 shows the Mediterranean rainfall trends for 1958-1998 based on NCEP reanalyses. These are grid data based on model and observations but global rain gauges analysis in recent 89 y [Dai et al., 1997] as well as regional analyses [Piervitalli et al., 1998, Trigo et al., 2000] fit well to the integrated picture of Fig. 1.6.1.

Rainfall predictions for the global warming scenarios also show dominant decreasing trends over the Mediterranean [IPCC, 2001]. Focusing over the Mediterranean the averaged 19 global model prediction show steep decreases are predicted over the central/west Mediterranean while mixed trends over the E/SE Mediterranean [Jouni Raisanen, personal communication]. This predicted Mediterranean drying is in accordance with the observed recent trends (Fig.1.6.1) and can be partly explained by the increased NAO since the 1970's [Thompson and Wallace, 2001]. Since, increased NAO is associated with stronger southwesterly flow over W. Europe and a drier Mediterranean. The recent increasing NAO index was also linked to the global warming [Thompson and Wallace, 2001; Steven Feldstein, personal communication] and hence with the drier Mediterranean. Here, we wish to address the mixed E. Mediterranean response in temperature and rainfall in view of the above teleconnections? The following arguments are that the unique mixed response over the E. Mediterranean can be directly explained by the global warming above-mentioned characteristics. First, the NAO increase in winter is negatively correlated with temperature and positively with the pressure, over the E. Mediterranean (Israel), Fig. 1.6.2. It is not clearly correlated with rainfall over the EM. Hence, NAO increases (linked to global warming) can explain the lower winter temperatures over Israel. Colder winter air temperatures over the EM were earlier found to correlate with more rainfall [Striem, 1979] but this may not be so simple here since surface pressure also increases along with the NAO increase (Fig. 1.6.2c). Surface pressure increases suggest reduced number of E. Mediterranean (Cyprus) Lows and weaker westerly flows which are the main agents for EM rainfall. Indeed, our objective analysis of daily EM synoptic systems based on NCEP reanalysis for 1948-2000 suggest a drop in the annual number of Cyprus lows as well as a significant increase in the frequency of Red Sea Trough (RST) days, Fig. 1.6.3, Alpert et al. (2003). The RST is associated with easterly drier and frequently colder flow over the EM [Saaroni et al., 1998]. At the same time active RST synoptic situations may yield severe storms with moisture from the tropics [Krichak et al., 1998] particularly over the SE region of the Mediterranean. Hence, less rainfall in the north EM and potential increase of rainfall over the SE region, particularly heavy rainfall.

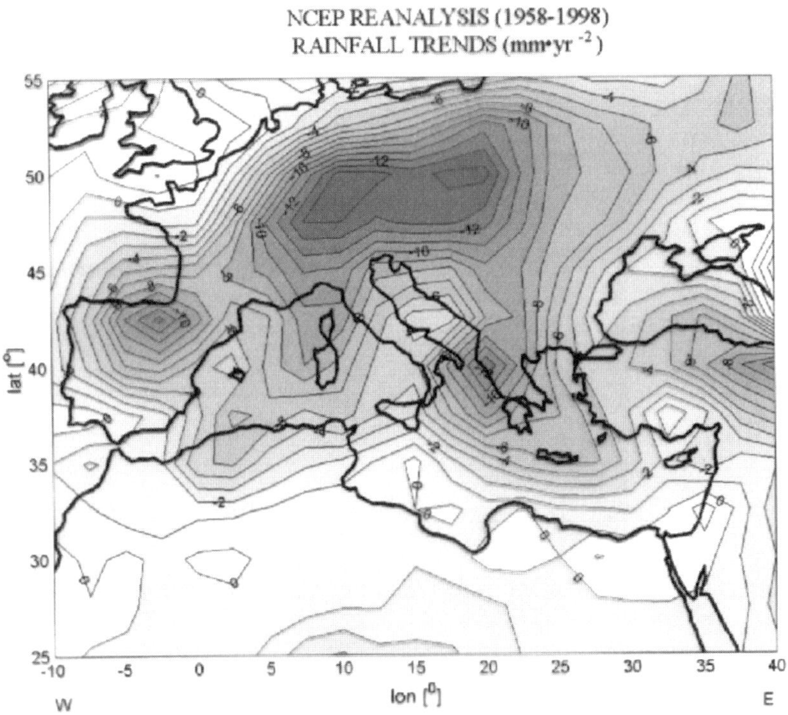

Fig. 1.6.1. Mediterranean rainfall trends for 1958-1998 based on NCEP reanalysis. Decreasing center of –12 mm/y over N. Italy is equivalent to 480 mm drop in annual rainfall over 40 y [see Piervitalli, 1998]. Additional decreasing centers are over Spain, Greece. These maxima are somewhat exaggerated compared to rain gauge data but agree well with the rain gauge analysis [Dai et al., 1997]. Over the SE and Israel some increases are noticed.

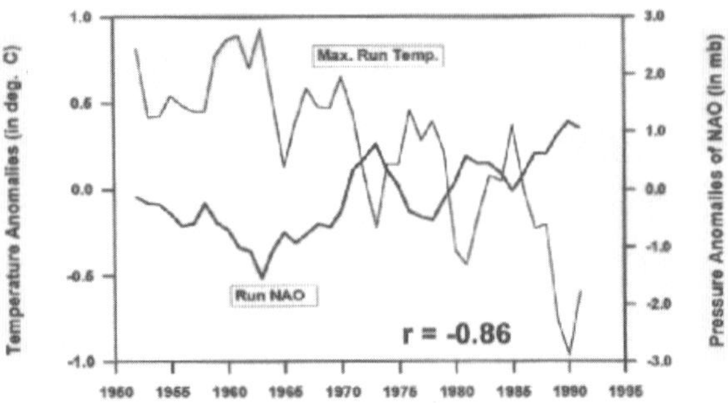

a. Winter Tmax & NAO

b. Winter Tmin & NAO

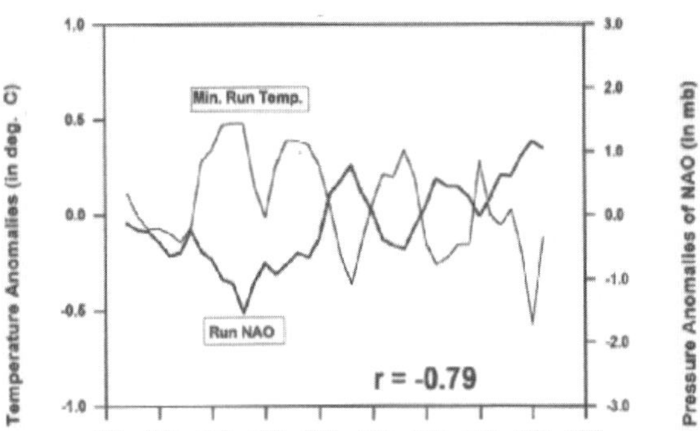

c. Sea level pressure & NAO

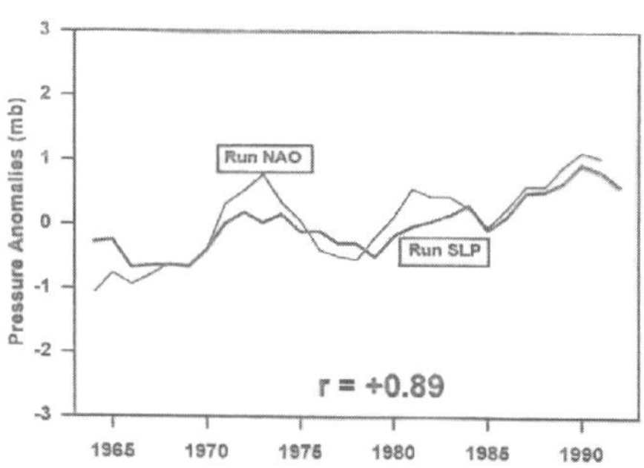

Fig. 1.6.2. Smoothed anomalies (5 y running mean) of Israel winter maximum and minimum winter temperature as well as sea-level pressure, SLP (December-March) and the NAO index. Coefficients of correlation between NAO index and anomalies of maximum and minimum temperatures and SLP are -0.86 (a), -0.79 (b), and +0.89 (c), respectively , from Ben-Gai et al (2001b).

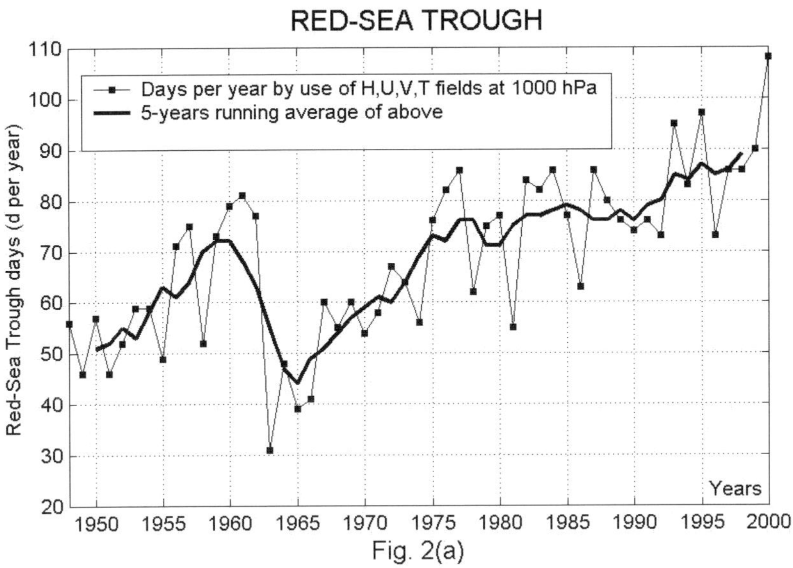

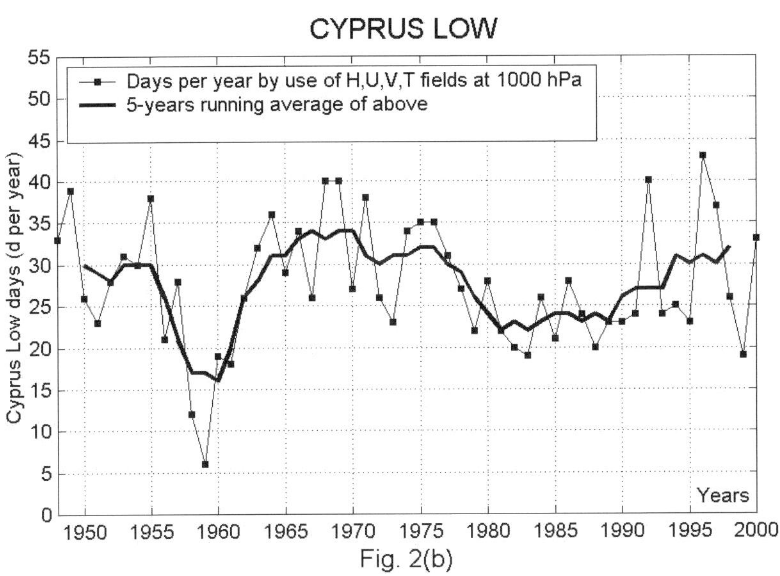

Fig. 1.6.3. The annual numbers of Red Sea Trough days (a) and Cyprus Low days (b) for 1948-2000. Thin lines with squares indicate results obtained by use of H, T, U, V fields at 1000 hPa level. Heavy lines are 5-y running means , from Alpert et al. (2003).

1.6.4 Summary

The outstanding features in the EM recent climate changes, i.e. decreasing winter temperatures, increased rainfall in the south EM in spite of dominant decreases over the whole Mediterranean and the increase in extreme daily rainfall along with general decrease in totals, can be all explained by global teleconnections and linked to global warming. The recent NAO index increase is directly related to cooler winters over the EM. The south EM does not suffer from the dominant rainfall decreases over the region thanks to the positive contribution of the EA/WR index which strongly affects the EM rainfall [Krichak et al., 2002]. Also, increases in El-Nino events were found to be positively correlated to rainfall in the region [Price et al., 1998].

Local land-use changes over central to south Israel were also shown to positively contribute to the rainfall increases there [Otterman et al., 1990, Ben-Gai et al., 1993]. The increase of extreme rainfall over Israel in spite of decrease in rainfall totals reflects a change in the rainfall distributions. The latter is suggested to be the result of increase in Red-Sea Trough synoptic situation on the account of other systems like the Cyprus Low.

References

Alpert P, T Ben-Gai, A Baharad, Y Benjamini, D Yekutieli, M Colacino, L Diodato, C Ramis, V Homar, R Romero, S Michaelides and A Manes (2002) The paradoxical increase of Mediterranean extreme daily rainfall in spite of decrease in total values. Geophys. Res. Lett. *29 11* 31-1 – 31-4 (June issue)

Alpert P, I Osetinsky, B Ziv, H Shafir (2003) Trends in Objectively-Classified E. Mediterranean Synoptic Systems. 1948-2000. Intern. J. of Climatol. (Submitted)

Ben-Gai T, A Bitan, A Manes and P Alpert (1993) Long-term change in October rainfall patterns in southern Israel. Theoretical and Applied Climatology <u>46</u> pp 209-217

Ben-Gai T, A Bitan, A Manes, P Alpert and S Rubin (1999) Temporal and spatial trends of temperature patterns in Israel. Theoretical and Appl. Climatology *64* pp 163-177

Ben-Gai T, Bitan A, Manes A and Alpert P. Climatic variations in the moisture and instability patterns of the atmospheric boundary layer at the east Mediterranean coastal plain of Israel, Bound. Lay. Meteor., 100, 363-371, 2001a.

Ben-Gai T, Bitan A, Manes A, Alpert P and Rubin S, "Temperature and surface pressure anomalies in Israel and the North Atlantic Oscillation", Theoret. And Appl. Climatology, 69, 171-177, 2001b.

Dai A, IY Fung and AD Del Genio (1997) Surface observed global land precipitation variations during 1900-1988. J. Climate *10* pp 2943-2962

IPCC, Climate Change, (2001) The Scientific Basis, Contribution of WG I to the Third Assessment Report of the Intergovernmental Panel on Climate Change Cambridge Univ. Press 2001

Krichak SO and P Alpert (1998) Role of large-scale moist dynamics in Nov 1-5, 1994, Hazard Mediterranean Weather. J. Geophys. Res. *103* D16 pp 19453-19468

Krichak SO, P Kischa and P Alpert (2002) Decadal trends of main Eurasian oscillations and the Mediterranean precipitation. Theor. and Applied Climatology *72* pp 209-220

Otterman J, A Manes, S Rubin, P Alpert and D Starr (1990) An increase of early rains in Israel following land use change? Bound. Lay. Meteorol. *53* pp 333-351

Piervitali E, M Colacino and M Conte (1998) Rainfall over the central-western Mediterranean basin in the period 1951-1995, Part I: Precipitation trends. Nuovo Climento C *21* pp 331-344

Price C, L Stone, A Huppert, B Rajagopalan and P Alpert (1998) A possible link between El-Nino and precipitation in Israel. Geophys. Res. Letters Vol 25 No 21 pp 3963-3966

Price C, S Michaelides, S Pashiardis and P Alpert (1999) Long term changes in diurnal temperature range in Cyprus. Atmospheric Research *51* pp 85-98

Saaroni H, B Ziv, A Bitan and P Alpert (1998) Easterly wind storms over Israel. Theoretical and Appl. Climatology *59* pp 61-77

Saaroni H, Ziv B and Alpert P, "Long-term variations in summer temperatures over the eastern Mediterranean", Geoph. Res. Lett., 2003.(in press)

Steinberger EH and N Gazitñ Yaari (1996) Recent changes in the spatial distribution of annual rainfall in Israel. J. Climate *9* pp 3328-3336

Striem HL, (1979) Some aspects of the relation between monthly temperatures and rainfall, and its use in evaluating early climates in the Middle East. Clim. Change *2* pp 69-74

Thompson DWJ and JM Wallace (2001) Regional climate impacts of the northern hemisphere annual mode. Science *293* pp 85-89

Trigo IF, TD Davies and GR Bigg (2000) Decline in Mediterranean rainfall caused by weakening of Mediterranean cyclones. Geoph. Res. Lett. *27* pp 2913-2916

Xoplaki E, JF Gonzalez-Rouce, J Luterbacher and H Warner (2003) Mediterranean summer air temperature variability and its connection to the large-scale atmospheric circulation and SSTs. Climate Dynamics *20* pp 723-739

1.7 Surface Water in Jenin Governorate-Palestine

Ghassan Abu Ju'ub & Kurt Schetelig

Department of Engineering Geology and Hydrogeology (LIH). Aachen University of Technology (RWTH) Aachen-Germany.

Key Words

Jenin Governorate, Palestine, West Bank, water resources, surface water, seasonal wadi, seasonal lake, Nahr Al-Muqatta', Nahr Al-Mufjir (Wadi Al-Khudairah), Marj Sanur, Wadi Shubash, Wadi Al-Maleh, Wadi Al-Fara'a, Nahr Jalut, Wadi Bardalah.

Abstract

Very little attention is normally given to surface water in Palestine. This paper focuses on this subject by studying surface water in Jenin Governorate, which is one of the major governorates in the country.

Water resources in Jenin Governorate are groundwater, surface water and cisterns. Even though groundwater is the main resource, still surface water can have a substantial contribution to water supply if it is used. Currently this resource is not used neither in Jenin Governorate nor in other parts of Palestine mainly because of the Israeli restrictions.

The area is discharged by several surface water bodies flowing either westwards to the Mediterranean Sea or eastwards to the Jordan River and the Dead Sea. The vast majority of the Governorate is discharged westwards.

In this paper surface water in Jenin Governorate is discussed. The results reached are preliminary and subject to modifications when more data are available.

Many researches and projects are still needed regarding surface water in Palestine, in order to develop this sector theoretically and practically.

1.7.1 General Information

Jenin Governorate is one of the major governorates in Palestine. It is located in the upper most part of the West Bank. The governorate has an area of about 600 km^2 and a total population of about 200,000 inhabitants living in 96 localities including the central Jenin City and the nearby Jenin Refugee Camp.

The topography in the Governorate varies between flat plains and low elevated hills. The lowest point is at 78 m amsl (above mean sea level) located in the northern Marj Bin Amer Plain. The highest one is at 764 m, which is the crest of the southern Hureish Mountain. Most of the Governorate is located at elevations less than 400 m amsl, around 23% of its area are flat plains.

The Governorate is a fertile area where many kinds of agriculture grow. It is normally called the "Basket of Food" for Palestine. Around 60% of the governorate's area is cultivated.

The climate is typical Mediterranean with rainy cool winters from October to May and dry warm summers in the other part of the year.

1.7.2 Surface Water Bodies (Catchment Areas)

Jenin Governorate has no access to any permanent surface water body, but there are several seasonal ones. In total, the governorate is drained by eight of them: seven wadis and one seasonal lake. Three belong to the western catchment (located west of the main surface water divide), these are: Nahr Al-Muqatta', Wadi Al-Khudairah (Nahr Al-Mufjir) and the seasonal lake of Marj Sanur. The other five wadis belong to the eastern catchment, they are: Wadi Shubash, Wadi Bardalah, Wadi Al-Maleh, Wadi Al-Fara'a, and Nahr Jalut.

Five of the wadis originating (totally or partially) in Jenin Governorate become perennial (permanent) in their final stages, these are: Nahr Al-Muqatta', Nahr Al-Mufjir (Wadi Al-Khudairah), Nahr Jalut, Wadi Al-Maleh and Wadi Al-Fara'a. The first two flow westwards to the Mediterranean, the last three eastwards to the Jordan river. Most of Jenin Governorate (93%) is drained westwards, only 7% is drained eastwards (Figures 1.7.1, 1.7.2, tables 1.7.1, 1.7.2).

1.7.3 Surface Water Quantities

Very rare, unreliable data exit about surface water quantities (also called flood water or surface runoff) in Jenin Governorate or other parts of Palestine. In this study water quantities in the different surface water bodies in Jenin Governorate are estimated. The calculations are based on the study of Husary et. al. [9], in which the water quantities in the Palestinian part of Wadi Al-Khudairah were estimated. Jenin Governorate makes around 50% of the total catchment of this wadi and around 74% of its catchment inside Palestine.

Table 1.7.1. Surface water bodies in Jenin Governorate and their direction of flow.

No.	Surface water body	Catchment area in J.G. (km^2)	% of catchment area from J.G. area	Basin (west or east of the surface water divide)	Direction of flow
1.	Nahr Al-Muqatta'	210	35.8%	Western	West (Med.)
2.	Wadi Al-Khudairah (Nahr Al-Mufjir)	282	48%	Western	West (Med.)
3.	Marj Sanur (Seasonal lake)	53	9%	Western	Closed basin
Total W.C.		545	92.8%		
4.	Wadi Shubash	33	5.6%	Eastern	East (J.R.)
5.	Wadi Bardalah	3	0.5%	Eastern	East (J.R.)
6.	Wadi Al-Maleh	1	0.2%	Eastern	East (J.R.)
7.	Wadi Al-Fara'a	5	0.9%	Eastern	East (J.R.)
8.	Nahr Jalut	0.2	0.03%	Eastern	East (J.R.)
Total E.C.		42	7.2%		
Total		587.2	100%		

No.: Number, J.G.: Jenin Governorate, Med.: Mediterranean Sea, J.R.: Jordan River, W.C.: Western catchment, E.C.: Eastern catchment.

The methodology used in the calculations is described next:

1. For every year between 1982/83 - 1991/92, annual runoff quantities were measured at a certain gauging station installed on the stream of Wadi Al- Khudairah inside Israel. This station measures all the wadi's water coming from Palestine (West Bank) plus some water coming from Israel.
2. The annual water quantities were divided over the catchment area of that part of the Wadi which the gauging station represents. This area is 458 km^2 according to [9], it includes all the wadi's catchment inside Palestine plus part of its catchment inside Israel.
3. The result is the annual water quantity per km^2.

This quantity was multiplied by the catchment area of the wadi inside Jenin Governorate to get the total quantity in the Governorate's part of the wadi.
The same was done to the other surface water bodies in Jenin Governorate.
In these calculations the following assumptions were made:

1. Runoff quantities per km^2 of Wadi Al-Khudairah are the same in Jenin Governorate as they are in the other parts of the wadi's catchment measured by the

gauging station. This assumption was made because the natural conditions in the two areas are almost similar.

2. Runoff quantities per km^2 are the same for the different catchments in Jenin Governorate. This assumption was also made because the natural conditions in the Governorate are almost the same. That includes precipitation, evapotranspiration and other climatic conditions, topography, landscape, land use, etc.

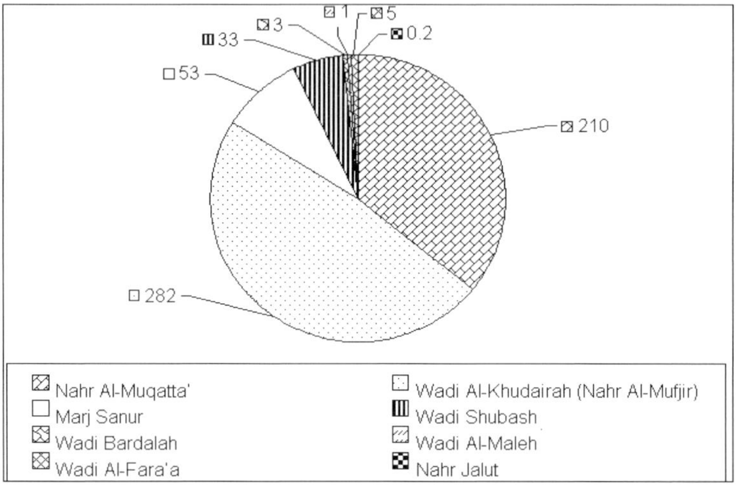

Fig. 1.7.1. Sizes of the catchment areas in Jenin Governorate (km^2).

Such assumptions may not be 100% correct, but they are not too far from reality, because indeed the differences in the natural conditions within Jenin Governorate are not so big. Also the northern part of Tulkarm Governorate through which Wadi Al-Khudairah flows later on is very similar to Jenin Governorate. But still there is a need for further studies about this subject. The results of the calculations are shown in tables 1.7.3, 1.7.4. Table 1.7.3 shows that the average total surface runoff in Jenin Governorate is around 23.6 Mm^3/yr. if 1991/92 rainy year is included and 11.7 Mm^3/yr. if it is excluded. The maximum runoff value is 119.2 Mm^3/yr. in 1991/92. See also figures 1.7.3, 1.7.4.

However, the period 1982/83-1991/92, for which the calculations were made belongs to the wettest period the area witnessed in the recorded 150 years. It includes the extraordinary rainy year of 1991/92. Therefore, it is very probable that the long term average runoff is less than these figures. According to the calculations here, surface runoff makes around 7.6% of rainfall (when 1991/92 is included), or 3.8% (when 1991/92 is excluded).

Table 1.7.2. Surface water bodies in Jenin Governorate and their distribution between Palestine and Israel.

No	Catchment name		Type of catchment	Catchment area (km^2)				
				Inside Palestine			Outside Palestine (in Israel)	Total area
	Arabic name	Hebrew name		Inside Jenin Governorate	Inside other Palestinian Governorates	Total inside Palestine (West Bank)		
1.	Nahr Al-Muqatta'	Nahal Qishon	Seasonal wadi, only last part is permanent	210 (19%)***	0.0 (0.0%)	210 (19%)	805 (81%)	1105 (100
2.	Nahr Al-Mufjir (Wadi Al-Khudairah)	Nahal Hadera	Seasonal wadi, only last part is permanent	282 (50%)	101 (18%)	383 (68%)	184 (32%)	567 (100
3.	Marj Sanur	Marj Sanur	Seasonal lake	53 (96%)	2 (4%)	55 (100%)	0.0 (0.0%)	55 (100
4.	Wadi Shubash	Nahal Bezeq	Seasonal wadi	33 (41%)	12 (15%)	45 (56%)	35 (44%)	80 (100
5.	Wadi Bardalah*	Nahal Bardalah*	Seasonal wadi	3 (5%)	50 (83%)	53 (88%)	7 (12%)	60 (100
6.	Wadi Al-Maleh	Nahal Malhah	Seasonal wadi, only last part is permanent	1 (1%)	119 (99%)	120 (100%)	0.0 (0.0%)	120 (100
7.	Wadi Al-Fara'a	Nahal Terssa	Seasonal wadi, only last part is permanent	5 (2%)	315 (98%)	320** (100%)	0.0 (0.0%)	320* (100
8.	Nahr Jalut	Nahal Harud	Seasonal wadi, only last part is permanent	0.2 (0.1%)	0.0 (0.0)	0.2 (0.1%)	219 (99.9%)	219.2 (100

This catchment includes several small wadis, some disappear before reaching the Jordan River. *Their individual names were not available, s catchment was given the name of the main Palestinian village through which the largest wadi passes (Bardalah).** This number is taken from all other numbers in the table are calculated in this study. *** Numbers between brackets are the percentages of the wadi's area in every locat from the total wadi's area.

Table1.7.3. Runoff per km^2 in Jenin Governorate in the period 1982/83-1991/92.

Year	Catchment area (A)* in (km^2)	Runoff (Q)* in (Million m^3/yr.)	Runoff per km^2 (Q/A) in (m^3/yr.)
1982/83	458	17.2	37600
1983/84	458	1.9	4200
1984/85	458	0.3	600
1985/86	458	Not available	Not available
1986/87	458	10.5	23000
1987/88	458	16.9	36900
1988/89	458	6.6	14400
1989/90	458	14.4	31500
1990/91	458	5.1	11100
1991/92	458	93.0	203100
Minimum (in any dry year)		0.0	0.00
Maximum		93.0	203100
Average (with 1991/92)		18.4	40300
Average (without 1991/92)		9.1	19900

* Q is the measured runoff quantity for the part of Wadi Al-Khudairah in Palestine (West Bank) plus some parts in Israel. A is the catchment area for which Q is measured. Data in the first 3 columns is taken from [9].

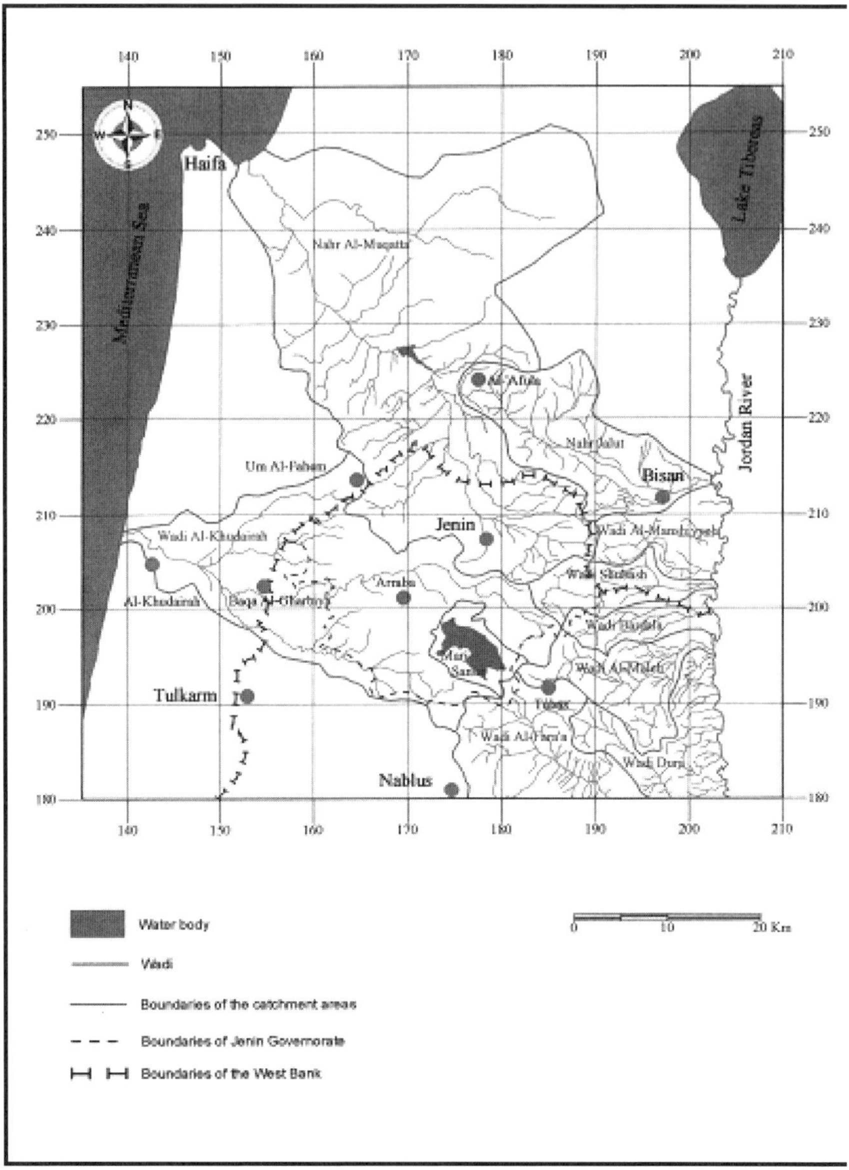

Fig. 1.7.2. Surface water bodies in Jenin Governorate and their catchments.

The four catchments of Nahr Al-Muqatta', Nahr Al-Mufjir (Wadi Al-Khudairah), Marj Sanur and Wadi Shubash make around 99% othe total area of Jenin Governorate. The first three belong to the western catchment, while the fourth belongs to the eastern. Any real thinking of using surface water in Jenin Governorate shall concentrate on these 4 catchments. Brief description of each of them is given next, more about them in [1].

1.7.4 The Four Main Catchments

Nahr Al-Muqatta'

Nahr Al-Muqatta' in Arabic means "the Disconnected River". It is one of the few perennial (permanent) rivers (wadis) in historical Palestine and one of the largest catchments there. The river is around 70 km long and becomes permanent in the last 11 km [12]. It originates mainly from the mountains of Jalboon (Israeli Gelboa) in the north east of Jenin Governorate. Other tributaries come from other parts of northern Jenin Governorate (north west and central north). The tributaries and the catchment of the river cover all the northern part of Jenin Governorate. More information about the catchment area and flow quantities of the river are in tables 1.7.1, 1.7.2, 1.7.4 and figures 1.7.1-1.7.4.

The tributaries flow later through Marj Bin Amer, which is now shared between Palestine and Israel. After crossing the borders to Israel, they join each other to form the main stream of the wadi not far from the Israeli town of Al-'Affulah, which is around 15 km north of Jenin City. Later on the river receives additional water from its tributaries originating in the Hills of Nazareth in Lower Galilee and from Marj Bin Amer itself.

The river continues to flow through the Marj in the north western direction until it reaches Mount Carmel Ridge near Haifa where it changes path and flows through the Plain of Akka (Acre). Finally it reaches the Mediterranean Sea and discharges in the Bay of Haifa, where the Israelis developed its mouth as part of the complex of Haifa Port and called it "Qishon Port" [7]

The Palestinians use nothing of the water of this wadi because of the Israelis restrictions. The Israelis have constructed a dam on its main stream soon after its first tributaries (which come mostly from Jenin Governorate) meet. The dam is located in Marj Bin Amer not far from the Israeli town of Al-'Affulah. A reservoir of about 5 km length stands behind the dam and is filled with the water of the wadi. It is used for several purposes, mainly irrigation, fishponds and flood control [7]. See figure 1.7.2.

When the river crosses the border of Jenin Governorate to Israel it is still made of 5 main tributaries. The largest is called by the name of the main river: Nahr Al-Muqatta'. It starts deep inside Jenin Governorate near the villages of Um At-Tut and Jalqamoos and also from Jalboon Mountains. One of its streams starts from Arraba Plain and passes through Jenin City.

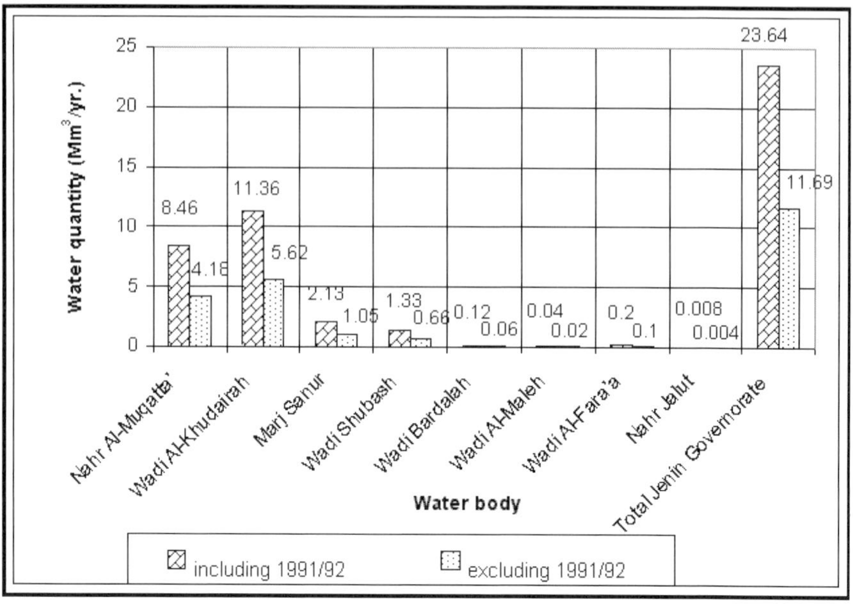

Fig. 1.7.3. Average annual surface water quantities in Jenin Governorate.

Table 1.7.4. Runoff quantities in the different surface water catchments in Jenin Governorate.

Catchment name	C.A inside J.G. (km²)	Surface runoff in different years (Million m³/yr.)													
		1982/83	1983/84	1984/85	1985/86	1986/87	1987/88	1988/89	1989/90	1990/91	1991/92	Min.*	Max.	Avr. with 1991/92	Avr Wi[t]h[out] 199[1/92]
Nahr Al-Muqatta'	210	7.89	0.88	0.12	NA	4.83	7.74	3.03	6.62	2.34	42.65	0.00	42.65	8.46	4.18
Nahr Al-Mudjir (Wadi Al-Khudairah)	282	10.59	1.19	0.17	NA	6.49	10.40	4.06	8.89	3.14	57.27	0.00	57.27	11.36	5.62
Marj Sanur	53	1.99	0.22	0.03	NA	1.22	1.95	0.76	1.67	0.59	10.76	0.00	10.76	2.13	1.05
Wadi Shubash	33	1.24	0.14	0.02	NA	0.76	1.22	0.48	1.04	0.37	6.70	0.00	6.70	1.33	0.66
Wadi Bardalah	3	0.11	0.01	0.00	NA	0.07	0.11	0.04	0.09	0.03	0.61	0.00	0.61	0.12	0.06
Wadi Al-Maleh	1	0.04	0.00	0.00	NA	0.02	0.04	0.01	0.03	0.01	0.20	0.00	0.20	0.04	0.02
Wadi Al-Fara'a	5	0.19	0.02	0.00	NA	0.12	0.18	0.07	0.16	0.06	1.02	0.00	1.02	0.20	0.10
Nahr Jalut	0.2	0.008	0.001	0.00	NA	0.005	0.007	0.003	0.006	0.002	0.041	0.00	0.041	0.008	0.00[4]
Total J.G.	587.2	22.05	2.47	0.35	NA	13.51	21.65	8.46	18.51	6.54	119.22	0.00	119.22	23.64	11.6[9]

Min.: minimum. Max.: maximum. Avr.: average. NA: not available. J.G.: Jenin Governorate. C.A.: catchment area. * The minimum value of ze[ro c]an happen in any dry year.

The second tributary is Wadi Al-Kaslan. It also starts from Jalboon Mountains and drains part of north eastern Jenin Governorate (north of the first one). The other three tributaries are: Wadi Al-Jamus, Wadi Ti'innik and Wadi Al-Bassa. These three drain the north western part of Jenin Governorate.

Nahr Al-Mufjir (Wadi Al-Khudairah)

It originates mainly from the eastern and south eastern parts of Jenin Governorate. It surrounds the catchment of Marj Sanur between its two main tributaries (Wadi Abu Nar in the north and Wadi Massin in the south). Its catchment makes most of the southern half of the governorate. The wadi flows through Jenin Governorate, then passes to Tulkarm Governorate, where it crosses the border of Palestine. In Israel it receives smaller tributaries and keeps flowing westwards through the coastal plain, until it reaches the Israeli town of Al-Khudairah (Hadera). Finally it discharges in the Mediterranean Sea.

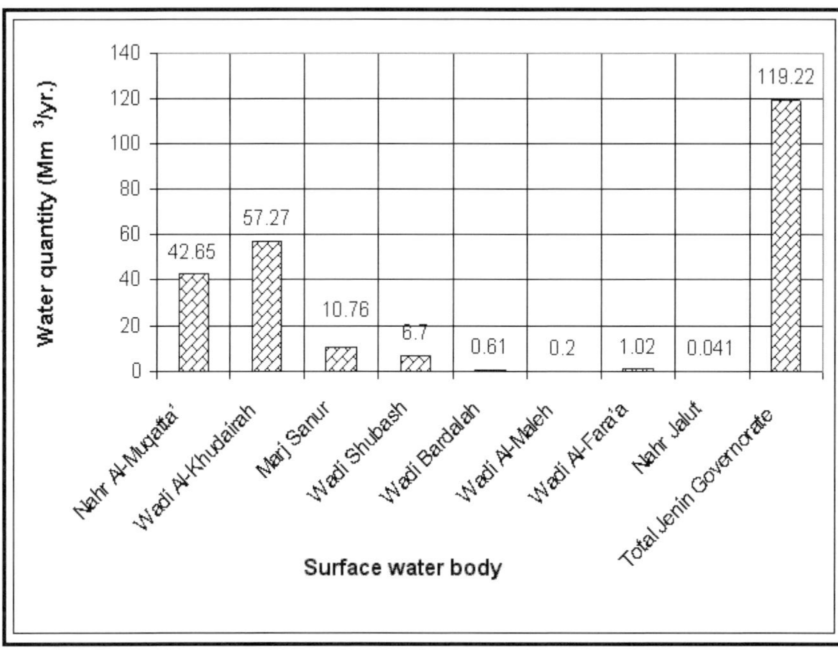

Fig. 1.7.4. Maximum annual surface water quantities in Jenin Governorate.

The length of the Wadi reaches up to 50 km, more than half of its stream is in Jenin Governorate. It is of middle to large size compared to other wadis in historical Palestine. More information about its catchment area and flow quantities are in tables 1.7.1, 1.7.2, 1.7.4 and figures 1.7.1-1.7.4.

The wadi's catchment is the largest in Jenin Governorate and the vast majority of its tributaries originate there. When it leaves Palestine it is made of 5 tributaries, three of them are small, two are large and make the main body of the Wadi.

The small tributaries are: Wadi Fudeil, Wadi Al-Yamuni, and Wadi Al-'Asal: They originate in the western part of Jenin Governorate near the towns of Barta'a, Ya'bad and Qiffin (This town is in Tulkarm Governorate).

The two large tributaries are: Wadi Abu Nar and Wadi Massin. Wadi Abu Nar is the largest and longest one. It originates in the eastern part of Jenin Governorate in the Plain of Sir (Al-Kufeir) just north of Marj Sanur.

Wadi Massin is the second largest tributary. It starts at the southern edge of Jenin Governorate on Mount Hureish just south of Marj Sanur.

The tributaries join each other in the Israeli territories to form the main stream of the Wadi.

Marj Sanur

It is a closed basin with about 20 km^2 area and a total catchment area of around 55 km^2. It has a maximum length of 7.5 km (extending north west-south east) and a maximum width of 3.5 km in the opposite direction. All the Marj is located inside Jenin Governorate. 53 km^2 of its total catchment area are in Jenin Governorate, the remaining 2 km^2 are in Tubas Area.

Elevations in the Marj range between 348 and 370 m amsl, but most of it has an elevation range of 350 to 360 m amsl. Mountains and hills with elevations from 450 up to 764 m amsl surround it from all directions. The highest one is Mount Hureish at the southern edge with 764 m amsl elevation. There are no natural outlets from the Marj.

Normally Marj Sanur gets flooded when precipitation exceeds 600 mm/yr. The probability of the flooding is around 47% and more than half of the Marj area can be covered with water when precipitation is extremely high as happened in 1991/92 rainy year when around 10.5 km^2 of the Marj were covered with several meters of water. The average flood quantities are about 2.2 Mm^3/yr. with a maximum of 11.2 Mm^3/yr. in 1991/92. More information about the Marj's catchment area and flow quantities are in tables 1.7.1, 1.7.2, 1.7.4 and figures 1.7.1-1.7.4. Additional information are in [1], [2], [3].

Wadi Shubash

It has a total catchment area of 80 km^2. It is small compared to other wadis in historical Palestine. It belongs to the eastern surface water catchment, where rainfall and runoff are less than in the western one.

The wadi is around 20 km long and it is the fourth largest catchment in Jenin Governorate. Its two main tributaries start in the eastern part of Jenin Governorate. The main stream of the wadi is formed in Jenin Governorate, then it passes through

Tubas Area before it crosses the borders to Israel. More information about the wadi's catchment area and flow quantities are in tables 1.7.1, 1.7.2, 1.7.4 and figures 1.7.1-1.7.4.

References

[1] Abu Ju'ub Gh (2002) Water Resources in Jenin Governorate / North of the West Bank – Palestine, A Part of the Water Problem in the Middle East. Ph.D. thesis Heft 83 Lehrstuhl fuer Ingenieurgeologie und Hydrogeologie-Aachen University of Technology (RWTH) Aachen

[2] Abu Ju'ub Gh, Schetelig K (2002) Using the Concept of Artificial Recharge to Solve the Flooding Problem of Marj Sanur in Jenin Governorate-Palestine. In: Dillon PJ (eds) Management of Aquifer Recharge for Sustainability. Balkema Lisse pp 419-423

[3] Abu Ju'ub Gh, Schetelig K (2002) Marj Sanur: A Harmful Seasonal Lake in Palestine Which Can Become Useful. In: Van Rooy JL, Jermy CA (eds) Engineering Geology for Developing Countries. SAIEG Pretoria pp 1691-1701

[4] Abu Ju'ub Gh, Schetelig K (2001) Water Resources in Jenin Governorate-North of the West Bank/Palestine. In: Seiler K-P, Wohnlich S (eds) New Approaches Characterising Groundwater Flow. Balkema Lisse volume 2 pp 1101-1105

[5] ARIJ-Applied Research Institute of Jerusalem (1998) Water Resources and Irrigated Agriculture in the West Bank. Bethlehem

[6] ARIJ-Applied Research Institute of Jerusalem (1996) Environmental Profile for the West Bank, Volume 7-Jenin District. Bethlehem

[7] Encyclopedia Britannica (1999-2000) Information about Nahal Qishon (Nahr Al-Muqatta'). Internet site:www.britannica.com

[8] Ghanem M (1999) Hydrogeology and Hydrochemistry of the Faria Drainage Basin/West Bank. Ph.D. thesis. TU Bergakademie Freiberg Freiberg/Saxony

[9] Husary SM, Aliewi AS (1996) The Determination of Wadi Runoff Volumes for the Palestinian Part of the Hadera Catchment. In: Haddad M, Feitelson E (eds) Joint management of Shared Aquifers, the Third Workshop Jerusalem

[10] PCBS-Palestinian Central Bureau of Statistics (1999) Population in the Palestinian Territory, 1997-2025 (in Arabic and English) Ramallah

[11] Shu'un Tanmawiyya (1988) Water in Palestine (in Arabic). Arab Thought Forum Publication Jerusalem

[12] The Columbia Encyclopaedia, Sixth Edition (2000) Information about Nahal Qishon (Nahr Al-Muqatta'). Internet site: www.bartleby.com/65

1.8 Herodion-Beit Fajjar Well Field of High Quality and Resource-Productivity Groundwater Endangered by Severe Over-Pumpage

Qannam, Z., Merkel, B. and Lange, T.

Abstract

Herodion Beit-Fajjar well field is located in a syncline area about 8 km to the east of the Surif Anticline, the main anticlinal structure in the southern West Bank. The selected well field lies within 35°9` and 35°16` longitude 31°35` and 31°43` latitude which is referenced on the Palestinian Grid as being between 110-125 N and 165-175 E. The area enjoys a Mediterranean climate, hot, dry summers and mild, wet winters, with 600 mm as the average annual rainfall and 1400-1600 mm as the annual potential evaporation. The elevation of the well field ranges between 750 masl at the well of PWAII and 455 masl at well of Azzariya 3. The main geological outcrops in the area are Turonian-Cenomanian limestones, dolomite and marls of Jerusalem Formation and to less extent the Bethlehem Formation. The wells in the area are discharging two main regional aquifers, the Turonian Cenomanian (Upper) and the Albian (Lower) aquifers separated by the marls and clays of the Yatta Formation. Both the surface and ground water flow eastward directed by the topography and the dip of the geological formations respectively. Studies about the groundwater quality conducted between 1996-2002 show that the water of the wells is of calcium carbonate type and of high drinking water quality from both the biological and chemical point of view. The content of ^{2}H and ^{18}O represents the local precipitation, while the content of ^{3}H dates it back to the 1950's. However, measurements of the static water levels at the different wells in the field show that the well field is subject to severe over-pumpage, lowering the water table e.g. at well Herodion 3 by 85 m between 1981 and 1997 and about 37 meters at well Herodion 2. Thus, Herodion Beit Fajjar well field is an important source of groundwater, but subjected to heavy over-pumping. Therefore sustainable management actions must be taken to protect these precious water resources.

1.8.1 Introduction

The boreholes of the Herodion-Beit Fajjar well field are located at 350-770 masl inside the Beit Sahour Syncline in the southern West Bank between 165-175 E and 110-125 N (Palestinian Grid). It is situated in the Jerusalem Desert Sub-Basin, a part of the Eastern Basin of the Mountain Aquifer (Fig. 1.8.1). Its potential yield was first recognized in 1963 by the Rofe and Raffety consultancy team, which was contracted by the Hashemite Kingdom of Jordan. Beit Fajjar well-1 was the first successfully drilled and tested well in this field. It is fitted with a submersible pump. Since being connected to the drinking water supply network it has contributed a drinking water amount of 1200 m³/day. Between 1971 and 1993 six more wells were drilled by the Israelis; Beit Fajjar well-1A at the same location of Beit Fajjar well-1 but tapping to greater depths and five wells having the names of Herodion 1 to 5 (Abed Rabbo et al. 1998). Since the signing of Oslo Agreement in 1994 additional wells were drilled, half way already set in operation while others are still in the drilling and construction state.

Based on the records of Al Arroub Meteorological Station, five kilometers to the southwest of the well field, the average annual rainfall for the period 1953 –2001 is 607 mm/yr. The maximum annual rainfall was 1200 mm during the precipitation period of 1991/1992, while the minimum was 212 mm in 1998/1999 (Fig. 1.8.2). The average annual number of rainy days between 1953 and 2001 is 46 day/yr. The maximum daily rainfall intensity was 148 mm/day on November 19th, 1953. Generally the wettest month is January. But January 1974 was the wettest month during the whole period with 491 mm. According to Qannam (2003), the annual rainfall in the area is distributed as 17.0 % surface runoff, 23 % recharge, 49 % actual evapotranspiration and 11 % as interception, depression storage and soil moisture.

1.8.2 Geology and hydrogeology

The geology of the Herodion-Beit Fajjar well field is composed of sedimentary carbonate rocks (limestone, dolomite chalk) and marls of Albian to Holocene age. It is structurally dominated by the Surif anticline and the Beit Fajjar-Beit Sahour syncline (Fig. 1.8.3). The general stratigraphy of the area is presented in Table (1.8.1).

All of the field's production wells discharge two regional aquifers; the Albian and the Turonian-Cenomanian, referred to as the Lower and Upper Aquifer respectively. The Albian aquifer is built-up from limestone and dolomite for both the Upper and Lower Beit Kahil formations (Fig. 1.8.3). This aquifer owes its high water bearing capacity and productivity to its great thickness (275-330 m). The geological formations of this aquifer are characterized by a steep gradient of 4-5 % (40-50 m/km) towards the east. Rainfall and infiltration on the outcropping aquifer formations is the major source of recharge. Despite its limited outcrops, most of the ground water abstraction in the area is from this aquifer, which could be attributed to additional feeding of this aquifer from the overlying Cenomanian–Turonian aquifer, mainly along the fault zones and fractures. Examples of the wells discharging this aquifer are the wells of Hebron 1, PWA I and II, Herodion 2, 3, and 4.

Jerusalem, Bethlehem and Hebron formations are the water bearing formations of the Cenomanian –Turonian aquifer, with a thickness ranging between 320 and 350 m. Generally, the formations of this aquifer are eroded at the Surif Anticline that forms the crest of the Hebron Mountains in this area, but they outcrop east and west of it. This aquifer is phreatic and therefore exposed to surface-derived pollution such as waste water infiltrating from the sewage conduit in wadi Al Arroub and widespread poorly designed cesspits. The principal wells extracting water from this aquifer are the wells of Hebron 2, Herodion 1 and 5 and Beit Fajjar 3.

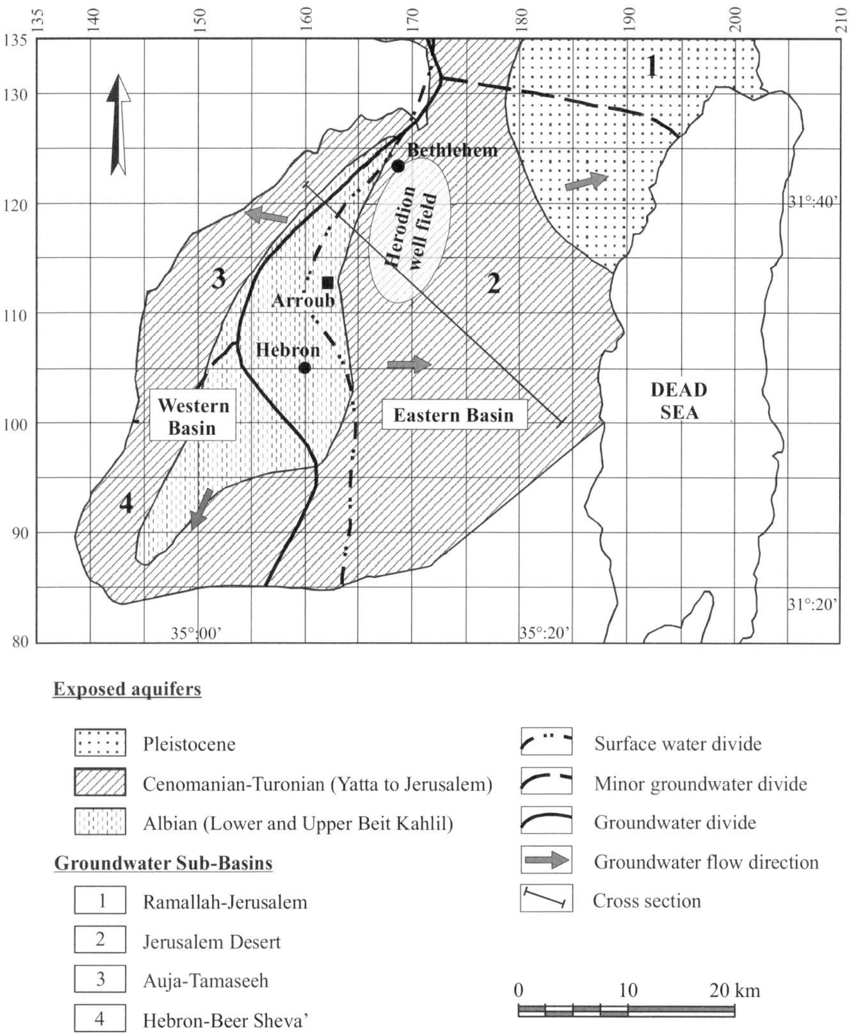

Fig. 8.1.1. Map of the Herodion-Beit Fajjar well field showing the geology, groundwater basin and aquifers exposed in the southern West Bank.

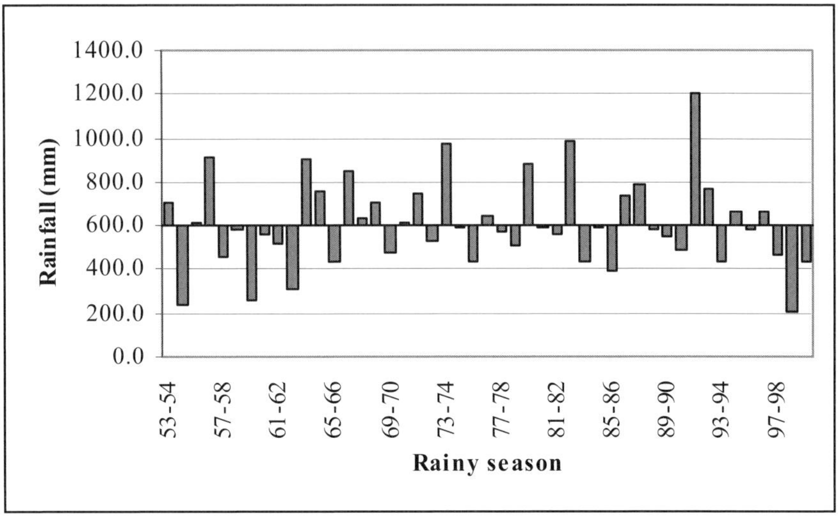

Fig. 1.8.2. Annual rainfall variations at the Arroub Meteorological Station during the period 1953-2001.

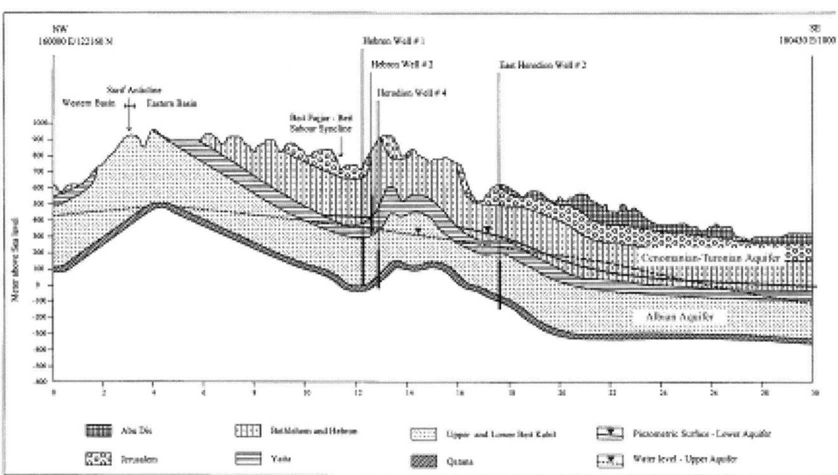

Fig. 1.8.3. NW-SE hydrogeological cross section showing the regional aquifers and aquicludes in the study area and its surrounding (after Millennium Engineering Group et al. 2000).

Table 8.1.1. General stratigraphy of the study area (Guttman and Gotlieb 1996).

Age	Formation		Thickness (m)	General lithological Description
	Palestinian	Israeli		
Turonian	Jerusalem	Bina	90-100	Hard limestone, some dolomite and marl
Cenomanian	Bethlehem	Weradim	90-100	Hard massive dolomite
		Kfar Shaul	30-40	Soft limestone, chalky limestone and marl
	Hebron	Aminadav	90-100	Hard porous, grey dolomite
	Yatta	Moza	10	Yellowish marl and limestone
		Bet Meir	80-110	Limestone, chalk, grey marl at bottom
Albian	Upper Beit Kahil	Kesalon	30	Hard dolomite, hard limestone, rich in fossils
		Soreq	100-110	Limy dolomite, marly dolomite, marl, at times some chert
	Lower Beit Kahil	Givat Yearim	60-70	Hard dolomite, hard massive limestone
		Kefira	150-180	Limestone with intermediate marl layers (marl increases in downward direction)

The major regional aquiclude that separates the Albian and the Cenomanian-Turonian aquifers is marl, clay, and chalk of the Yatta Formation. The separation is manifested in the difference of the regional water levels in the two aquifers. The water level in the Cenomanian-Turonian aquifer is at about 50-80 m higher than that in the Albian aquifer indicated from data derived from the wells of Herodion 4 and 5 and at about 40 m from the wells of Hebron 1 and 2.

Herodion area, was identified to have relatively moderate flow gradients compared to a steeper gradient in the areas around the wells of Jerusalem 5 and Azzariya 1 at the northern border of this well field. Based on that, it was assumed that the ground water flow is influenced by transmissivity barriers (low transmissivity zones).

1.8.3 Hydrochemistry and water quality

The evaluation of the water quality of this well field is based on the analysis of samples collected from the wells of Herodion 1, 2, 3, 4, 5, PWA I, PWA II, Hebron 1 and Hebron 2 during the period 1996 and 2002. During this study the samples were analyzed for the major elements, metals as well as for total and fecal coli forms. Water chemistry was investigated by the WSERU labs at Bethlehem University. The results (summarized in Table 1.8.2) show that the water of these wells is free of coli form bacteria. All analyzed chemical parameters were found to be below the WHO (1995) guidelines for drinking water. In many cases trace metals were found to be below the detection limits. Thus the water of these wells is suitable for drinking from both the biological and chemical aspect. Seasonal variations in the chemistry of water were detected and supposed to be attributed mainly to alternating dry and wet seasons as well as to the varying quantity and distribution of rainfall. The water of the wells evinces calcium carbonate type as shown by Durov diagram (Fig. 1.8.4).

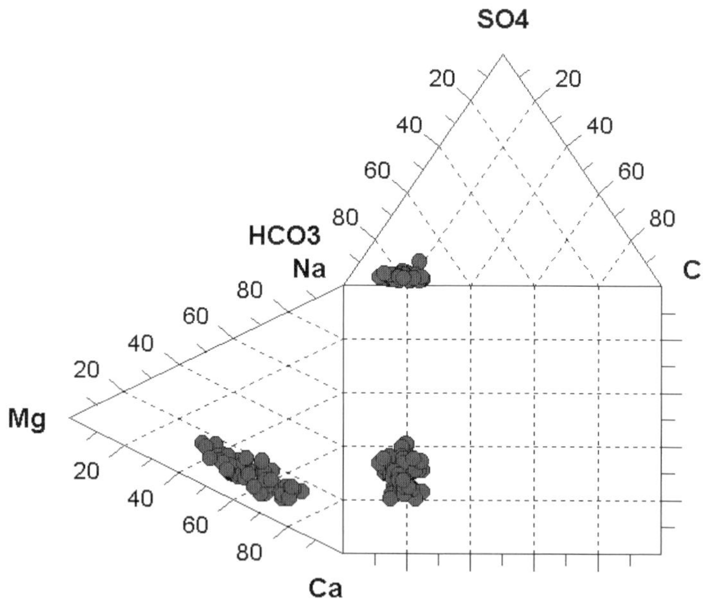

Fig. 1.8.4. Plot of wells sampled from the Herodion well field between 1996 and 2002 on the Durov diagram.

The isotopic content of water samples collected from the wells of Herodion 4 and Beit Fajjar 3 on April 30th, 1998 being analysed for ^{2}H and ^{18}O, plot on the Mediterranean Meteoric Water Line proving its meteoric origin (Fig. 1.8.5). Taking into consideration that the ^{3}H content in precipitation during the study was about 6 TU, then 0.4 TU in the water of Herodion well # 4 tapping the Albian aquifer and 1.5 TU in the water of Beit Fajjar well tapping the Turonian-Cenomanian aquifer date them back to the early and late 1950's respectively, assuming the Piston flow model.

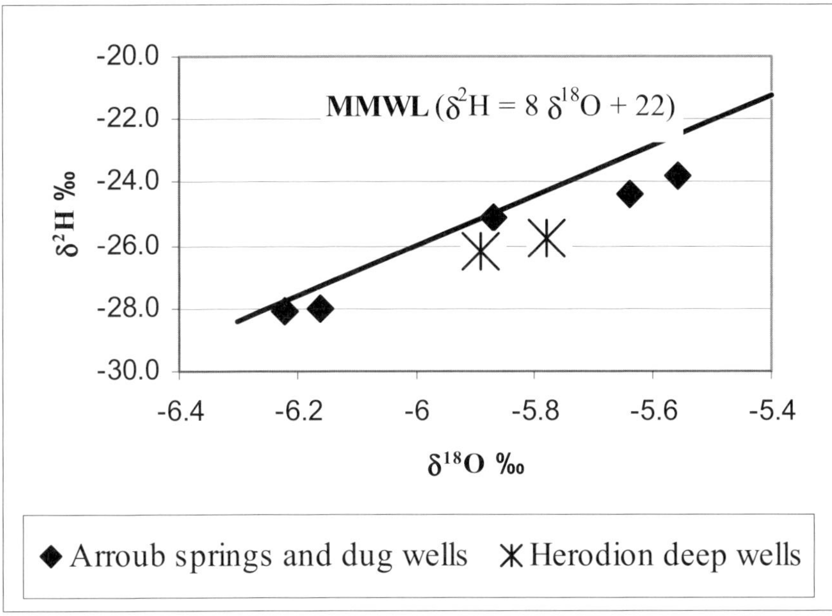

Fig. 1.8.5. Relation of δ^2H versus $\delta^{18}O$ from samples collected at Herodion deep wells, springs and dug wells from Wadi Al Arroub drainage basin compared to the MMWL.

1.8.4 Overexploitation of the Ground water resources

The aquifers of this area, especially in the Herodion well field, are being heavily overexploited indicated by the serious drop in the static water levels overall the area. Table (1.8.3) shows that the existing pumping scheme is causing a continuous drop in the water tables of the aquifers. Continuing this exploitation policy leading to serious drops in more wells in the area will endanger the prospect of safe water supply in the near future. Aliewi and Jarrar (2000) quoted from a modeling study delivered to the PWA by the CDM that, if the newly drilled wells are put in operation, which already happens, then the simulated draw-down is about 120 m over a four-year period in the Herodion field, and this will significantly dewater the utilized aquifers. Fig. 1.8.6 and 1.8.7 graphically present the changes in water level with time at the wells of Herodion wells 2 and 4.

Table 1.8.2. Summary of the characteristics of the wells sampled.

Parameter	Unit	Concentration	
		Average	Std. Dev.
EC	µS/cm	580	60
Ca^{2+}	mg/L	70.1	14.3
Mg^{2+}	mg/L	23.2	6.9
Na^{+}	mg/L	17.2	3.2
K^{+}	mg/L	1.9	0.9
HCO_3^-	mg/L	295	40
SO_4^{2-}	mg/L	10	2.4
Cl^-	mg/L	34	7.1
NO_3^-	mg/L	10.6	4.2
SiO_2	mg/L	20.5	6.7
Total coli form	# of colonies/100ml sample	0	0
Fecal coli form	# of colonies/100ml sample	0	0

Table 1.8.3. The drop in the static water level of selected wells in the study area and its surroundings (after PWA, 2002).

Well	Aquifer	Static water level in masl				Drop in W.L. (m)
		Date	W.L.	Date	W.L.	
Beit Fajjar (substitute)	Cen.-Tur.	15.04.87	582.8	20.11.88	571.2	11.6
Herodion 5	Cen.	28.01.97	430.0	09.11.98	426.6	3.4
Herodion 4	Albian	10.08.86	363.5	07.11.99	292.1	71.4
PWA II	Albian	12.05.99	335.6	26.11.99	325.5	10.1
PWA I	Albian+Cen.	12.05.99	450.4	20.12.99	446.6	3.8
Herodion 2	Albian	27.12.72	416.2	08.12.99	370.3	45.9
Herodion 2a	Cen.	28.01.97	346.0	07.12.99	339.5	6.4
Herodion 3	Albian	04.11.81	400.2	21.08.97	314.9	85.3
Hebron 1	Albian	15.02.98	359.0	15.07.01	325.5	33.5

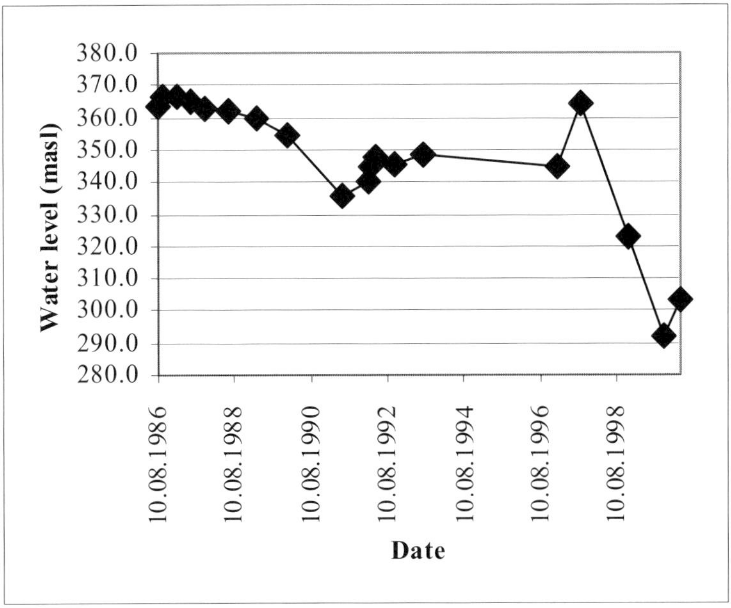

Fig. 1.8.6. Water level at Herodion well 4.

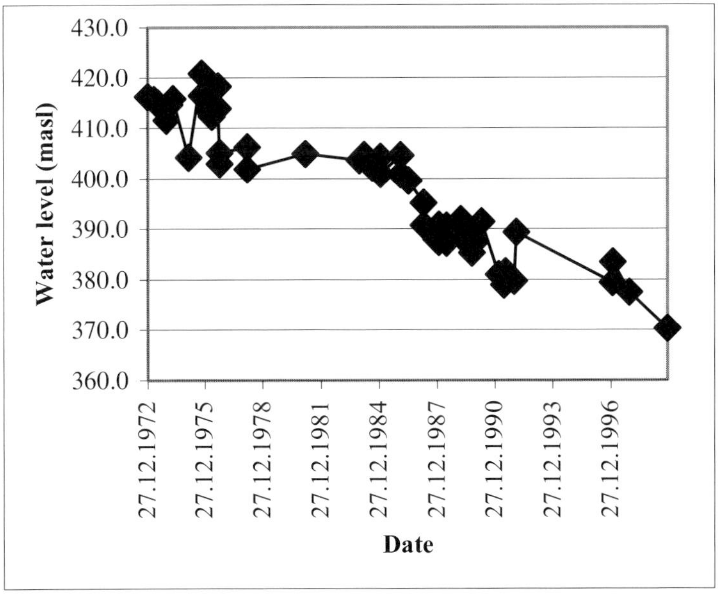

Fig. 1.8.7. Water level at Herodion well 2.

Both figures show a general drop in the water level at the production wells of Herodion 2 and 4, which can be attributed to over-pumping. Despite partial rehabilitation of the aquifer through wet rainy seasons especially that of 1991-1992, it could be noticed that the drop in the water level after 1996 was much sharper than before, especially at Herodion well 4. This could be explained by the installation and putting into operation of new wells, which the 1994 Oslo agreement allows the Palestinian Authority to install. The start of continuously production of these newly installed wells is more affecting the water level of Herodion 4 than Herodion 2. Obviously this is a consequence of the much closer distance of the newly installed production wells Hebron 1 and 2, as well as PWA 1 to Herodion 4. On the other hand the only new well close to Herodion 2 is PWA 3 (Fig. 1.8.8).

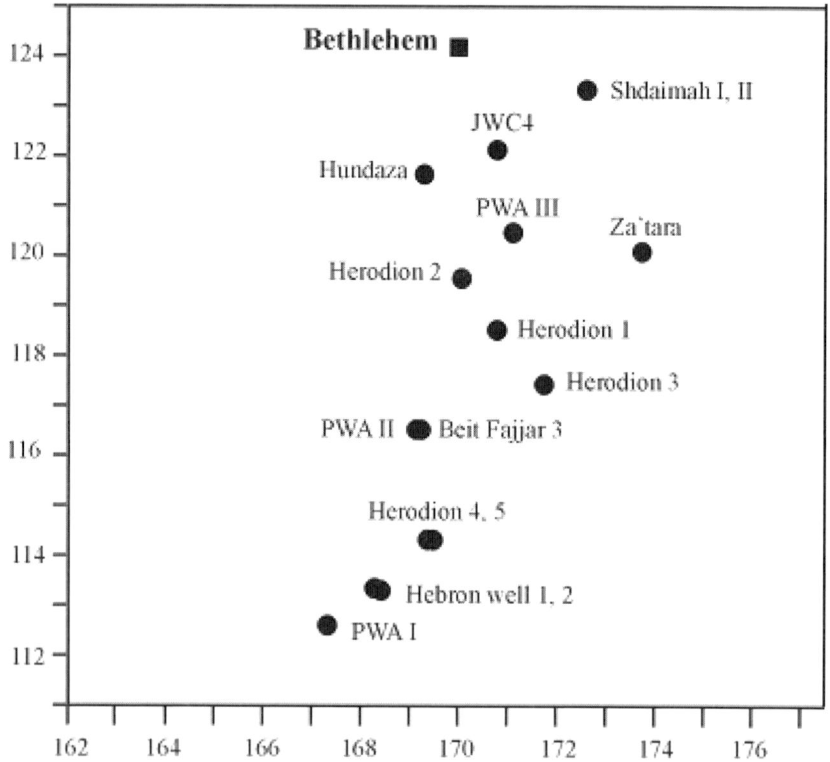

Fig. 1.8.8. Map showing the production wells of Herodion-Beit Fajjar well field (Palestinian grid).

1.8.5 Summary and recommendations

Generally, the Herodion-Beit Fajjar well field has a distinct ground water production potential as it is located within a syncline structure with high aquifer transmissivity. So the well field delivers ground water of high drinking water quality, but is subjected to heavy over-pumping. To protect this precious resource against overexploitation, sustainable management and policy actions must be taken. Examples of such actions could be:

Reduce the abstraction from these aquifers, while compensating this reduction through rain water harvesting using house cisterns and runoff storage by small dams or even more sophisticated artificial recharge methods.

Stop installing new wells in the well field.

Focus on other well fields such as Bani Naim.

Limit urban expansions at the recharge areas of these aquifers

References

Abed Rabbo A, Scarpa D and Qannam Z (1998) A study of the water quality and hydrochemistry of the Herodion - Beit Fajjar Wells, West Bank. Bethlehem University Journal 17: 11-28.

Aliewi A and Jarrar A (2000) Technical assessment of the potentiality of the Herodion well field against additional well development programs. Palestinian Water Authority, Ramallah, Palestine. (unpublished report).

Guttman J and Gotlieb M (1996) Hebron boreholes 1 and 2, final report, 5477-R96.253(E). Tahal Consulting Engineering LTD. Tel Aviv. (unpublished report).

Qannam Z (2003) A hydrogeological, hydrochemical and environmental study in Wadi Al Arroub drainage basin, south West Bank, Palestine. Wissenschaftliche Mitteilungen, Heft 22/2003, Institut für Geologie - Technische Universität Bergakademie Freiberg, Freiberg, Germany. ISSN: 1433-1284.

Millennium Engineering Group - CH2M Hill / Montgomery Watson / Arabtech Jordaneh (2000) West Bank water resources, program 2 and Bethlehem 2000 project – Ground water management modeling, Task 7 – the Hebron model, final report. (unpublished).

PWA - PALESTINIAN WATER AUTHORITY (2002) Wells of the southern West Bank – water table data. Palestinian Water Authority- Water Data Bank Section. Personal communication.

WHO - World Health Organization (1995) Drinking water guidelines, Amman.

1.9 Water Resources, Protection and Management in Palestine

Abu Safieh, Yousef

Minister of Environmental Quality Authority, Palestine

Abstract

Water is the most precious and valuable natural resource in the Middle East and North Africa in general and in Palestine in particular. It is vital for socio-economic growth and sustainability of the environment. Water scarcity has been a source of social and economic stress all over the ages, but water quality is a new problem emerging to dominate the crisis in the area. Preservation of the quantity and quality of Palestinian water resources may be the most important environmental challenge facing the Palestinian Authority. Demand for fresh water in the area exceeds the naturally occurring, renewable supply. Most of the available limited water is being polluted due to Israeli practices, industrial, agricultural wastes and lack of adequate infrastructure of sanitary wastes disposal. Palestine is in critical situation that requires immediate intervention to improve the water situation in terms of quality and quantity

Groundwater is so far the largest source of water for the people of Palestine. Other minor sources are springs, surface water occurrences and collected rainwater. Already the water availability for Palestinians has been decreased to the low average of 25-30 m^3 per year per capita.. Some studies show the water deficit to be in the range of 50 MCM/yr in the Gaza Strip and approximately 70 MCM/year in the West Bank.

This paper presents water resources, protection and management in Palestine. Also it will describe water stress in the region, which is not limited to scarcity but stems from three interacting crises: quantity, quality, and equity. In this region, the geopolitics of water is probably of greater concern than anywhere else in the world. Moreover, because of these crises, any solution must deal with all of the three water problem dimensions (quantity, quality, and equity), and at the same time it has to be economically sound, ecologically sustainable, and politically acceptable. The Palestinian population now living in Palestine is expected to grow to over five million by the year 2020, and the demand for water will far exceeds the sustainable capacity of the aquifer. Continuous urban and industrial growth will place additional stress on the aquifers system, unless appropriate integrated planning and management actions are instituted immediately. It is evident that immediate action must be taken to save and sustain the aquifers system to allow Palestine to support its people and continues agricultural development in the future to meet the needs of the growing population and the needs of the future generations.

1.9.1 Geography and historical background

Palestine has a strategic location, as the gate between Africa and Asia. It is situated on the eastern coast of the Mediterranean Sea; in the transitional zone between the arid desert climate of the Sinai Peninsula and the temperate and the semi humid Mediterranean climate. The total area of the historic Palestine is about 27,000 km^2. After the first Israeli-Arab war in 1948, Israel occupied about 78% of the Palestinian land and left two separated segments under the Jordanian and the Egyptian administrations in the West Bank and the Gaza Strip respectively with a total area of 6,020 km^2. After the 1967 war, which was waged by Israel on the Arab neighboring countries, Israel occupied the West Bank, Gaza Strip and Jerusalem. A peace interim agreement was signed between the Palestinian Liberation Organization PLO and the State of Israel in 1993. The Palestinian National Authority was established in Gaza Strip and fragmented parts of the West Bank in 1994. The West Bank is a term used to mean the lands located west of the Jordan River. It is an area of about 5655km^2 divided into four geographical regions. Up to sixty percent of the population lives in approximately 400 villages, and the remainder in urban areas, refugee camps and cities.

Gaza Strip is a narrow piece of land located on the south eastern coast of the Mediterranean Sea. Its position on the crossroads from Africa to Asia made it a target for occupiers and conquerors over the centuries. Israel occupied the Gaza strip in 1967 after 19 years of Egyptian administration. The Gaza Strip is a very crowded place with an area of 365 km^2. The population is mainly concentrated in the cities, small villages and eight refugee camps.

1.9.2 Introduction

Water Scarcity may be the most crucial environmental problem facing Palestinians today. Groundwater is so far the largest source of water for the people of Palestine. Other minor sources are springs, surface water occurrences and collected rainwater. Palestinians have also water rights in the surface water of River Jordan but they are denied access to this resource. Depletion of the (fresh) groundwater resources is already a severe problem in Gaza and parts of the West Bank. Water demands in Palestine are expected to grow further, aggravating the problem of depletion of groundwater as the major water source. Already the water availability for Palestinians has been decreased to the low average of 70 m^3 per year per capita. For comparison: the average Israeli per capita water availability is about 400 m^3 per year.

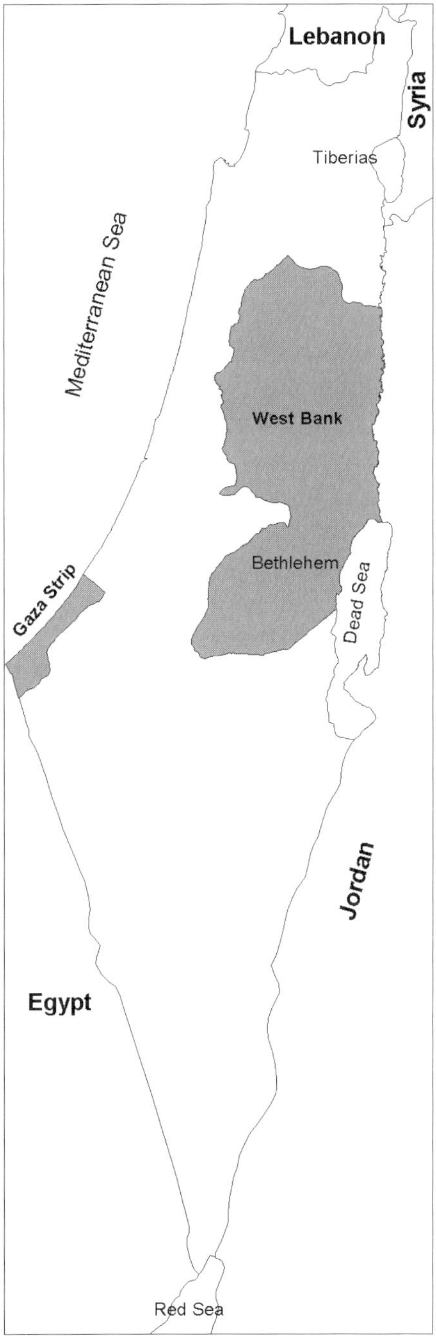

Fig. 1.9.1. Map of Palestine

1.9.3 Water Quantities

Water Quantities in the West Bank

Water Resources is one of the most crucial political issues in the region. Currently, Palestinian use of the West Bank water resources is highly restricted by the Israelis. The Palestinians are not allowed to drill wells in the Western and North Eastern basins, while only limited drilling is permitted in the Eastern Basin. Large parts of Palestinian water demands are obtained through the Israeli controlled well fields in the West Bank, while the largest share of the water from these fields is utilised by the Israelis themselves. The total unconstrained water demands for the year 2000 have been estimated to be approximately 193 MCM/year, leaving a deficit for the West Bank in the year 2000 of approximately 70 MCM/year.

Total Groundwater Quantities

The total annual available ground water quantities in the West Bank Governorates from the existing groundwater sources are around 123.6 MCM out of that 62.8 MCM come from wells and 60.8 MCM from springs. Table 1.9.1 below shows the distribution of water sources between the three basins of the west bank.

Table 1.9.1. Distribution of water sources between the three basins of the West Bank (M3)

Basin	Agricultural wells		Domestic wells		Springs		Total discharge
	No.	Annual yield	No.	Annual yield	No.	Annual yield	
Western	124	16.18	21	6.45	34	1.98	24.61
North eastern	61	3.50	6	4.91	24	3.26	11.67
Eastern	145	14.58	13	17.24	56	55.54	87.36
Total	330	34.26	40	28.6	114	60.78	123.64

Surface water (River, Wadis and seasonal lakes)

The only permanent river in the West Bank is the Jordan River. The Jordan River satisfies around 50% (MOPIC, 1999) of Israel's and Jordan's demand, Lebanon and Syria are minor users, gaining 5% of their water demand through the river water. Israel is forbidding the Palestinians their share in the river. Mostly the wadis are of seasonal type; they flow only in the winter season during the flood periods. The estimated quantities of lost flooded surface water are about 70 MCM/yr (MOPIC, 1998). In addition to these wadis there are seasonal lakes, which can provide the West Bank with additional annual millions of cubic meters of water.

Other water resources

Among the other resources of water, cisterns are of major importance in the West Bank. Most of them are of a volume ranging from 60 to 100 cubic meters. The total water quantity collection through these cisterns in the West Bank is estimated to be about 6.6 MCM.

Water Quantities in the Gaza Strip

The main source of water in Gaza Strip is the groundwater from the coastal aquifer. The Coastal Aquifer Management Program (CAMP) estimates the total Palestinian municipal and agricultural abstraction as 127-147 MCM and the Israeli Mekrot abstraction as 5-8 MCM and the natural groundwater discharge as 10-15 MCM. The estimated total inflow is 131-160.2 MCM. These figures give a water deficit in rang of 9.8-28.9 MCM/yr. Other studies show the deficit to be in the range of 40 to 60 MCM/yr. (PES, 1999).

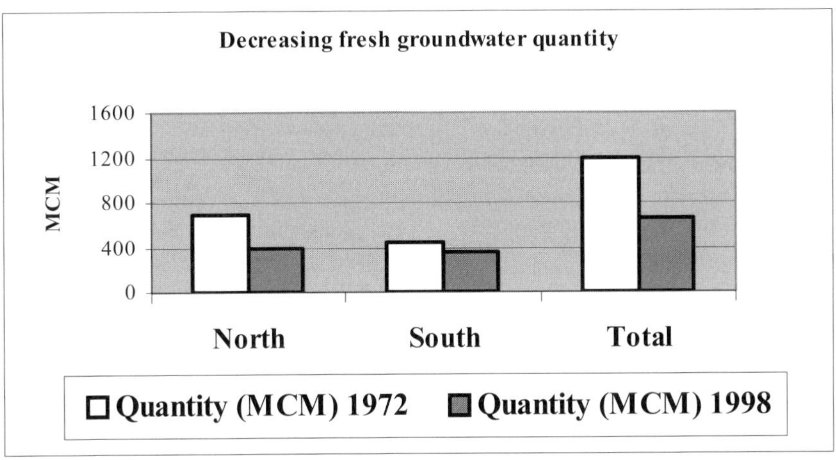

Fig. 1.9.2. Decreasing fresh ground water quantities. *Adapted from PES, 1999*

1.9.4 Main causes of depletion of water resources

The gap between available water resource supplies and future demands is continuously exacerbating due to fast growing population and development in the Palestinian territories, the over-pumping of the Israeli controlled wells and the Israeli domination on the Palestinian water resources, which is not in harmony with limited water resources. The following points summarized the main causes of depletion of water resources:

- Palestinian access to the West Bank water resources are highly restricted by the Israelis,
- Limitation for funding of environmental actions in water sector,
- Insufficient monitoring systems and networks for data collection,
- The geographical discontinuity of areas in the West Bank and Gaza Strip,
- Mismanagement, over utilization of natural resources and lack of sovereignty over land use strategies,
- The volume of the groundwater abstracted by the Israeli settlers, and by the Israeli wells around the border of Gaza Strip,
- Palestinians have water rights in the surface water of River Jordan but they are denied access to this resource,
- Absence of legislation, rules, and regulations to control, protect, and sustain all aspects of the water sector.

1.9.5 Water Quality in Palestine

Water Quality in the West Bank

Excluding some areas where saline groundwater is encountered, the groundwater quality in the West Bank is still considered good. However, the shallow and unconfined aquifers are susceptible to wastewater pollution. Also, discharge of untreated wastewater and dumping of solid and hazardous waste by the Israeli settlements pose serious risks for the quality of the groundwater resources. It is estimated that about 34% of the Israeli settlements' wastewater is discharged without treatment in the West Bank wadi systems. Infiltration of leachate from solid waste dumping sites comprises another large and uncontrolled risk of contamination. Also, the excessive use of fertilizers and pesticides poses a serious threat to the water quality of aquifers and springs. Information on these sources of pollution is very scarce indeed. The figure below gives the classification of domestic wells according to nitrate concentration in different West Bank Governorates. The world health organization guideline for non-polluted water is 50 mg-NO_3^-/liter for drinking purposes.

Water Quality In Gaza

Pollution of the groundwater is a major problem in Gaza. Not only there are numerous sources of pollution, but also the aquifer of the Gaza Strip is highly vulnerable to pollution. Figures 1.9.4, and 1.9.5 presents chloride and Nitrate pollution in Gaza. The domestic water is becoming more saline every year and average chloride concentrations of 500 mg/l or more is no longer an exception. The permissible limits for nitrate are exceeded by a factor of 8 times for a number of public wells; most of the public water supply wells don't comply with the drinking water quality standards.

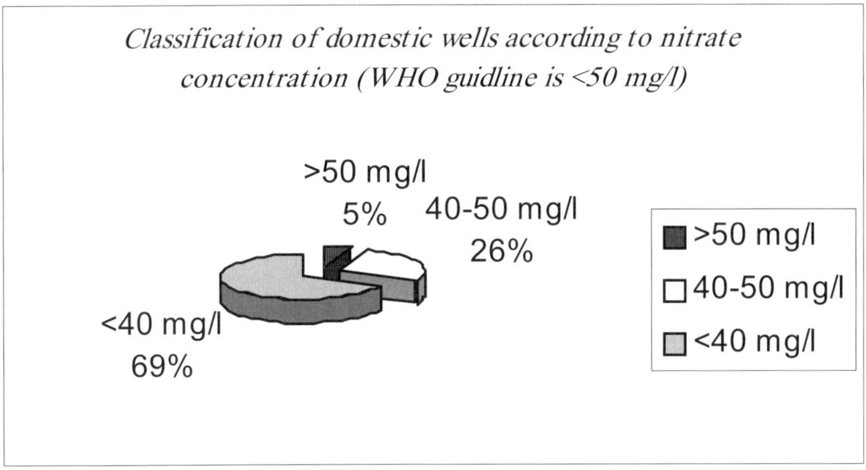

Fig. 1.9.3. Domestic wells Nitrate content in the West Bank. *MOPIC, 1998*

The main causes of deterioration of groundwater quality are:

- Infiltration of untreated domestic and industrial wastewater;
- Discharge of poorly operating wastewater treatment plants;
- Infiltration of wastewater from cess-pits;
- Mismanagement and extensive use and disposal of pesticides and fertilisers;
- Leakage from unprotected solid and hazardous waste dump sites;
- Seawater intrusion into coastal aquifers (due to over-pumping);
- Saline water up-coning due to over pumping;
- Discharge of industrial waste and waste water generated by the Israeli settlers;

1.9.6 Water Management

Existing Strategies and Policies

The real water resources management has started only a few years ago especially after the establishment of the Palestinian Water Authority (PWA) and the Palestin-ian

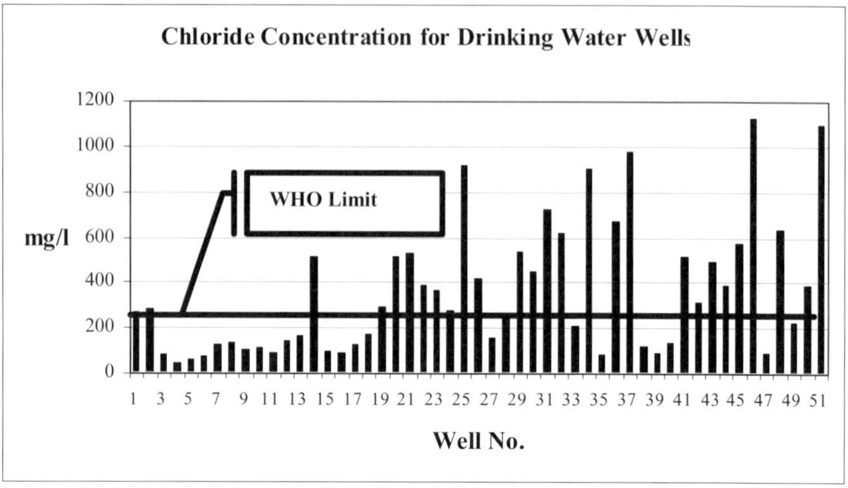

Fig. 1.9.4. Domestic wells Chloride content in Gaza

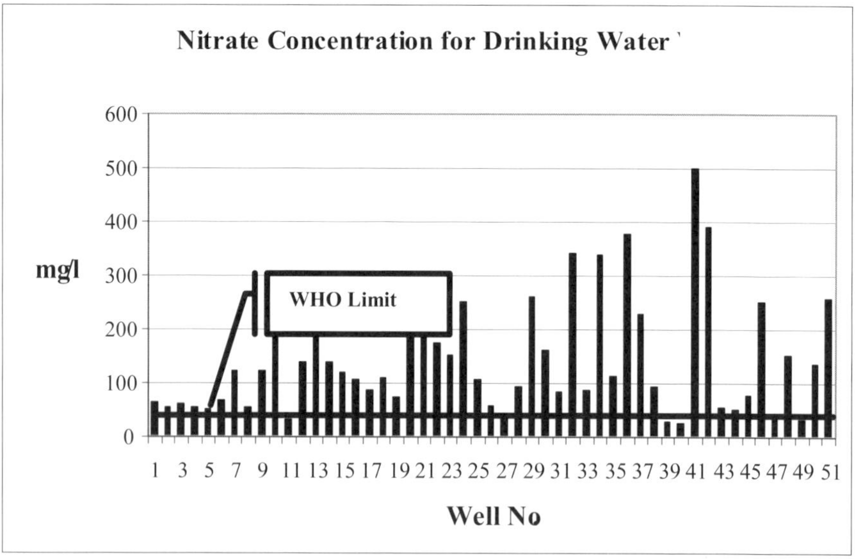

Fig. 1.9.5. Domestic wells Nitrate content in the Gaza Strip

Environmental Quality Authority (EQA). The PWA has the authority and mandate over water resources management, policy and strategy. In 1999 the PWA has published the Palestinian Water Policy and Strategy. The EQA has developed the Palestinian environmental strategy, which provides a systematic analysis of the causes of current problems; it identifies the most urgent environmental themes that needed to be addressed in the coming years. Depletion of water resources and deterioration of water quality are considered as the highest priority. In order to achieve the goals of water resources management two environmental targets were stated to minimize the depletion of water resources and to safeguard the water resources system from any form of serious pollution. The environmental targets are formulated as:

"To achieve a balanced situation in which the Palestinian water rights are effectuated and the total Palestinian water demand is met under the condition that annually the total abstractions (Palestinian and Israeli) from the water resources system do not exceed the natural and artificial recharge of the water resources system''.

"To protect the quality of the water resources in order to be suitable for the desired or designated uses of water. The suitability is expressed as maximum allowable concentrations of various substances or groups of substances for particular uses of water resources''.

Furthermore the PWA has formulated its strategy, which is based on the following activities:

- Pursue Palestinian water rights
- Strengthen national policies and regulations
- Build institutional capacity and development of human resources
- Improve information services and assessment of water resources
- Govern water and wastewater investments and operations
- Enforce pollution control and protection of water resources
- Promote public awareness and participation

Institutional Arrangement

Several ministries and authorities are involved in water quality management including planning, development, source protection, monitoring, research and regulation. The following institutions are the main actors that either produce regulations or follow them regarding the water and its management and development. The main ministries and authorities are:

The Palestinian Water Authority (PWA) was established by Decree no. 90/1995, and its power and responsibilities were defined in Law no. 2 of 1996. the mandate of PWA is to ensure the right of water for each citizen within the framework of sustainable development of the available resources. The responsibilities of PWA cover water resources, wastewater management as well as sanitation. The authority has a policy making and a regulatory role and is in-charge of:

1. Strategic planning of water resources,
2. Monitoring and protecting water resources, and
3. Regulating the water sector including licensing of water abstraction, disposal of wastewater and reuse of treated wastewater

The Ministry of Planning and International Cooperation was established in 1994. The Ministry is concerned with the physical and strategic planning for rural and urban areas. It is involved in national development plans for medium and long term and with international cooperation and coordination.

Environmental Quality Authority (EQA): The prime responsibility of EQA is to promote sustainable development of the Palestinian society. Its main task is to protect the environment, including its water, soil, air, natural resources, nature and bio-diversity, and to prevent public health risks related to environmental issues. EQA is responsible for giving environmental approval for water projects. EQA is also responsible for following up the Environmental Impact Assessment or Initial Environmental Examinations of these projects. In addition, EQA, in cooperation with other concerned institutions, is responsible for setting water quality standards, increasing public awareness in the water sector, protection of water resources, and setting the required regulations.

Ministry of Agriculture (MOA) is in charge of developing water resources for agriculture, controlling the quality of water for irrigation, as well as monitoring the import, marketing and usage of agrochemical. The ministry has a water laboratory for monitoring of the quality of irrigation water and recently the ministry has established a specialized laboratory for analysis of pesticides.

Ministry of Local Government (MoLG) is assigned responsibility for the local government system and has been actively engaged in defining the structure of local government, the institutional arrangements and the key organizations at the various levels and the role and functions at these levels. The ministry also supervises the works of the municipalities for all infrastructure works including water, sewage and drainage.

The Municipalities and the village councils are responsible for providing the public with water supply. In the areas where there is a collection system, the municipalities are also responsible for the wastewater services as well.

Ministry of Health (MOH): The development objective of Ministry of Health is to improve preventive and curative health care for the entire Palestinian population. It is responsible for domestic sources of pollution. Monitoring drinking water quality and its health impact is a major responsibility of the Ministry of Health in water sector.

In addition to the above organizations, there are several councils and institutes that are in charge with water or have responsibilities related to water. Among these is the national water council (NWC) that was nominated under article 8 of law No. 2 of 1996.

Legal and regulatory framework

The Water Law, confirms and assures the optimal utilization of water resources and addresses the protection of public water from pollution, the reuse of treated water and identifies sanctions to be applied in case of violating the conditions of protecting water quality and /or using polluted water for irrigation. The Environmental Law is clear in assigning tasks and responsibilities to EQA and other agencies, without creating duplication of tasks, or interfering with other activities. A number of principles and guidelines have been defined in the Environmental Law, which form the basis for environmental legislation, decisions, and structures of environmental institutions. Accordingly the EQA has prepared regulations and standards in many different themes regards water pollution control. Regulations and standards prepared pursuant to the environmental law and related to water pollution control include: Drinking water quality standards, wastewater reuse standers and its executive list, standards of wastewater discharge to the sea, discharge of industrial sewage into the sewage system.

Ongoing and planned water projects

After the establishment of PWA (1996) and EQA (1998) several water and wastewater project has been implemented. These projects have direct or indirect positive impacts on water quality improvement in Palestine. On the other hand several projects are ongoing or planned. Also large efforts have already been taken to improve the wastewater situation in the West Bank and Gaza. Various feasibility studies and design of wastewater treatment plants have been completed, and construction of several plants is soon to be expected. Still a large effort is needed to realise full collection and treatment of all domestic and industrial wastewater in Palestine. Reuse of treated wastewater for irrigation purposes has to be further developed.

1.9.7 Main water resources management issues

Water Resources Policy

Water resources must be developed and managed efficiently in order to meet present and future water needs, in an environmentally sustainable way. The PWA has considered the following three principal objectives for sustainable water resources management:

- Providing quantity and quality of water for domestic purposes in compliance with WHO standards.
- Supplying adequate quality and sufficient quantity of water that is required for the planned agricultural production.
- Managing the Groundwater Aquifer at its safe yield and preventing further deterioration of the aquifer water quality.

Institutional setup

Most of the environmentally related institutions are still facing some major institutional deficiencies in addition to the other technical or other issues. The lack of clear mandate for some of the national level governmental institutions is still resulting in some duplication of efforts as well as causing confusion to the public and even to donors. The absence of clear task description to the different departments and personnel is resulting in internal confusion and delays in performing the required tasks. The lack of defined approved strategies for some institutions is resulting in short term action plans based on external interference and support.

Legal and Regulatory Framework

The legal system in Palestine is derived from several past and present legal systems. The legal system in the Gaza Strip is derived mostly from Egyptian legislation whilst that of the West Bank has been strongly derived from Jordanian legislation. During the Israeli occupation there was insufficient developmental and environmental legislation in Palestine. The many military orders were only enforced to serve the occupation and the Israeli security objectives with no regards to sustainable development. The Palestinian National Authority has established a legal and institutional infrastructure such as the Palestinian Legislative Council (PLC) and the civil institutions.

Information and Awareness

Collection of information and data on water quality and quantity is the basis for water management and water policy. There are some available data concerning water, but the method for distribution and dissemination of information to concerned stakeholders and to the public is poor. Public awareness of water pollution has been raised and has stimulated a common concern over the water resources management. Awareness regarding water quality management, option and alternative for improvement, clean technologies, and proper agricultural practices require considerable strengthening.

1.9.8 Options for improvement

The situation of water sector especially the water quality and quality management issues indicate that much work is still needed to meet the challenges and tackle the problems in water and its management. The actions and developments regarding water management should be done without duplication to increase the benefit and effort and to ensure that the actions are taken for specified objectives without delay. Required actions to abate depletion of water resources and deterioration of water quality are:

- Management of domestic and industrial waste water

- Implementation of water projects for maximization of water availability
- Setting water quality standards and norms, wastewater emission standards and regulations, and solid waste disposal regulations
- Minimization of water demands and losses
- Encouragement of capacity building and improvement of institutional set-up
- Establishment of a data base system on water quality and pollution sources
- Management of fertilisers and pesticides
- Management of water resources and irrigation practices
- Issuing of lists of hazardous substances and waste
- Control of waste import
- Enforcement of (transboundary) groundwater protection zones
- Development of an integrated solid and hazardous waste management plan. Development of a management plan for agrochemical practices
- Introducing an economic regulatory instruments and economic measures;
- Development of a monitoring system for groundwater quality

References

Al-Jamal and Shoblack (2000) Institutional reforms towards sustainable water sector.
Ministry of Environmental Affair, Palestinian Environmental Strategy, (2000)
Ministry of Environmental Affairs, Land base Pollution Sources, (2001)
Ministry of Environmental Affairs, State of Environment, (2000)
Ministry of Environmental Affairs, Environmental Action Plan, (2000)
Palestinian Water Authority, water Policy (1996)
Ministry of Heath, Yearly reports related to water quality monitoring and heath
The Coastal Aquifer Management Program (CAMP) reports

1.10 The Role of Transient Flow and Drainage on Regional Groundwater Flow Pattern in Semi-Arid and Arid Climates

Klaus-Peter Seiler

GSF-Forschungszentrum, D-85758 Neuherberg

1.10.1 Introduction

Subsurface water, the oceans and the atmosphere form the largest reservoirs on earth. The main link between them is the hydrologic cycle, which provides renewable water for beings, ecosystem functions, weathering and the sediment transport and, which is co-responsible for the temperature equilibration on earth as well. The driving force of the water cycle is solar radiation.

Water resources are horizontally uneven distributed on continents. This is due to the distribution of the incoming solar radiation, the albedo and heat capacities of the interface atmosphere, lithosphere, biosphere, as well as the global atmospheric and ocean circulation patterns and its modifications by ocean floor and continental topography. According to the most recent world water balance (UNESCO 1999), the yearly discharge from the continents to the oceans amounts to 44,800 km³ and the distribution of discharge by continents is shown in Table 1.10.1.

Table 1.10.1 Distribution of discharges in world regions (1922-1985) as related to the mean discharge of 44,800 km³ and the availability of water per capita and year (UNESCO 1999).

Continent	Percentage of discharge	Available water in m³/capita and year
Asia	31.5	3,920
Europe	6.7	4,200
Africa	9.8	5,720
N-America	18.4	17,400
S-America	28.0	38,200
Australia & Oceania	5.6	83,700

The annual water demand of humans amounts

- To a total of about 1,000 to 1,500m³,
- For households to about 50m³ and
- For the minimum to survive to about 5m³.

As compared to the averages in Table 1.10.1, Asia, Europe and Africa might soon reached the limits of water availability by quantity means or already did in special areas. By quality means, it should be considered that the use of blue water produces grey water and by experience,

- $1m^3$ of untreated grey water needs $9m^3$ and
- $1m^3$ of treated grey water $3m^3$ of blue water

to efficiently support natural attenuation before the continental discharge joins the ocean in exoreic and salt pans in endoreic hydrologic systems. These numbers and relations in mind, it gets obvious that in the near future by quantity and quality means, only N-America, S-America and Autralia & Oceania will not suffer from water shortage; in many other areas water shortage already exists.

The Middle East is so short in renewable water resources, that most of the nations of this region already overexploit the existing water resources; by nature, this has meteorological and hydro geological reasons and gets aggravated by the rate of population growth, sometimes an economic development in disagreement with nature and will get worse in the future by the expected global warming, as far as a transient behaviour of the subsurface system does not buffer the immediate consequences of global changes.

From the many facets of water shortage, this paper emphasis to draw attention to two facts, which are mostly disregarded by hydro geologists, but quite important in semi-arid and arid climates for the development of appropriate strategies of water management. These facts are the inherited transient hydraulic behaviour of many water resources in arid and any deep drainage in all climate zones.

1.10.2 General flow pattern in groundwater resources

On continents, the groundwater volume amounts to $8,000,000km^3$; as compared to this the volume of lake and surface water, soil and air humidity is negligible. In comparison with $8,000,000km^3$ of groundwater, the overall discharge from continents of $44,800km^3$ is small. Supposing that this discharge entirely recharged groundwater and was evenly distributed in the subsurface, the mean residence time of the groundwater resource was minimum 178 years. This assumption, however, is in obvious disagreement with all field observations on pollution and radioactive environmental isotopes in groundwater.

Groundwater flows in aquifer systems, in which each aquifer has its individual hydraulic properties. To demonstrate the quantitative turn-over of groundwater recharge in the subsurface, numerical modelling has been performed with commonly known distributions of hydraulic parameters; these numerical results have been compared with field studies (Seiler, Lindner 1995). According to the numerical studies, the turn-over quantities as compared to the groundwater recharge in individual layers is in near surface aquifers (active turn-over zone) more than 85% and beneath (passive turn-over zone) less than 15% (Fig. 1.10.1). Related to these results, groundwater in near surface aquifers is young (<50 years) and in deep aquifers old (>100 years). Both these turn-over zones occur world wide (Seiler, Lindner 1995, Alvarado et al. 1996) and are underlain by connate water (Fig. 1.10.2), which originates from sedimentation times and did not join the biosphere since. Thus, there is not only a horizontal but also a vertical uneven distribution of groundwater recharge.

In dry and hyper-dry areas, groundwater recharge is mostly of a patchy type; here it is often difficult to differentiate all over the catchment area between an active and passive turn-over zone in groundwater. Contrary, in semi-arid areas the active recharge zone always exists and has a thickness of meters to a few decameters, in tropics of some decameters and in temperate climates of maximum 100 m. Generally spoken, the thickness of the active recharge zone depends on effective groundwater recharge and the drainage and storage properties of the respective aquifer system.

Since only a small portion of continental groundwater is actually recharged and most of the present recharge does not penetrate to deep aquifers, the majority of groundwater resources belongs to a long-term reserve (Freeze,Witherspoon 1967) respectively to historic or even fossil groundwater (> 10,000 years old).

As a golden rule in water management, the long-term availability of water for beeings and ecosystems should not exceed the excess water of the water cycle (precipitation minus actual evapotranspiration). All use in excess is called groundwater mining or overexploitation and leads on a long run of time to a decline of water tables and of ecologic functions of landscapes and special cases even to a splitting of the hydraulic heads in the aquifer systems, often favouring deep drainage of aquifer systems. This golden rule, however, can be broken, if unproductive losses of groundwater occur (chapters 2.1, 2.2).

On a very long run of time only groundwater recharge, on a short run of time, also the transient behaviour of aquifer systems (chapter 2.1) and an adapted economic development are the sustainable regulating factors of the welfare of a region.

Related to groundwater recharge, there exists outstanding differences between water resources in temperate/tropic climates as compared to semiarid/arid climates:

- As an average temperate/tropical climates receive a significant groundwater recharge (> 30 mm/year), the catchment areas are essentially small in size, and the discharge system is predominantly of the exoreic type.
- Opposite, in semi-arid to arid climates groundwater recharge is small (< 30mm/year) and feeds groundwater resources of very large catchment areas, which mostly belong to the endoreic type.

Small recharge numbers, which are representative for arid to semi-arid climates, thus can sum up in huge catchment areas to a significant discharge as it is known from desert springs (see chapter 2.1).

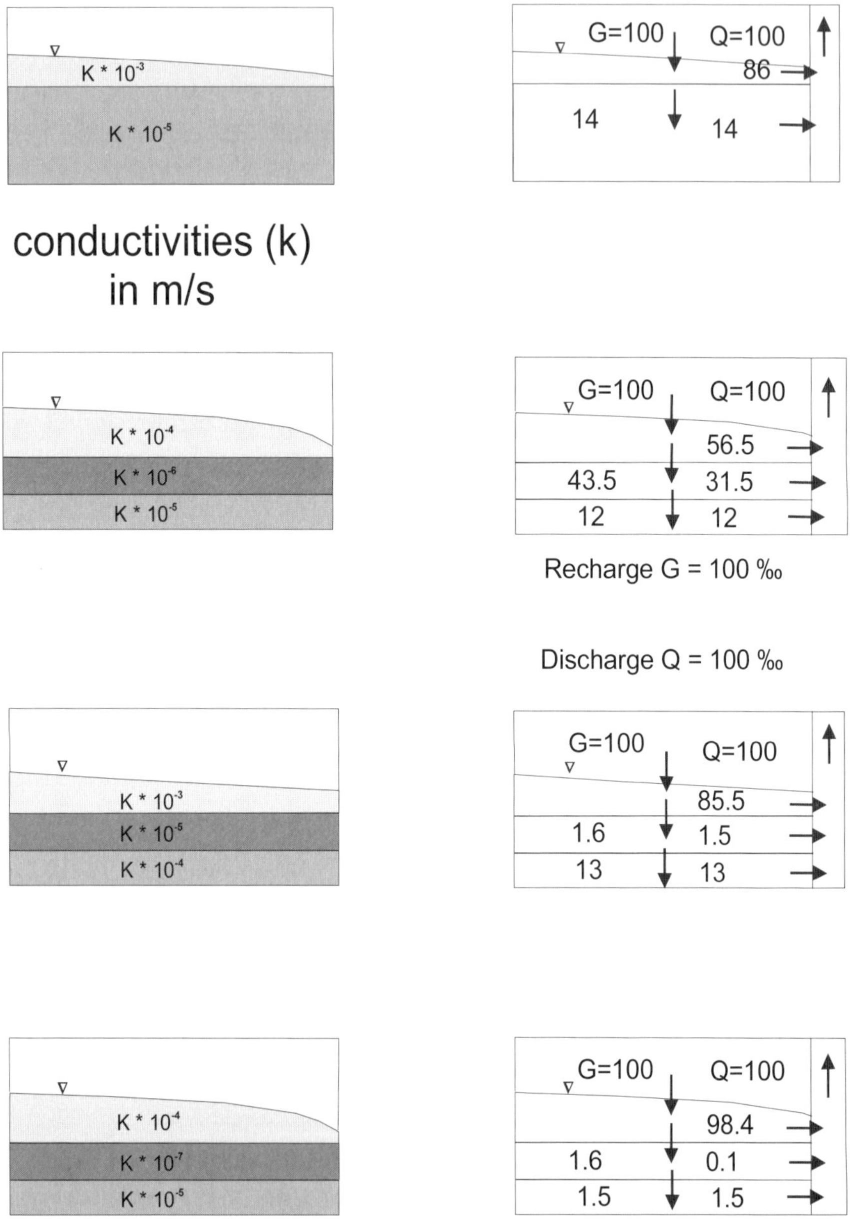

Fig. 1.10.1. Numerical results on the distribution of groundwater recharge in individual strata of aquifer systems. Left site: the distribution of hydraulic conductivities. Right site: the turn-over quantities relative to recharge (≅ 100%).

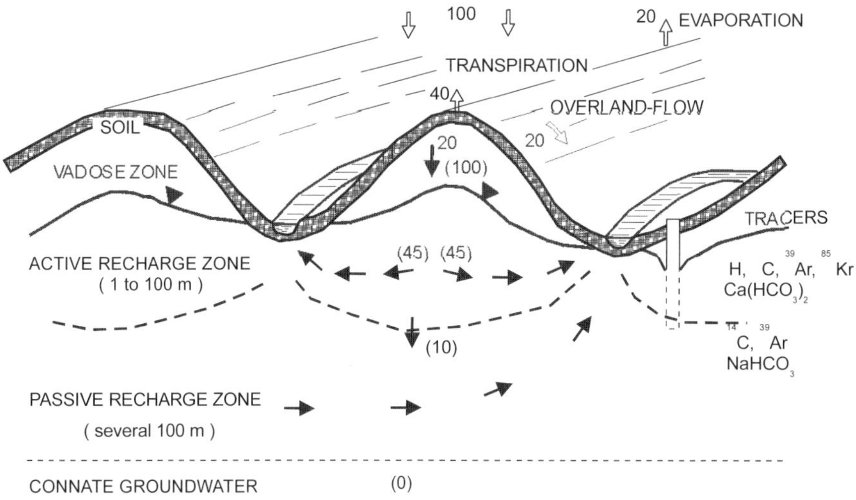

Fig. 1.10.2. Scheme of the three main turn-over zones of subsurface waters.

1.10.3 Transient behaviour of groundwater flow

Most of the desert areas of the world belong to endoreic discharge systems; only in the case that extended wet regions, which are separated by desserts from the oceans (Egypt, Punjab), have to crosses the dessert an exoreic discharge exists and the recharge of aquifer systems in the desert area is locally enhanced by river bank infiltration.

Many desert areas underwent within the last 50,000 years repeatedly climate changes from wet to dry or hot to cold (Sonntag et al. 1979, Edmonds et al. 2003). Coupled with this climate changes, the discharge system turned from exoreic into endoreic and vice versa, which led in the case of diminishing precipitations to an enlargement of the sizes of the catchment area, by pulling down former subsurface water divides; this enlargement of the size of watersheds is a slow process, coupled with changes in the groundwater flow directions and thus introduces to the subsurface system a long-term transient, hydraulic behaviour.

Taking ^{14}C-dating of groundwater in the Near East and north Africa (Fröhlich et al. 1987), most data group either close to 25,000 years or to 5,000 years; this is interpreted as times of high recharge intensities as compared to present. These large turn-over times in spring water of deserts and an exponential age distribution in mind, results in an actual groundwater recharge contribution of 0,0004 with respect to 5,000 years and 0,00004 with respect to 25,000 years. In other words, the actual groundwater recharge contributes to less than 1‰ to the observed spring discharge.

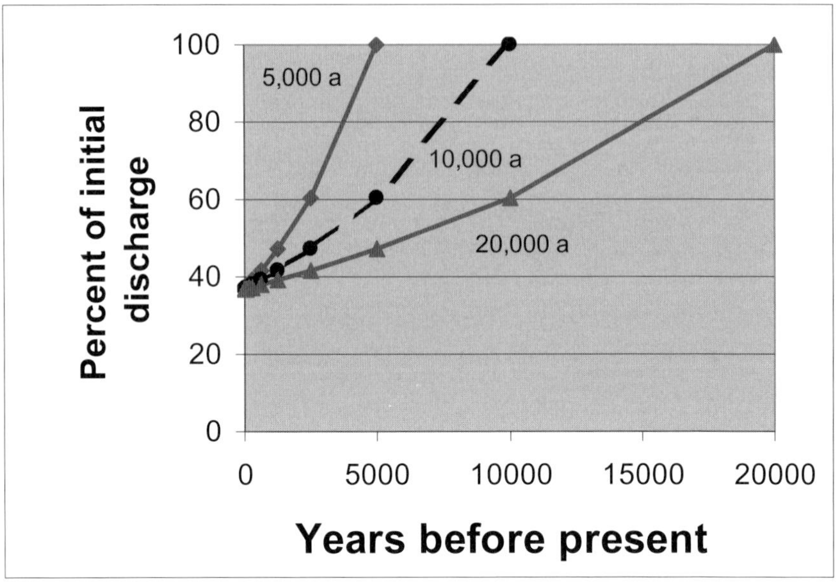

Fig. 1.10.3. Outflow of a spring according to an exponential age distribution in groundwater. 20,000a = mean residence time equals to 20,000 years.

Characterizing the outflow (D) of a subsurface reservoir without groundwater recharge by an exponential equation

$$D_{(t)} = D_{(T)}\, e^{-t/T}$$

D = discharge, T = turn-over time, t = time variable,

it can be seen from fig. 1.10.3 that within 5,000 to 20,000 years the original discharge reduces to about 1/3. This actual discharge, however, does not reflect the present groundwater recharge conditions, but expresses the transient behaviour of a groundwater system, which gets even more transient, if a change from an exoreic into an endoreic system changes the extend of the catchment.

The Azraq spring in the Jordan Badia region discharged about 1.2m³/s before systematic groundwater abstraction by tub-wells started close to the spring area. The spring draines an orographic area of 12,700km²; for this catchment and with the actual discharge, a recharge rate of 0.1L/(s km²) or 3mm/year is calculated; this however was incorrect, because the water age is about 5,000 years and the contribution of actual recharge was calculated to be less than 1‰. This interpretation is supported by Tritium analysis of samples from shallow wells: only few samples out of dug wells close to Wadis yielded detectable ^{3}H concentrations all other wells were free of Tritium (see chapter 1.10.2) or recently recharged groundwater.

Referring to a three times higher discharge 5,000years ago and the same size of the catchment resulted in 0.3 L/(s km²) or 10mm/year for the past and considering a smaller size of the catchment would approach the groundwater recharge to 20 to 30mm/year, which was close to the groundwater recharge under actual semiarid climate conditions that probably prevailed 5,000 years ago in this area.

All these data contribute to assess the actual system as a transient one, which is not representative for present, but for past climate conditions.

Since groundwater from actual till historic times discharges through the Azraq springs and gets lost by evaporation, it was convenient to use it for water supply, although it is actually not measurably renewable; long-term groundwater abstraction under these conditions, however, should be somewhat lower than the natural spring discharge (chapter 2.2). As can be seen from fig. 1.10.3 the actual discharge recession is smoothly slowing down; under these transient conditions, groundwater management can still be assessed as sustainable or available for more than three generations.

In the case of the Azraq springs no real water balance can be established, because it belongs to a transient system. For management purposes the unproductive water losses through evaporation can be reduced and transferred to humans. This, however, should be well elaborated and monitored, to avoid return any access of saline water from the Sabkah area or from very deep groundwater (chapter 2.2) by an excess draw down in single wells or by well fields.

1.10.4 Drainage through the Jordan rift

Groundwater flow in a catchment is governed by

- Actual and past groundwater recharge and
- The altitude of the discharge area.

It is modified by the petrographic and tectonic properties of the aquifer system.

Israel and Jordan are located on the unstable shelf of the Near East and hydraulics of the groundwater system are dominated by

Near surface discharge, because of the rock mechanic properties of the consolidated sediment and

Deep drainage because of the Jordan graben with a discharge level of –490m asl.

The tectonic development was studied by Neev,Emery (1967) and Bender (1968), the hydrodynamic consequences of this tectonic development have been discussed by Salameh, Udluft (1983).

- Till the late Miocen the axis Aqaba-Hula was uplifted and the uplift axis corresponded to a continental subsurface water devide; groundwater west of this axis discharged to the Mediterranean Sea, east of it in direction Saudi Arabia, Indian Ocean.
- This uplift was followed by the graben formation, creating a new discharge base for surface and groundwater and allowing the Mediterranean Sea to get

in temporary connection with the graben; this gave rise to the Plio-Pleistocen salt deposits in the graben area, lasting till about 100,000 years b.p..

- From 100,000 b.p. to about 20,000 b.p. no connection to the Mediterranean Sea existed and the remaining lake level dropped by climatic and tectonic reasons to about –400m asl.

From Miocen on the axis Aqaba-Hula changed from a continental subsurface water divide into an endoreioc discharge area. Linked to this tectonic development

- Groundwater flow directions and hydraulic gradients changed, and
- The pressure distribution in the aquifer system of the region was lowered.

These changes in pressure distribution within the aquifer system express in a remobilisation of petrol, once caught in sedimentary or tectonic traps. Mobilized petrol reappears as asphalt on the floor and sometimes at the level of the Dead Sea; this asphalt in the abiotic environment of the Dead Sea originates from flushed petrol resources.

Since the aquifer system in the Near East is characterized by fissure flow and fissure permeability decreases statistically with depth (Seiler 1968), groundwater flow also diminishes with depth. As compared to the graben formation, hydrodynamic changes are slow and therefore proceed in this special case from depth to the surface and express in regional dimensions; aside from the slow depletion of groundwater reservoirs (chapter 2.1), also tectonics makes groundwater flow transient for a long run of time.

An isotope-hydrochemistry study of the Azraq groundwater, about 130 km east of the Dead Sea, showed (Almomeni, Seiler 1996) that

- Groundwater close to the surface flows to the two Azraq springs and the Azraq Sabkah and
- Chloride concentrations in the Azraq area are unexpetedly low till 1,000m depth (<1g/L) and than rise to a maximum of 86g/L, which exceeds by far the salt content of ocean water, presumably because of diagenetic processes.

Since water of such huge salinity is of high specific weight, fresh water on saline water cannot reach a thickness of 1,000m, unless the fresh water system is hydraulically disturbed, these observations are also interpreted as a drainage, triggered by the Dead Sea, which pulls down fresh water to great depth.

This interpretation is in good agreement with Salameh, Udluft (1983), reporting from groundwater flow directions in Jordan, directed from the graben shoulder to the east in near surface aquifers and in the same area to the west in deep aquifers: Near surface groundwater still follows the inherited flow direction, deep groundwater is already influenced by the graben drainage and gets fed by lekage of near surface groundwater.

1.10.5 Conclusions

In desert areas, springs in the neighbourhood of Sabkahs belong very frequently to transient groundwater flow systems in extended catchment areas: as a rule, any contribution of present groundwater recharge is negligibly low. Under these conditions, the traditional rules of groundwater management do not apply any more. Since this transient groundwater recharge gets lost by evaporation, the unproductive losses was better used for beings.

The management of a transient groundwater flow system has to adapt to the transient behaviour of the discharge system and should be controlled hydro-chemically and isotopically; this can be realized by monitoring the movement of the fresh-/saltwater interface along the Sabkah and beneath the freshwater area; if this interface moves laterally or upward to the exploitation wells, groundwater abstraction is in disequilibria with natural subsurface discharge.

As in many other areas of the world, there exits in the Azraq area, Jordan additionally a deep drainage, which originates from the development of the Jordan graben, and enhances the transient behaviour of the aquifer system in the Badia desert.

It was wise to explore and exploit the water resources close to the Azraq springs and to use unproductive water losses for the water supply; it was, however also of interest to better explore and exploit the water resources from deep drainage before entering the Dead Sea and then being lost. For the management of these deep resources the same rules apply as in the near surface groundwater around Azraq.

References

Almomani, M, Seiler, K-P (1996) Environmental isotope study of the shallow and deep groundwater in the Azraq basin, Jordan. - Proc on Isotopes in Water Res. Management, vol. 1, 433-434, (IAEA) Vienna

Alvarado, J, Seiler, K-P, Trimborn, P (1996) Investigación hidrogeologicá, isotopica and hidrichimica de la cuenca del Lago de Valencia, Venezuela. – Proc on Isotopes in Water Res. Management, vol. 2, 281-299, (IAEA) Vienna

Bender, F (1968) Geologie von Jordanien. Beitrag zur regionalen Geologie der Erde. – 235 p., (Bornträger) Berlin

Edmonds, WM, Dodo, A, Djoret, D, Gasse, F, Gaye, CB, Croni, IB, Travi, Y, Zouari, K, Zuppi, GM (2003) Groundwater as an archive of climatic and environmental changes. – In Battarbee, R.W., Gasse, F., Stickley, C.E. (Eds.): Past climate variability Through Europe and Africa, (Kluwer) Dordrecht (in press)

Freeze, RA, Witherspoon, PA (1967) Theoretical analysis of regional groundwater flow: 2. Effect of water table configuration and subsurface permeability variations. - Water Resources Res.3, 623-634

Fröhlich, K, Geyh, MA, Verhagen, BT, Wirth, K (1987) Isotopenhydrologische Methoden zur Begutachtung von Grundwasser in Trockengebieten. - Res. Rep. of the Federal Ministry for Economic Cooperation 85, 179 p., (Weltforum Verlag) München, Köln, London

Neec, D, Emery, KO (1967) The Dead Sea, depositional processes and environments of evaporites. – Geol. Survey Israel 41, 147 p.

Salameh, E, Udluft, P (1984) Hydrodynamic pattern of the central part of Jordan. - Geol. Jahrbuch; Hannover

Seiler, K-P (1968) Kluft- und Porenwasser im Mittleren Buntsandstein des südlichen Saarlands. – Geol. Mitt. 9, 75-96

Seiler, K-P, Lindner, U (1995) Near surface and deep ground waters . - J. Hydrol. 165, 33-44, Amsterdam

Sonntag, C, Klitsch, E, Loehnert, EP, El-Shazly, EM, Munnich, KO, Junghans, Ch, Weistroffer, K, Swailem, FM (1979) Paleoclimatic information from deuterium and oxigen-18 in carbon-14 dated north Saharan groundwaters. – Proc. on Isotope Hydrology, 569-581, (IAEA) Vienna

UNESCO (1999) World map of hydrogeological conditions and groundwater flow. - Hydro-Science Press

1.11 Why no Middle East Water Wars: Global Solutions to Local Deficits?

Tony Allan

King's/SOAS London Water Research Group

Summary

The purpose of the chapter will be to show how processes in the international political economy have so ameliorated the water resource scarcity of the Middle East and North Africa (MENA) that there have been none of the predicted water wars. The chapter provides a narrative on MENA water policy and analytical frameworks for understanding the changing approaches to the valuation, the allocation and the management of water resources in the water scarce Middle East and North Africa (MENA). The study will mainly address the experience of the past half-century. The useful role of the concept of sustainability will be demonstrated. It will be shown how the discursive allocative politics of those contending different versions of sustainability helps an understanding of water policy-making and conflict avoiding processes.

Discursive hydro-politics are shown to mediate the contested claims on the fresh-water resources required by society, the economy and of the environment. The main insight provided by the analysis is that the MENA region solves its serious water deficit problems by resorting to a 'virtual water' remedy. This economically invisible and politically silent economic process, enabled by international trade in water intensive commodities such as wheat, addresses the big water problem [90 per cent of needs] of not being able to produce the region's food needs. It will also be shown that the recent fall in the costs of manufacturing desalinated water makes the future challenge of providing freshwater for the small water problem [10 per cent of needs] of providing freshwater for domestic and industrial water, easily addressable for the majority of the region's population living near the coast or major rivers.

The political economy of the virtual water solution will be shown to have important consequences for water policy making. Its existence de-emphasises the problem and as a result delays the introduction of measures to improve water use efficiency and the recognition of the environmental services provided by water in the environment. It will be argued throughout that a hydro-centric approach to the analysis of water policy-making is unsafe, as is the simplistic linking of single disciplines, such as economics, sociology, politics, international relations and law, to hydrology.

Key words

Middle East and North Africa, freshwater, soil water, sustainability, virtual water, political economy, social adaptive capacity, security, politics, sanctioned discourse, avoiding the hydro-centric.

'Pessimists are wrong but useful; optimists are right but dangerous.'

1.11.1 Introduction: dangers of the hydro-centric

The first purpose of this study is to analyse the water resource challenges facing the economies of the Middle East and North Africa (MENA). A second purpose it to show that it is dangerous for policy analysts and policy-makers to focus only on the freshwater resources of the region located in the watersheds of its major rivers and political economies. The region's water resource deficits cannot be successfully addressed in these watersheds. It will be shown that for the past three decades they have been very successfully ameliorated in the problemshed outside the region via the global trading system.

The political economy of the water resources of the MENA region is subordinate to the political economy of the global trade in staple foods. The political economy of this global trade in grain is in turn subordinate to the global hydrological system. Fortunately the global hydrological system has the capacity to meet the needs of the world's current and future populations for water intensive commodities, including those of the MENA region. To make sense of the policy-making context as well as the way water policy makers operate it is necessary to look outside the MENA region. It is necessary to integrate global systems into the analysis of the MENA water resources problems. It will also be shown that hydro-centricity is not a safe starting point. Linking water resources to theory deriving from disciplines such as politics and international relations is shown to be especially unsatisfactory. The theoretical linkages must be much more comprehensive, with culture, society and political economy being included as essential in a more comprehensive analytical framework.

1.11.2 Neo-Malthusian temptations

Linking water and political and international relations analyses via environmental determinism has proved to be a fatally attractive intuitive explanation of how communities react to resource scarcity. Environmental determinism was totally discredited over six decades ago in the discipline of geography. But disciplines ranging from hydrology, politics, international relations to law have in the recent past been tempted by the beguiling simplicity of environmental determinism. The argument runs, if you run out of water you reach for a kalshnikov or summon an air-strike. Societies in practice have choices and they can, and do, substitute for scarce economic factors. But it is a remarkable feature of the analysis of the political and legal analysis of water that the preferred interpretation is that there is something of a straight line trajectory with shortage, anxiety, conflict and violence figuring sequentially in a linear mode.

This tendency to deduce violent outcomes is one of the dangers of hydro-centricity. When politics, international relations and legal analysts focus on water resources they are unable properly to address the water scarcity policy options. Their theorising scope is too restricted and observation tends to ignore processes operating in the political economy. This is a particularly dangerous approach in relation to water resources as there are economically invisible and politically silent processes, which enable water scarce economies very effectively to escape the challenges of their water deficits. The MENA region, the first region to run out of freshwater has more than its share of misguided deterministic analysis. (Bulloch and Darwish 1992)

The most striking recent evidence of the temptation to devote misguided intellectual energy and misallocated funds to research inspired by environmental determinism was a project launched at the end of the Cold War. The four decades of Cold War confrontation had ended with an economic victory and not with a reckoning based directly on the enumeration of missiles. The security community obsessed for four decades by the precautionary posture of threatened mutual destruction quickly recognized that national security had important economic, social and environmental dimensions beyond those of conventional military security. The reactive shift of focus on the parts of the security and the science communities was predictable. One project that attracted a strong coalition of related funding foundations – MacArthur, Ford, Rockefeller – was a review of water resource security. The endeavour spanned the first half of the 1990s and sustained itself to the end of the decade. The intellectual basis of these studies was shallow environmental determinism. The outcome was a series of publications with titles shouting the banality of the assumptions driving the research such as, 'On the threshold: environmental changes as causes of acute conflict' (Homer –Dixon 1991), and 'Environmental scarcities and violent conflict: evidence from cases' (Homer-Dixon 1994). Homer-Dixon even edited a publication entitled, *Environmental scarcity and violent conflict: briefing book* (Homer-Dixon and Percival 1996). In the event the cases did not fit the theory and Homer-Dixon had to conjure the concept of 'ingenuity' to explain why determinism did not work as explanation. The social decision-making processes underpinning the notion of 'ingenuity' had been identified in the 1920s and called 'possibilism'. He then defaulted to making a virtue out of the utility of introducing 'useful pessimism' in policy debates and de-emphasised the unsatisfactory intellectual outcome of the project. (Hartmann 2003). Ohlsson (1999) and Ohlsson and Turton (1999) have very successfully demolished the utility of a Malthusian inspired analysis of water resource use and have introduced the very useful concept of social adaptive capacity. The shortage of this second order scarcity is much more determining of socio-economic outcomes than the absence of a first order scarcity, such as a shortage of freshwater.

Lowi (1993) had provided a very well researched international relations interpretation of the hydro-politics of the Jordan Basin. She identified a relationship, which she captured in the term hegemonic cooperation. This relationship reflected the coercive context in which the riparians contested the inadequate water resources of a river basin. The collective action outcome was predictably unfair but enduring. While useful this analysis did not explain why the outcome endured despite the radical worsening of the resource circumstances. Her mentor, John Waterbury, also a realist in inclination has provided fine syntheses of the hydro-politics of international shared waters and of the Nile Basin, in particular, in a recent book. (Waterbury 2002) Neither author satisfactorily explains why the unfair outcome has endured despite the dramatic worsening of the status of the shared water resources. Nor do they explain why the interesting but incomplete discourses that they have identified have been able to remain current and inspired by the same partial ideas despite the seriously worsening resource context.

1.11.3 The central role of hydro-politics in combining water and other capitals

Students of political economy have shown that societies can combine and substitute natural resources, such as water and other factors of production, to avoid apparently very challenging resource constraints. The process can be usefully illustrated in a diagram showing the five capitals and the central role of politics in marshalling these capitals according to endowment and the level of diversity achieved by the political economy.

Figure 1.11.1 conceptualises the five capitals and the sixth central process of hydro-politics. The extent to which a political economy can address the scarcity of one of the capitals, such as water, depends on its diversity (Karshenas 1994) and its social adaptive capacity (Ohlssson 199). All societies, including pre-modern societies, use environmental capital and human capital and evolve some form of social capital. As economies modernise via industrial modernity they develop manufactured capital and financial capital.

Over half of the economies of the Middle Eastern and North African economies and over half of the populations of the region are oil rich or partially enriched by oil derived financial capital. This advantage has made it possible for most of the region's population to solve the water scarcity problem by importing water intensive commodities. It will increasingly enable the political economies of the region to address the demand for freshwater for domestic and industrial purposes by manufacturing desalinated water.

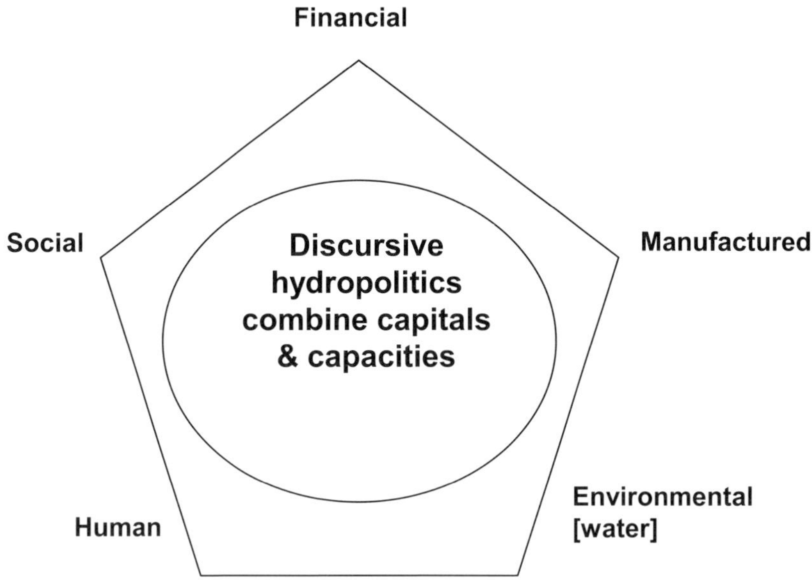

Fig. 1.11.1. The five capitals - land [water], human [labour], social of pre-modernity. and manufactured and financial capital especially associated with modernity. Hydro-political processes combine these capitals more or less effectively according to the social adaptive capacity of the political economy.

1.11.4 Virtual water in the problemshed

The explanation of why hegemonic cooperation endures lies in the political economy. Much else is explained by processes operating in the political economy; for example, the simultaneous existence of contradictory discourses. Insiders contend that there is no water shortage; while outsiders insist there are serious shortages of freshwater and that these deficiencies are addressable. (Allan 1994, 2001, 2003a) The economies in question, in the Jordan Basin, have been able to access volumes of water equivalent to between seventy and ninety per cent of their national needs via trade in food. Every tonne of grain imported allows the importing economy to avoid the costly environmental, economic and political stress of mobilising one thousand tonnes (cubic metres) of freshwater. The 1000 tonnes of water associated with the production of each tonne of grain has been termed – virtual water. (Allan 1996, 2001a, 2003a, Hoekstra and Hung 2002, Hoekstra 2003)) The importers also enjoy the advantage of half cost commodities as the exporting economies – the USA and those in the EU – set prices as low as half the production cost, of for example wheat, through subsidies. It is rather like an individual troubled by credit card debt in a Northern economy having access to long-term interest-free debt– with the added blessing of subsidised re-payments. Such perverse economic solutions are nowhere gainsaid by governments that can avail of them without making if public that they are. Unrecognised, these invisible processes, have spectacular impacts on the capacity of political scientists to identify reasons for the enduring and unfair water allocation at the same time as the non-conflictual collective action outcome.

The powerful insight of the 'problemshed' forces us to shift the analysis from a hydro-centric focus to a comprehensive approach embracing the political economy and other relationships that are part of operational water allocation and use. The solution provided by the economically invisible and politically silent water, food and trade nexus (McCalla 1997) solves the water resource problem so spectacularly that the long-time players such as water policy makers – forgive-ably, and academic analysts – unforgivably, can pretend that familiar and reassuring discourses are still relevant. A very important example of the impact of the virtual water solution on political processes is found in the Middle East. Here discursive coalitions in individual economies can reinforce the politically acceptable 'sanctioned discourse' prevalent across the Middle East region, namely that 'all we need is a little more water, then we shall manage it more carefully and everything will be all right' (Allan 2001b).

The problemshed explanation, based on the movement of global virtual water to solve local watershed problems, is also consistent with the analytical approach of political ecology with its emphasis on discursive coalitions which may be well informed on under-lying fundamentals, but even if well informed, they are capable of constructing knowledge to address historically familiar rather than new and dangerously destabilising risks. (Hajer 1996, Allan 2000) The no go area in such discourses is the topic of water and food insecurity. The global economy solves the problem but this is not an admissible truth in the Middle East region. It is politically feasible on the other hand to construct an acceptable discourse that particular economies are secure, on the basis of the sanctioned discourse, rather than on the basis of the security provided by the problemshed of global systems. (Allan 2002) Back-grounding and foregrounding risk (Douglas and Wildavsky 1982) in this way is the natural language of politicians if not of environmental, social and political scientists. Goleman (1996) has provided the helpful notions that keep relationships in place. He suggests that enduring relationships require a mutual acceptance of 'simple truths' – 'all we need is a little more water'; and 'vital lies' – 'there is no water shortage'. Allan (2001b) has elaborated the exemplification for the Middle East. The water users and water policy makers are in denial in the Middle East and North Africa.

1.11.5 Deniable underlying fundamentals: types of water, types of economy

Any analysis of the water resources of a region facing water resource shortages could be based on awareness of environmental and economic fundamentals. In practice it is easy for those developing water policy in such water scarce regions of the Middle East to overwhelm the observed science of water scarcity with constructed knowledge. It has even been possible for resources to take on the role of national fantasies. (Allan 1983) Fantasies are easier to accommodate than concepts and information that fail to gain a place in the policy making discourse because if publicly debated they would be associated with very serious political costs.

Even defining the analytical domain of water policy making is politically subversive. Here are some important fundamental concepts that relate to the nature of the water resources themselves that if kept out of the discourse make it easier to sustain the familiar and comfortable, but blind, current regional hydro-politics. There are two types of water – small water and big water. Small water is the water needed for drinking, domestic uses and the water needed by industry and services. The small water is about ten per cent of the water needed by an individual as well as the ten per cent of water needed for the self-sufficiency of an economy as a whole. This small water must come from freshwater sources – rivers, lakes, reservoirs and groundwater. Small water can command quite high prices and is commonly delivered for about one US$ per cubic metre. The big water is the 90 per cent of water needed by an individual and economy to be self-sufficient in food. The water to raise food can come from freshwater sources in which case it competes with the provision of water for domestic and industrial livelihood uses as well as for the security of environmental services. The water to raise food can also come from the soil profile. Soil water, or effective rainfall, is the majority water in economies located in humid temperate and humid topical regions. Semi-arid regions have to endure their very poor endowment in soil water. The huge volumes of big water are expected to be free if available in a soil profile. And much more worryingly they are expected to be nearly free when delivered by an irrigation system from surface or groundwater sources.

Figure 1.11.2 conceptualises the big water and small water idea. It also illustrates how an economy satisfies its water needs by accessing freshwater and soil water available locally. Where there is a deficit then an economy has to make up the deficit by importing water intensive commodities to meet the big water element of national water needs. With the fall in the cost of desalination an economy has the option of manufacturing water for the population that lives near the sea.

A second de-emphasised water resource fundamental is soil water. It is very regrettable that the existence of soil water is denied by those who draw up the water budgets of national economies. Economists and engineers are also blind to its existence. Soil water accounts for at least 50 per cent of the water used to raise field crops worldwide. Soil water also accounts for most of the feed used to rear the world's livestock and almost all the timber and wood products that enter the world's economic systems. It is soil water in the humid regions, which enjoy water surplus circumstances and can 'export' virtual water, that solves the water deficits of poorly endowed economies across the arid world. Hoekstra and Hung (2002) have estimated that 15 per cent of the water used to raise crops goes to raise commodities that enter international trade. The number is much higher if the livestock trade is taken into account (Hoekstra 2003).

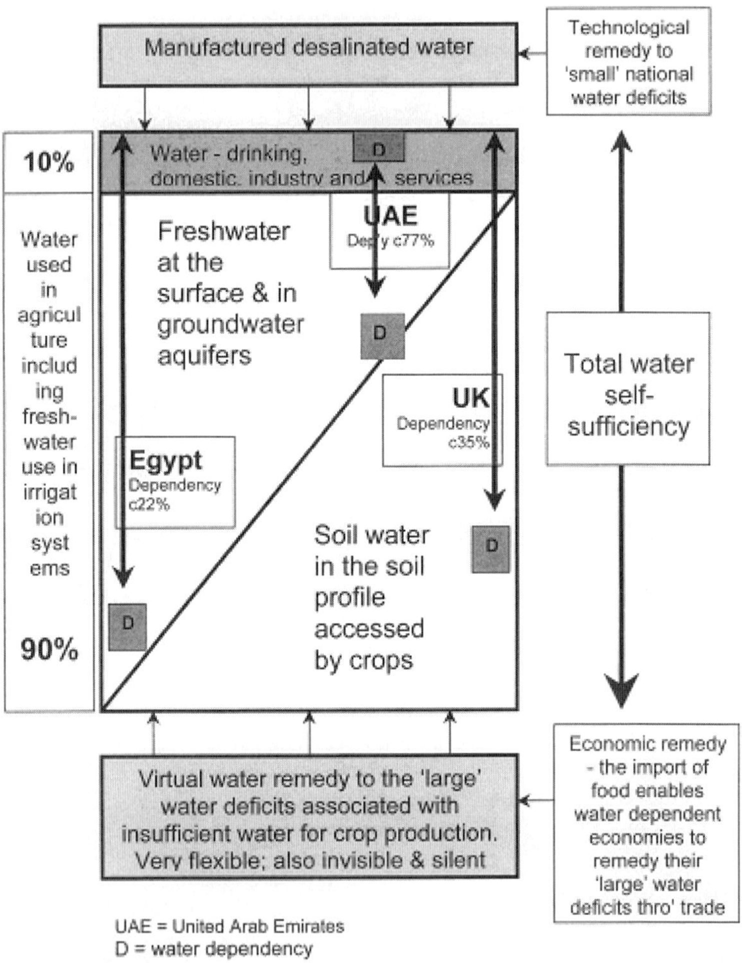

Fig. 1.11.2. The extent of the water dependency (D)of different water deficit economies (Hoekstra & Hung 2002 p55-59). 2. How local water can come from freshwater sources and soil water sources. 3. How manufactured water and virtual water can remedy the water deficits.

The global phenomenon of about 1600 cubic kilometres of water (Hoekstra and Hung 2002) being marshaled annually in international trade to meet the needs of water deficit economies dwarfs the achievements of engineers to store and deliver water. More important it addresses something that the much less flexible engineering solutions cannot, namely the constantly variable demand for food across the global system. Trade is an ideal solution because of its flexibility as well as its extraordinary reach. Virtual water embedded in traded staple commodities can readily address the variable demands. It can also address the periodic emergencies brought about by drought and famine.

1.11.6 Modernity theory and water allocation and management

Any analysis of water resource management at the global and the local levels is enhanced by an awareness of how water policy-making has been affected in the North by the ideas and technologies associated with modernity. An element of such analysis is also helpful in explaining why trajectories of investment and development have diverged between the North and the South with the onset of late modernity in the North since the late 1970s. (Carter 1983, Reisner 1984, Swngedouw 1998, Allan 2001b)

The engineering of water has figured in the histories of all civilisations and especially in the ancient civilisations, which were substantially located in the arid and semi-arid Middle East and North Africa. Water management has also absorbed the driving ideas and technologies associated with the modernity of the past two centuries. Engineers solve problems and engineers showed themselves to be very competent in solving water problems in early modernity. They came to be essential allies of the state in achieving economic goals such as food self-sufficiency. Politicians, engineers, farmers and food consumers were all certain that the progressively larger withdrawals of water, associated with what has come to be called the 'hydraulic mission' (Swngedouw 1998) of 'industrial modernity' (Beck 1992 and 1995, Giddens 1990), was good.

Figure 1.11.3 shows the trajectories of freshwater use in the Northern and the Southern economies over the past two centuries. The constantly rising trajectory reflects the process of taking more and more water out of the environment to produce food in the agricultural sector. Before about 1980 'certainty' prevailed that developing more water for food and fibre production was sound. The trenchant messages constructed by the green movement, especially in the western states of the United States, shifted the emphasis of the water discourse from 'certainty' to 'uncertainty'. By 1976 the US President Jimmy Carter wanted to raid the budget devoted to the building of more structures to store and divert water to fund his ambitious social programme. In the event the institutions that built the structures, the Corps of Engineers and the Bureau of Reclamation in coalition with senators who had been waiting for years for heir next dam frustrated his intent. (Carter 1983) But his successor Reagan, to the surprise of many, pushed through the Carter policies. The notion of uncertainty about the soundness of the hydraulic mission prevailed had gained such significance by the beginning of late modernity for it to become the preferred idea underpinning water resource allocation in the North.

The neo-liberal North comprised only about one billion out of the 6.5 billions of the world's population at the millennium. The message of the uncertainty of late modernity accepted in the North proved to be rejectable in the economies of the South.

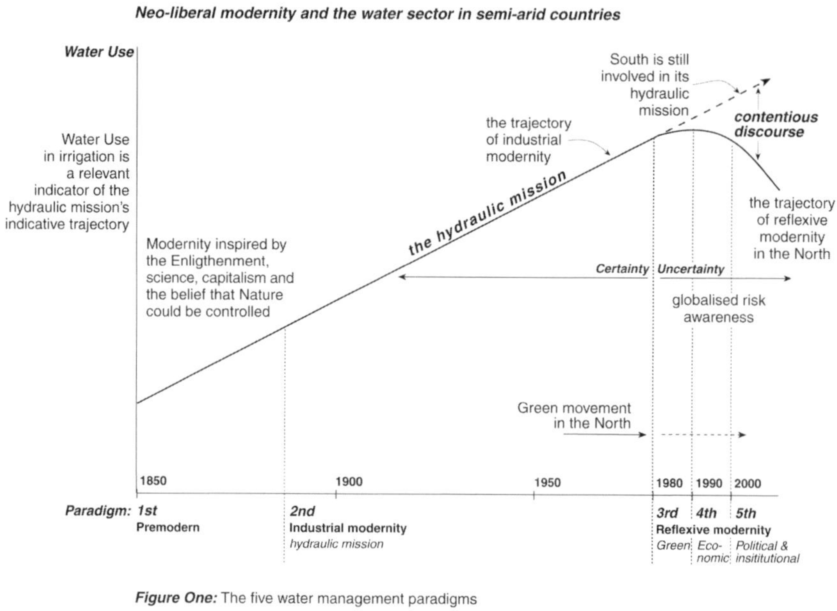

Fig. 1.11.3. The five capitals and illustrating the central the role of discursive politics in combining factors of production and compensating for scarce factors.

It was especially rejectable in the major economies of China in East Asia and in those of South Asia. The three most populous economies in the Middle East – Egypt, Iran and Turkey – comprising 75 per cent of the region's population, also rejected the idea that there should be an end to engineering interventions in their water environments.

Governments, engineering establishments and major water users in agriculture of over half of the world's population, including those of the MENA region, created almost unassailable coalitions against the arguments of the international green social movement and the greened international agencies such as the World Bank. Many Northern bi-lateral donors, emulating the World Bank, also extended their green, and later their economic efficiency, principles to their aid policies. Figure 1.11.3 illustrates the different trajectories of freshwater use in the arid and semi-arid North and the semi-arid Middle East and North Afrcia since 1980. In the neo-liberal North there was a shift towards putting water back into the environment. In the South, including the economies of the MENA Region, on the other hand, there remains a predictable commitment to taking more water out of the environment in order to further increase the output of food to meet rising food demands. Mobilising more freshwater is still seen as the way to avoid dependence on imports, and to increase the wealth of the respective economies as a whole. It should be observed that some of the major Southern economies also achieved spectacular increases in production. Production of staple grains increased by four and five times, between 1961 and the end of century, by a mix of increased freshwater use, increased efficiencies in the use of land and water and the combination of other inputs such as energy and fertilizers. Egypt doubled its wheat production after 1986. (FAO crop production data for all the above production figures 1961-2002) Again hydro-centricity would lead to only a partial explanation of these problem-solving increases in crop production.

The purpose of this amplification of the relationship between water resource use and the ideas in currency in the North and the South is to emphasise both the changes in approach that can and have taken place through time and how these differ predictably between the North and the South. Any analysis of the allocative hydro-politics of a particular economy or region can only be understood in terms of the discursive politics of the region in question. In order to illustrate the nature of such discursive processes it is necessary to achieve some consensus on the meaning of sustainability.

1.11.7 Water security: sustainability as a useful discursive hydro-political concept

The concept of sustainability is very relevant to any analysis of water policy. But it can lead to a very sterile analysis if its analytical scope is restricted to the water environment. If the notion of sustainability is captured in the larger context of the sustainability of society, the economy as well as the maintenance of the environmental services provided by water in the environment the concept of sustainability is very rich indeed. It is especially rich if discursive hydro-politics are seen to be integral to the achievement of sustainability.

Figure 1.11.3 illustrates the three dimensions of water sustainability – social, economic and environmental - and the central mediating role of discursive hydro-politics. Sustainability is a discursive outcome of the contending articulated concerns of society, those involved in the economy and those anxious about the status of the environment. There is no market and no perfect regulatory system that brings about an optimum outcome. The outcome is second best in economic terms, less than satisfactory in social terms as well as inadequate in pleasing those who want to reconstruct pristine environments. Political processes are also the only processes that can comprehensively, albeit imperfectly, respond to the driving forces of demography, technology and especially reflect the changing ideas in currency such as the innovations advocated by social movements. The diagram also shows how the level of emphasis given to social, economic and environmental principles in hydro-political processes changes through time. The diagram indicates how the emphasis in the water policy discourse is responsive to different actors – namely to the concerns of civil society, hierarchy (government), to civil movement (NGOs) and to entrepreneurs (the private sector). Identifying the actors involved in discursive hydro-politics is another useful service, which social scientists can contribute.

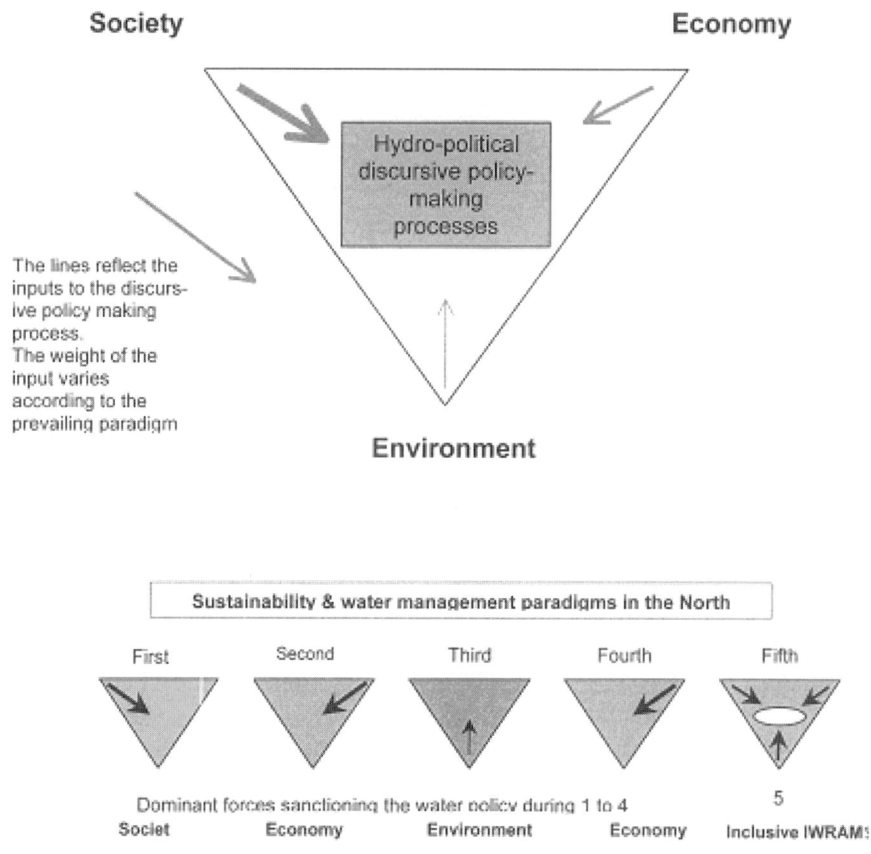

Fig. 1.11.4. The concept of sustainability and the water sector; water management as a political process and determining perceptions of the diverse values of water in the North. The third, fourth and fifth paradigms have only been very partially adopted in the South

1.11.8 Conclusion: resources in global problemsheds ameliorate the deficits of watersheds and nations

'Political analysis is not useful in explaining why water scarce regions can be economically secure or the extent of their involvement in water related conflicts. Politics is useful in the water sector in its capacity to explain what 'simple truths' and 'vital lies' are in currency in particular water policy discourses.'

The relations over scarce water resources are likely to be highly politicised. The purpose of this study has been to show that the it is dangerous to predict hydro-political outcomes on the basis of a defined area – on for example, a river basin or a nation, or on the assumption that the amelioration of water scarcity depends on capturing new water resources. Social science and law theories can be deployed to address such relationships but where such theorising is hydro-centric it will always come up with the wrong answer.

It has been shown that political economies combine – to a greater or lesser extent according to their adaptive capacity - the five capitals of land (water), human capital, social capital, manufactured capital and financial capital. In the pentagon which presents diagrammatically these five capitals it is the social adaptive capacity (Ohlsson 1999, Ohlsson and Turton 1999) deployed via political processes from the center of the diagram that achieves, or not, outcomes that address the water scarcity. Focusing on water is misleading a theoretical point of departure, which leads to incomplete and even banal explanation.

The central role of hydro-politics is also shown diagrammatically (Figure 1.11.4) in the demonstrating the importance of understanding the three dimensional nature of secure and sustainable water policy. Hydro-political discourse provides the necessary mediating process. This discourse enables the voices of society, the economy and the water environment to interact and develop a broadly based discursive power, rather than a narrowly based hydro-centric position.

The study has drawn on modernity theory to situate shifts in approach by those managing the political economies of the North and the South over the past two centuries and especially during the past 50 years. New ideas on the value of the environmental resources, such as water and their role in securing the environmental services of water, eventually overwhelmed a century of more of technological and economic certainty. Such certainties were reflected in the progressive mobilisation of water resources to address immediate economic and social goals. Five water management paradigms have been identified for the water scarce economies of the semi-arid industrialised North. The reflexive water using trajectory of late modernity (Figure 1.11.1) of these Northern political economies has not been implemented in the economies located in the semi-arid less industrialised South. The notion of discursive hydro-politics is a recurring explanation in this study. Here again it is useful in explaining first, the responses to the new ideas introduced by the green social movement and then by the epistemic water resource management community who saw water as an economic resource, and secondly, the divergence in the trajectories of water use in the North and the South.

Finally, the study shows that no single discipline has a suitable set of analytical tools to address issues of water resource security, water resource sustainability and the consequences of encountering water resource scarcity. The underlying observed science on first, the hydrology and secondly, the economics of the balance between water availability and water demand of a river basin or nation, are inadequate foundations on which to analyse or predict water policy. It is much safer to use the concepts of a combination of politics and ecology – political-ecology, and of politics and economics – political economy. Political ecology helps identify the discursive coalitions that influence policy outcomes. Political economy encourages the analyst to look beyond the narrow regional focus of the river basin or the national boundary. Political economy provides the framework, which frees us from the hydro-centric watershed, and allows us to identify the economically invisible and politically silent virtual water that brings a version of water security and the possibility of local sustainable water management regimes. Political analysis is not useful in explaining why water scarce regions can be economically secure or the extent of their involvement in water related conflicts. Politics is useful in the water sector in its capacity to explain what 'simple truths' and 'vital lies' (Goleman 1996) are in currency in particular constructed water policy discourses.

References

Allan JA (1983) Natural resources as national fantasies, Geoforum, pp 243-247.

Allan JA (1994) Overall perspectives on countries and regions, in Rogers, P. and Lydon, P. (editors) Water in the Arab World: perspectives and prognoses. Cambridge, Massachusetts: Harvard University Press. pp 65-100.

Allan JA (1996) The political economy of water [in the Jordan Basin]: reasons for optimism but long term caution, in Allan, J. A., Water, peace and the Middle East: negotiating resources in the Jordan Basin, London: Tauris Academic Studies, pp 75-120.

Allan JA (2000) Contending environmental knowledge on water in the Middle East: global, regional and national contexts, in Stott, P. A. and Sullivan, S., Political ecology, London: Edward Arnold, pp xx-xx.

Allan JA (2001a) Virtual Water – economically invisible and politically silent: a way to solve strategic water problems, International Water and Irrigation Journal, November 2001, pp 4-11.

Allan JA (2001b) The Middle East water question: hydro-politics and the global economy, London: I B Tauris.

Allan JA (2002) Hydro-peace in the Middle East: why no water wars? A case study of the Jordan River Basin, SAIS Journal, Vol XXII, No. 2, pp 255-272.

Beck U (1992) From industrial to risk society, Theory, Culture and Society, Vol 9, pp 97-123.

Beck U (1995) Ecological politics in an age of risk, Cambridge: Polity Press.

Bulloch J and Darwish A (1993) Water wars, London: Gollancz.

Carter J (1982) Keeping faith: memories of a president, New York: Bantam Books, 622 pp.

FAO (1961-2002) Crop production data, FAO: Rome. www,fao.org

Giddens A (1990) The consequences of modernity, Cambridge: Polity Press.

Goleman D (1997) Vital lies, simple truths, London: Bloomsbury.

Hajer M (1996) The politics of environmental discourse: ecological modernization and the policy process, Oxford: Clarendon Press.

Hoekstra AY and Hung PQ (2002) Virtual water trade: a quantification of virtual water xflows between nations in relation to international crop trade, Value of Water Research Report Series No *11*, Delft: IHE Delft.

Hoekstra AY (ed) (2003) Virtual water trade: Proceedings of the International Expert Meeting on Virtual Water Trade, IHE Delft, the Netherlands, 12-13 December 2002', Value of Water Research Report Series No.*12*, IHE, Delft, The Netherlands

Homer-Dixon TF (1991) On the threshold: environmental changes as causes of acute conflict, International Security, Vol *16*.1, pp 76-116.

Homer-Dixon TF (1994) Environmental scarcities and violent conflict: evidence from cases, International Security, Vol *19*.2, pp 5-40.

Homer-Dixon TF and Percival V (1996) Environmental scarcity and violent conflict: briefing book. Washington DC: American Association for the Advancement of Science.

Lowi M (1993) Water and power: the politics of a scarce resource in the Jordan River basin, Cambridge: Cambridge University Press.

Hartmann E (2002) Strategic scarcity, Unpublished PhD, London: London School of Economics

Karshenas M (1994) Environment, technology and employment: towards a new definition of sustainable development, Development and Change. Vol. *25*. 2: 723-757.

McCalla A (1997) The water, food and trade nexus, Paper delivered at the MENA-MED conference convened by the World Bank in Marrakesh, May 1997.

Ohlsson Leif (1999) Environment, scarcity, and conflict - a study of Malthusian concerns, PhD dissertation (12 February), Dept. of Peace and Development Research, University of Göteborg.

Ohlsson L and Turton AR (1999) The turning of a screw: social resource scarcity as a bottle-neck in adaptation to water scarcity, Occasional Paper No *19* , SOAS Water Issues Group. www.soas.ac.uk/geography/waterissues/

Reisner M Cadillac desert, New York: Penguin Books

Swyngedouw E (1999a) Modernity and hybridity - the production of nature: water and modernisation in Spain. Paper presented to the SOAS Water Issues Study Group, University of London, 25 January 1999.

Waterbury J (2002) The Nile: national determinants of collective action, Newhaven: Yale University Press

2 Water Protection

2.1 Management, Protection and Sustainable Use of Groundwater – Results of Long-Term Technical Co-operation in the Middle East

Schelkes, K., Hobler, M., Schmidt, G. & Steinbach, V.

2.1.1 Introduction

Groundwater represents the main water resource in most parts of the arid and semi-arid regions of the Middle East and Northern Africa. Adequate groundwater management is a prerequisite for the sustainable use of the scarce water resources. Groundwater protection measures have to be incorporated in integrated water management activities as an important feature for sustainable development. To fulfil these requirements, the German Federal Institute for Geosciences and Natural Resources (BGR) supports since several decades some national ministries and institutes as well as international organisations which are operating in the groundwater sector in this region.

During the seventies and eighties, work was mostly concentrated on the assessment, exploration and use of groundwater resources. In the following years the interest in groundwater protection issues grew continuously. This shift in interest was induced by the improved knowledge of the processes and events influencing the groundwater regime and the quality of the groundwater. Better understanding of groundwater issues made it obvious that groundwater protection is one of the main pillars for sustainable use of the scarce water resources in the Near East. It was not acceptable any more to diminish the potable groundwater resources by often irreversible contamination. The trend from quantity oriented to more quality oriented groundwater studies can easily be seen in the changes in the focal points of the projects as shown in this paper.

This improved knowledge led to changes in the people's mind. Sustainable and careful use of the scarce groundwater resources became a key issue. Consequently, project activities were not only concentrated on advice for the use of modern methods, e.g. for the preparation of planning documents, but also for the preparation of groundwater-related guidelines and policies as part of an overall strategy for better groundwater resources management. Based on the understanding that the guiding principles for sustainable groundwater use have to be implemented by governments and their administrations, this last topic forms now a fundamental part of German Technical Co-operation and is an important component of the ongoing groundwater related projects.

The following examples of bilateral and international co-operation show the changes in the types and goals of technical co-operation projects. The presented results highlight the groundwater situation in the Near East and should provide understanding for the wide range of problems related to the groundwater situation.

2.1.2 Bilateral co-operation: Jordan

In the early projects, the assessment of the groundwater resources was based on the results of geophysical investigations and hydrogeological field work. These investigations and the interpretation of the results were mostly carried out by foreign organisations and their contractors. The partner institution often played a minor role because of insufficient knowledge and experience especially in modern geophysical, mathematical and hydrogeological methods and lack of adequate equipment. To overcome this situation, improvement of the infrastructure and the qualification of the personnel of the partner institutions got high priority in all co-operation projects and formed an essential part of the overall project strategy. In turn, the role of the project partners in the Near Eastern countries changed with improving qualification. More and more, investigations and data interpretations were carried out by the specialists of the partner institution. Scientists and technical specialists of the foreign organisations assisted as advisors only. Support and advice in learning and using modern methods in groundwater management and groundwater protection are today well to the fore.

Groundwater cooperation projects with Jordan and groundwater models and modelling results in particular, are examples for this development as shown in the following.

Support for the preparation of the National Water Master Plan (NWMP) in 1977 was one of the first major groundwater activities of the BGR in Jordan (GTZ 1977). Data collection was mostly done by the Natural Resources Authority of Jordan (NRA), data interpretation and map preparation by German specialists. This NWMP was the first comprehensive study of the water resources of the entire country. It gave an overview of the water situation and served as a general planning basis for about twenty years. Thematic maps show the hydrogeological situation, e. g. favourable areas for ongoing or potential groundwater exploitation and available volumes of stored groundwater. For the new NWMP the same boundaries of water basins have been used as they had been defined during that survey.

In the following years, co-operation of BGR with Jordan was concentrated on other sectors like geology and geophysics. Studies in the water sector continued on a quite different level with an investigation of the possibilities to supply an envisaged oil shale processing plant at El Lajjun with water. Fig.2.1.1 shows the location of the study area in the Mujib basin and Fig. 2.1.2 an impression of the landscape and the base flow in the upper Wadi Mujib. The data collection and the field work for this hydrogeological study was supported by the Water Authority of Jordan (WAJ). Most of the hydrogeological interpretation as well as the numerical modelling were still done by BGR but on-the-job-training was already an important issue.

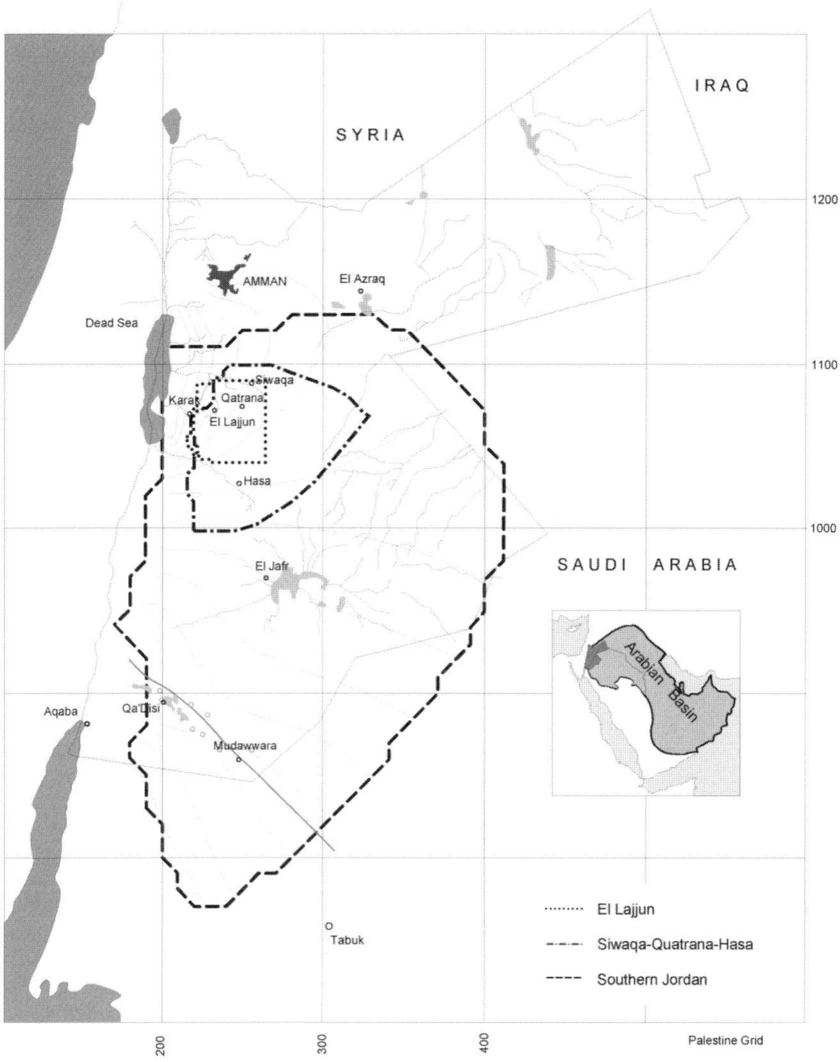

Fig. 2.1.1. Areas of the numerical model investigations 1985 – 1990

Based on the hydrogeological set-up, it was possible to calculate the water balance of the Wadi Mujib basin with the help of a numerical model (Fig. 2.1.3). The results showed very clearly that the quantities of recoverable groundwater from the upper aquifer have been widely overestimated in the past. The water abstraction from this aquifer for drinking water (including e.g. a well field for the supply of Amman) and industrial purposes already exceeded the recharge in that area. Any future demand has to be extracted from aquifer storage (Bender et al. 1989).

Fig. 2.1.2. Baseflow in the Upper Wadi Mujib

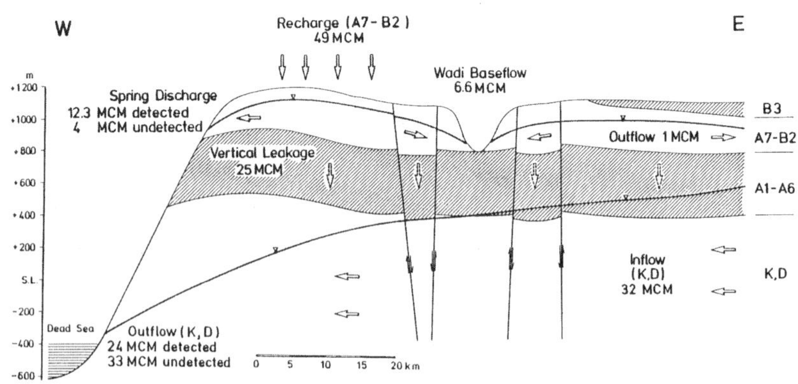

Fig. 2.1.3. Groundwater balance for the Mujib basin

The result of this "El Lajjun Oilshale Feasibility Study" was one of the reasons to investigate the groundwater situation in much more detail. It was the first of four technical co-operation projects in Jordan, which included numerical groundwater flow modelling as a major component.

The following list gives an overview of the various numerical models for the time period from 1985 to 2002 (for locations see Fig. 2.1.1 and 2.1.6):

- El Lajjun Model, (Schelkes, 1986)
- Siwaqa-Qatrana-Hasa Model, (Schmidt, 1989; validated 2001)
- South Jordan - Natural Aquifer Depletion Model, (Schmidt, 1989)
- Numerical model for the Muwaqqar well field, (Brunke, Khalifeh and Al Mahamid, 1996)
- 3D-Groundwater Model of Northern Jordan, (Brunke, 1995; revised 1998)
- Numerical model of the Siwaqa-Qatrana-Hasa well field (Khalifeh and Al Mahamid, 1998)
- Well field models (Corridor well field, Musaitbeh well field), (Khalifeh and Al Mahamid, 1999/2000)
- 3D-Groundwater Model of Jordan (ongoing development)

As it can be seen, in the beginning modelling was mostly done by specialists from BGR. As a result of on-the-job training and of advanced training courses, modelling was more and more accompanied and carried out by the Jordanian counterparts. This effect confirms the importance of capacity building as an integral part of co-operation projects.

The next co-operation project on "Groundwater Resources of Southern Jordan" started in 1987 and lasted until 1990. It included capacity building, the establishment of a hydrogeological data bank and the preparation of a map of the groundwater resources of Southern Jordan as well as the preparation and installation of two specific groundwater models (Schmidt and Hobler 1994).

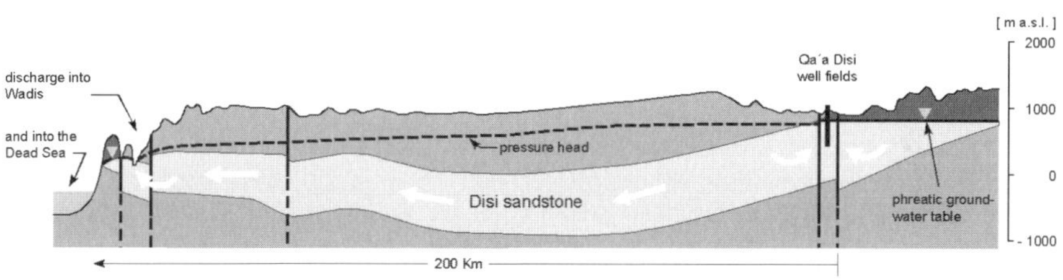

Fig. 2.1.4. Sketch of the flow system in the deep sandstone aquifer

The first of these models, the "South Jordan - Natural Aquifer Depletion Model" was planned as an overall evaluation of the groundwater potential of Southern Jordan. The model takes the natural time-dependent behaviour of the aquifer system on a long term into account. The calculations are based on the assumption that the whole system was filled with water at the end of the last humid period, approximately 5000 years ago. The model (location see Fig. 2.1.1) was intended to simulate the natural long-term depletion of the aquifer system during these 5.000 years. Fig. 2.1.4 shows a two-dimensional sketch of the flow system with the high groundwater potential in the southern part of Jordan and the deep depression of the Dead Sea which forms the base level for the flow system in the deeper aquifer.

It could be shown that the aquifer system in Southern Jordan is continuously under depleting conditions with a time-dependent decreasing outflow rate. The groundwater system is affected by the Wadi Sirhan depression at shallow to medium depths and by the Dead Sea at greater depths. The model gives a rough picture of the time-dependent behaviour of the flow system and the available groundwater resources in Southern Jordan. The calculated present distribution of groundwater levels compares fairly well with recent measurements of the groundwater level in the deep aquifer (Fig. 2.1.5). Since groundwater level observations are missing in large areas of the model domain, a real calibration is not yet possible. The model is therefore only one possible, even plausible representation of the system (Schmidt et al. 2002).

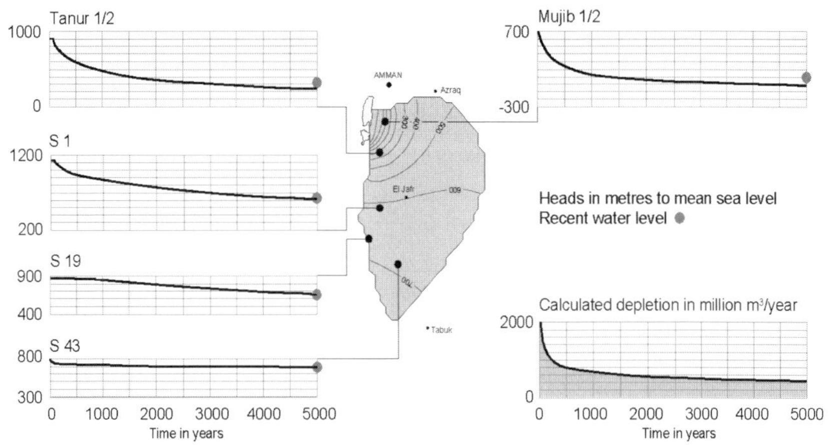

Fig. 2.1.5. Calculated groundwater-head contours of the deep sandstone complex as results of the three-dimensional South-Jordan model. After 5000 years of continuous depletion the calculated heads are close to the monitored water levels. During the recent time the calculated groundwater flow to the Dead Sea amounts to about 400 MCM annually.

The second model, a detailed flow model of the Siwaqa-Qatrana-Hasa well fields (see model area in Fig. 2.1.1), was planned as a contribution to an optimal management of the well fields for the water supply of Amman. The groundwater situation in part of the model area was already studied in the earlier "El Lajjun model". Due to overpumping, modelling results showed a high yearly drawdown. With the new model it was possible to study the reaction of the aquifer in more detail and to calculate various groundwater abstraction scenarios. The main model results confirmed that water table drawdown continued since 1984 and reached 20-30 m after 10 years and 30-40 m after 20 years in the main well fields. It was obvious that groundwater withdrawal created a mining situation. Based on these results a strategy for a more sustainable groundwater abstraction was developed

(Schmidt and Hobler 1994). Due to severe water shortages in Amman, the Water Authority of Jordan was not able to follow these recommendations during the next few years.

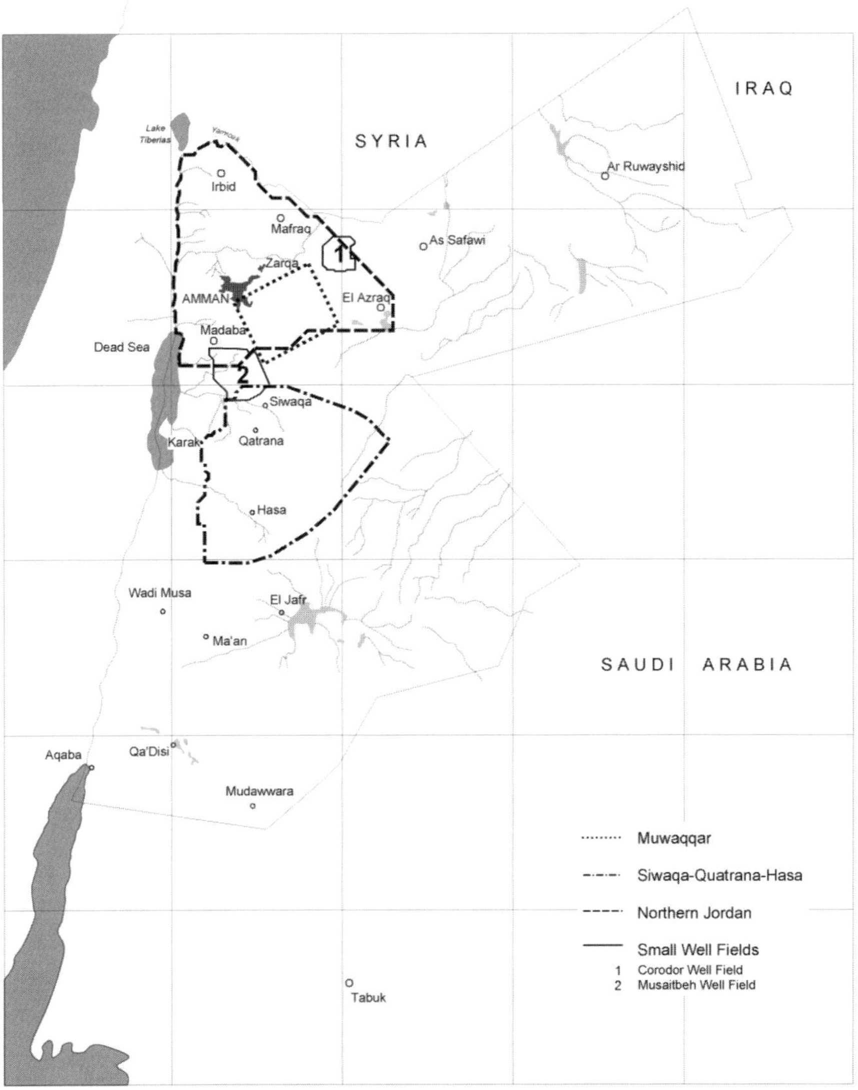

Fig. 2.1.6. Areas of the numerical model investigations in the Northern Jordan project

The next project, "Groundwater Resources of Northern Jordan", started mid 1991 and can be seen as a complement to the South Jordan Project. The main objectives included capacity building as the central topic, development of a data bank and creation of a map of the groundwater resources potential (Margane et al. 2002). As part of the assessment of the groundwater resources, a flow model was developed that takes all of the main aquifers of northern Jordan into consideration (see Fig. 2.1.6). Other than with the more generic South Jordan model, the intention was to create the model as an instrument for advising the Water Authority in the safeguarding and improvement of the drinking water supply of Jordan.

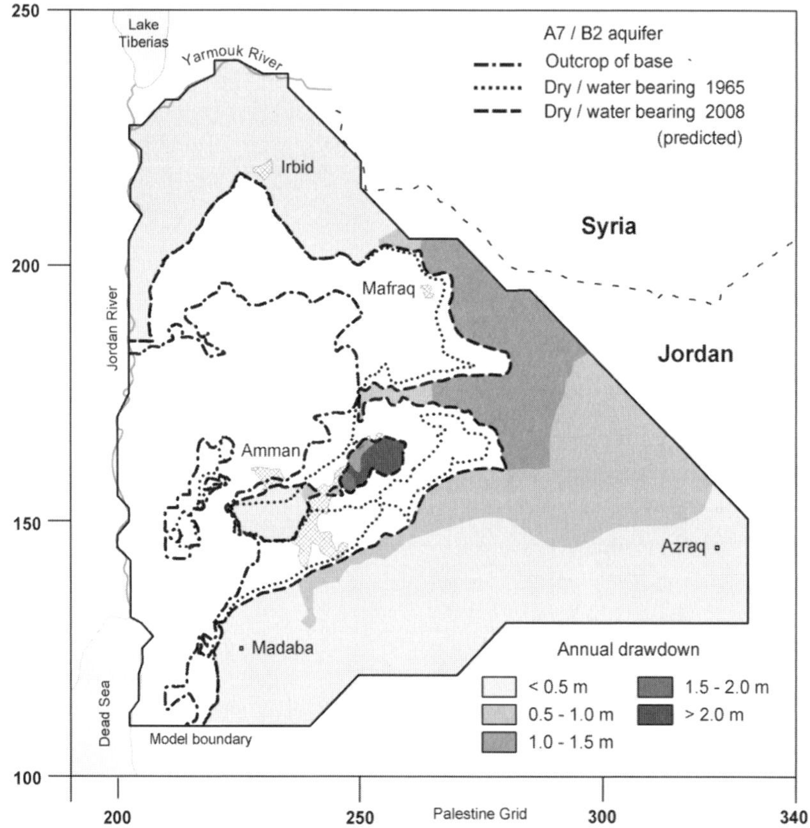

Fig. 2.1.7. Predicted annual drawdown and shift of the limit of dry / water bearing part of the A7/B2 aquifer (Groundwater model of Northern Jordan)

The data base for this model was much better than for the South Jordan model. Model reliability was proved by the good agreement of calculated discharge with spring discharge data. The model results illustrated again the severe consequences of an annual groundwater withdrawal rate of more than 200 MCM, an amount in excess of the recharge rate. Groundwater mining has now become common and the drawdown of the water table reached a typical mean value of about 1m/a in the main abstraction areas. Some springs dried up after approximately 30 years of groundwater development (see Fig. 2.1.7).

As part of the advanced training, the Jordanian counterparts also developed a new model of the Siwaqa-Qatrana-Hasa well fields with a different computer code. The model takes all available new data like abstraction and water level monitoring into account (Fig 2.1.6). The results confirmed the findings of the earlier model of 1989. Within the framework of a technical co-operation project, the model of 1989 has also been validated in 2001. This project dealt with the validation of numerical groundwater models developed and implemented by the BGR during the last 25 years in various countries.

In the last years of the North Jordan Project more emphasis was given to groundwater protection problems. It became obvious that aspects of groundwater quality are often more important than quantitative aspects. Methods for the prevention of contamination of the scarce groundwater resources by agriculture, industry, private users, waste dumps etc. were introduced as project topics. Groundwater vulnerability maps (Fig. 2.1.8) and maps of hazards to groundwater were developed for the first time in Jordan for the Irbid area in the north and an area near Amman (Margane et al. 1999).

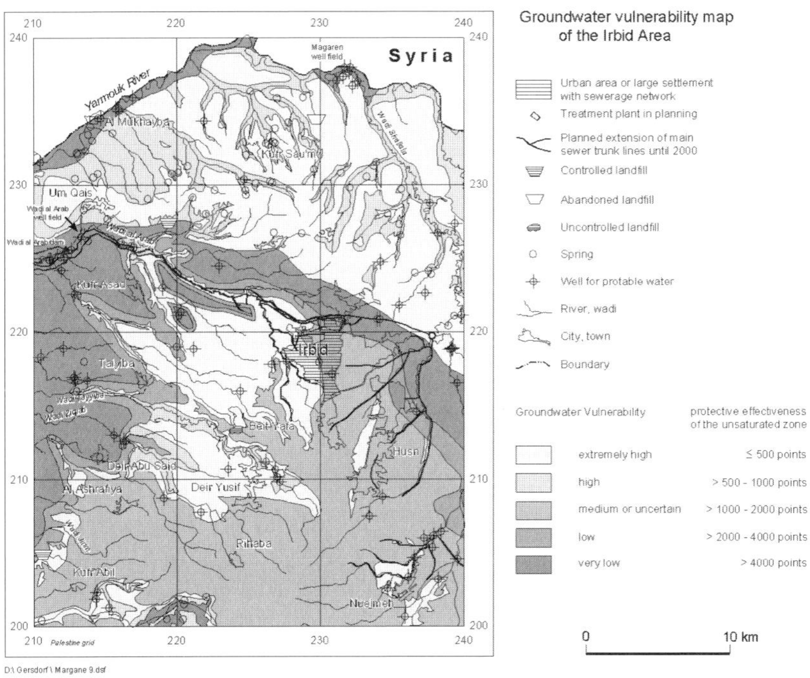

Fig. 2.1.8. Groundwater vulnerability map of the Irbid area

The maps were prepared as tools for land-use planning, which takes aspects of groundwater protection into account. The vulnerability map of the Irbid area showed for example, that the "Wadi el Arab" well field is at risk to become polluted. The establishment of a groundwater protection zone for this well field is therefore very important and will be a major topic of the ongoing cooperation project (Margane et al. 2003). Another step in the improvement of groundwater quality measures was the establishment of a groundwater protection zone for the "Pella spring (Tabaqat Fahel)". The spring is used for the water supply of Irbid. For the first time a groundwater protection zone was officially installed in Jordan and local authorities were instructed on the necessary measures (Fig. 2.1.9).

Fig. 2.1.9. Groundwater protection zone at the Pella spring (Tabaqat Fahel)

The main goals of the new cooperation project on "Groundwater Resources Management" include the elaboration of measures for groundwater protection and their implementation by the Ministry of Water and Irrigation (MWI). This coincides with the German Technical Co-operation strategy to advice governmental organisations in the policy sector as a prerequisite for optimal water management. The activities include the elaboration of guidelines for the delineation and implementation of groundwater protection zones at real sites. These activities serve as examples for the application of concepts which prevent groundwater contamination by the analysis and evaluation of groundwater vulnerability. Other topics are the examination of salt water intrusion problems and the evaluation of the effects of anthropogenic contamination by groundwater modelling including contaminant transport. The recommendations for preventive groundwater protection and the guidelines will be disseminated by the Ministry of Water and Irrigation in co-operation with other Jordanian organisations, ministries and stakeholders in the environmental and water sectors. In addition, support will be given for the hydrogeological part of the new National Water Master Plan. This also includes the development of a groundwater model for Jordan.

2.1.3 International co-operation: ACSAD and ESCWA

Since many years BGR co-operates with two international organisations in the Middle East, namely ACSAD ("The Arab Center for the Studies of Arid Zones and Dry Lands") and ESCWA ("United Nations Economical and Social Commission for Western Asia"). The role of these organisations as multipliers of knowledge to other Arab countries is an important aspect for the co-operation. This implies that ACSAD and ESCWA personnel are qualified to advice specialists of the member countries in modern methods of groundwater management and protection. The organization of advanced training courses, workshops and expert group meetings and the compilation of guidelines are therefore some of the most important topics of the cooperation.

These guidelines take general aspects of the situation in the whole region into account and can be used as a basis for national guidelines of the different Arab countries.

Since important groundwater systems are often shared between different countries, management on the national level has its limitations. Examples of shared aquifer systems in the Arab region are the Nubian-Sandstone aquifer in northern Africa and the Disi, the Paleogene and the Basalt aquifers in the Arabian peninsula. Shared groundwater resources bear an immanent potential for conflicts between countries, especially if the groundwater resources are scarce. This also holds if groundwater contamination affects the groundwater quality in a neighbouring country. Therefore, not only surface water basins but also shared groundwater resources need to be managed in bilateral or multilateral co-operation. International organisations can initiate the dialog between the involved countries. They can assist and support the process of a co-operative management of the shared resources and help in the prevention of conflicts between the different parties. Strengthening of this role is another important aspect of the co-operation with ACSAD and ESCWA.

Advanced training of the staff of ACSAD was therefore one of the main topics in all projects during the long time of co-operation. From 1988 to 1996 project tasks included methods for groundwater protection and groundwater management in general. Special emphasis was given to modern methods of geohydraulics, groundwater modelling and isotope hydrology. ACSAD specialists were prepared to carry out training courses in these fields. As an example, a groundwater management plan was developed for the Palmyra basin in Syria. Fig. 2.1.10 shows the groundwater situation in that basin.

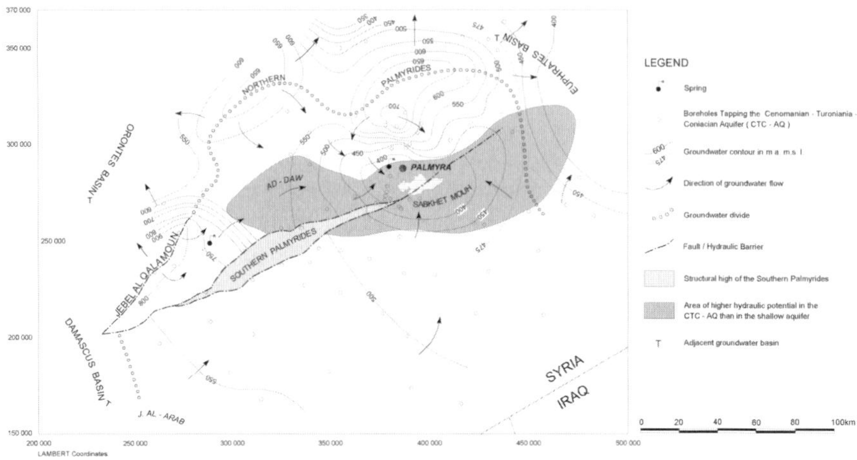

Fig. 2.1.10. Regional hydraulic setting of the Palmyra basin

The goal of the next project was the management, protection and sustainable use of groundwater and soil resources. During the first project phase (1997-2000) the main emphasis was on capacity building in hydrogeological and soil science studies and the dissemination of knowledge in specific workshops. In the ongoing second phase, the strengthening of the national and regional competence in the groundwater sector and the elaboration and implementation of guidelines were the main topics. New or improved methodological approaches are being applied and tested in pilot areas. These studies also serve for on-the-job training of specialists from participating national institutions. Maps of soil and groundwater vulnerability, hazards and risks of contamination were compiled, groundwater and transport models developed and a groundwater and soil information system installed. Based on the results from the pilot areas, guidelines for monitoring, protection and sustainable development of groundwater and soil resources were prepared. Awareness of the consequences of careless use of the scarce water resources is a precondition for the sustainable development of the resources. Dissemination of project results in regional workshops is therefore an important project task. Awareness and information days for stakeholders as well as information of the public in any possible way are essential for the success of such projects.

Three areas have been chosen as pilot areas in three different countries of the ACSAD region: the Damascus Ghouta plain in Syria, the Beka'a valley in Lebanon and the coastal area of Ras El-Jebel in Tunesia. General hydrogeological and soil science surveys were carried out for the Ghouta and the Beka'a valley. Sets of thematic maps including groundwater vulnerability maps and maps of heavy metal distribution in soil were compiled (Margane et al. 2003). In a second step, more specific studies were selected for each of the three pilot areas. Studies on heavy metals were carried out in the Ghouta Plain and on nitrate fluxes in the Beka'a valley. This included model calculations on contaminant transport in general. The effects of overexploitation of groundwater, salt water intrusion and contaminant transport are topics of the work in the Ras El-Jebel coastal area. The main purpose of these studies was advanced training in modelling contaminant transport including chemical reactions and density effects and the use of models in decision making processes. One result of the contaminant transport calculations is shown in Fig. 2.1.11 (Altfelder 2003). Contaminants spread widely from three waste dumps (two existing and one planned) in the central Beka'a Valley. The predicted concentration of pollutants will be highest near the newly planned waste dump, but the spreading will be very low as a result of an appropriate construction method of the dump. The northernmost waste dump seems to be the most dangerous one with respect to groundwater contamination.

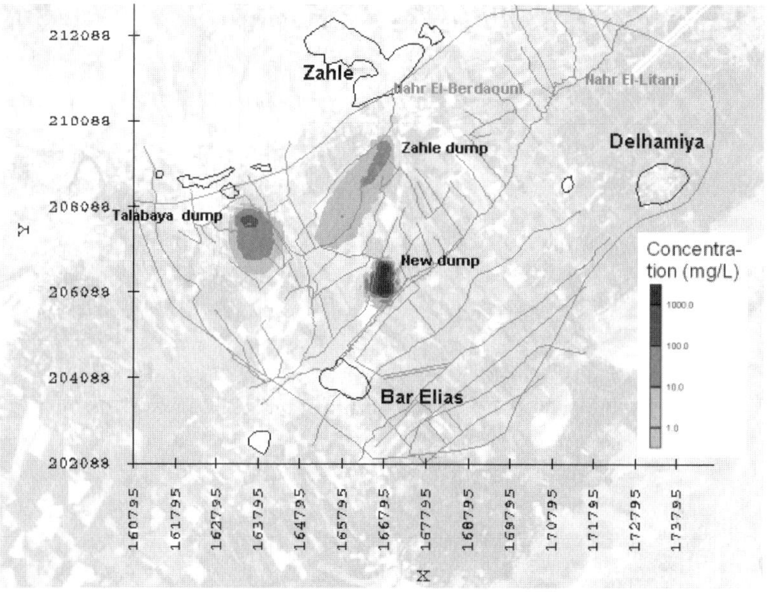

Fig. 2.1.11. Calculated plumes of nonsorbing nondegrading contaminants in the upper aquifer of the Beka'a valley 13 years after an assumed contaminant input

Cooperation with ESCWA and its member states in the field of water resources is going on since about ten years. During the first phases of this project, the main emphasis was concentrated on two topics. The first one was knowledge transfer of new methodologies in the groundwater sector mostly in form of workshops and advanced training courses. The new methods include remote sensing, use of environmental isotopes and application of transport models in hydrogeology. Reports were prepared on project results and distributed to the member countries.

The second focal point of the project activities was the evaluation of regional studies on groundwater systems in the ESCWA region. The hydrogeology of the Basalt aquifer of Syria and Jordan and of the Paleogene aquifer in the Arabian peninsula have been described in two comprehensive reports (ESCWA 1996 and 1999). For the storage of the data, a GIS based data bank on regional aquifers was developed. The project tasks also included advice on groundwater related problems for ministries in the various countries.

In the ongoing phase, the management of internationally shared aquifer systems has the highest priority. This has to be seen in light of the above mentioned conflict potential of shared water resources. Therefore, the project supports the ESCWA member states in questions of an integrated management of such shared groundwater resources. Bilateral committees have been established to improve the regional co-operation. Compilation of guidelines for an integrated water resources management, for example priorities for water allocation, water resources protection, use of non-conventional water resources, water rights and reduction of water consumption is also an important task of the project team. Naturally, the guidelines have to be modified for the purpose of the respective countries. As in earlier phases, results of all activities will be presented and discussed in workshops and disseminated in the member countries.

2.1.4 Conclusions

In general, the groundwater resources of the Near Eastern and North African countries are fairly well known. The knowledge on groundwater protection issues is also increasing continuously. Many new tools like GIS and data bank systems, groundwater and transport models, remote sensing methods and hydrochemical methodologies for the management of the groundwater resources and their protection are available. Important next steps are the elaboration of guidelines and policies and the implementation of groundwater protection measures e.g. in the national laws. At the same time, the information on groundwater protection and sustainable use of the resources has to be presented to the public and disseminated to stakeholders using appropriate methods.

A very important issue is the use and management of shared groundwater resources. The problem is not new, but the significance of the problem was not noticed for a long time. Therefore, it is a ‚new‘ task for many governments. Advice in handling this task, in finding a consensus between different countries and in preventing them from conflicts could be provided by organisations working on an international level. This is another good reason to support organisations like ACSAD and ESCWA.

All of the BGR groundwater projects aim to ensure a safe supply of good quality drinking water for the population. The interests of the various competing users have to be taken into account on national and also international levels. To fulfil these requirements integrated groundwater management systems have to be established and guidelines and policies for various aspects of groundwater protection and groundwater use including e.g. wastewater treatment and reuse or water price structure have to be elaborated. These complex tasks can only be worked out successfully in co-operation with other scientific disciplines such as geosciences, soil and environmental sciences and disciplines dealing with water rights and water supply.

Acknowledgements

The examples presented here are based on the results of various technical co-operation projects. Many scientists from BGR, the WAJ and MWI in Jordan, ACSAD in Damascus and ESCWA in Beirut have worked together with the authors in these studies. The authors wish to thank all these colleagues. They are mentioned in the text and the respective publications.

References

Altfelder S (2003) pers. comm., see: Tätigkeitsbericht BGR 2001/2002, 65 S, Hannover

Bender H, Giesel W, Knoop RM, Schelkes K, Rashdan J (1989) Feasibility study on the groundwater development for the water supply of an oil shale processing plant in Central Jordan. Proc. Int. Symp. on Groundwater Mangement: Quantity and Quality, IAHS Publications No. 188, pp 571-579, Wallingford, UK.

Bender H, Hobler M, Klinge H, Schelkes K (1989) Investigations of Groundwater Resources in Central Jordan. Desalination, 72, pp 161-170, Amsterdam, The Netherlands

ESCWA (1996) Investigation of the Regional Basalt Aquifer System in Jordan and the Syrian Arab Republic. E/ESCWA/ENR/1996/11, United Nations Publication, ESCWA, Amman

ESCWA (1999) Groundwater Resources in Paleogene Carbonate Aquifers in the ESCWA Region: Preliminary Evaluation. E/ESCWA/ENR/1999/6, United Nations, New York

GTZ (1977) National Water Master Plan of Jordan. Vol. IV (Authors: Vierhuff, H. & Trippler, K.), Hannover

Margane A, Hobler M, Almomani M, Subah A (2002) Contributions to the Hydrogeology of Northern and Central Jordan. Geologisches Jahrbuch, Reihe C, Heft 68, 52 S.

Margane A, Hobler M, Droubi A, Rajab R, Subah A, Khater AR (2003) Groundwater Vulnerability Mapping in the Arab Region. in: Water in the Middle East and in North Africa: Resources, Protection, and Management (ed. Zereini F)

Margane A, Hobler M, Subah A (1999) Mapping of Groundwater Vulnerability and Hazards to Groundwater in the Irbid Area, N Jordan. Z. angew. Geol., 45 (4), pp 175-187, Hannover

Schmidt G, Hobler M (1994) Groundwater modelling in Central and Southern Jordan. In: Proceedings of the First Jordanian Mining Conference (23.-25.04.1994), pp 239-252.

Schmidt G, Hobler M, Söfner B (2002) Investigations on Regional Groundwater Systems in North-East Africa and West-Asia. Proc. Int. Conf. on Aquifer Systems in Arid Zones: Managing Non-Renewable Resources.- Tripoli, Libya, Nov. 1999, UNESCO 2002, pp 195-203, (on CD), to be published in 2003 as Technical Documents in Hydrology, No. 42

2.2 Groundwater Vulnerability Mapping in the Arab Region

Margane, A., Hobler, M., Droubi, A., Rajab, R., Subah, A. & Khater, A.R.

2.2.1 Introduction

Within the framework of technical cooperation projects several groundwater vulnerability studies have been carried out by scientists from the Federal Institute for Geosciences and Natural Resources (BGR) and partner organizations in the Arab region over the past decade. Land use planners and decision makers are usually not trained to take aspects of groundwater protection into consideration. Groundwater vulnerability mapping is a suitable tool for this purpose. It also facilitates groundwater quality control measures, such as the delineation of groundwater protection zones and the development of groundwater quality monitoring schemes. The mapping scale depends on the intended use, the available data, the time frame and the availability of funds. For land use planning the scale is normally 1:100,000, whereas for outlining of groundwater protection zones the mapping scale should not be smaller than 1:25,000.

The first map of this type was developed for the Irbid region in Jordan (MARGANE et al., 1997, 1999). It is based on the GLA-method, developed and used by the German geological surveys (HÖLTING et al., 1995). Figure 2.2.1 shows the groundwater vulnerability map for the Irbid area. Other maps of this type were developed for the area south of Amman/Jordan (HIJAZI et al., 1999), the Ghouta area in Syria (HOBLER & RAJAB, 2002) and the Beka'a Valley in Lebanon (HOBLER & RAJAB, in prep.). The basic data needed for the GLA-method are: the effective field capacity of the soil, the percolation rate (groundwater recharge rate), the rock type, the thickness of the soil and rock cover above the saturated aquifer as well as information about perched aquifer systems and hydraulic pressure conditions. In order to make recommendation for land use, potential hazards to groundwater quality should be inventoried and mapped. The comparison of both maps gives an indication of areas and locations where the groundwater resources may be endangered. The next steps could involve recommendations for more detailed studies and /or groundwater monitoring. In addition, groundwater vulnerability maps can be used in the selection of locations where land use does not present a potentially high risk to groundwater quality. For example, areas of low groundwater vulnerability may be suitable locations for waste disposal sites, sewage treatment plants and industrial zones. However, in most cases more detailed studies will be necessary before decisions can be made on final locations. Further, if maps of the groundwater exploitation potential are available, recommendations could be made for the selection of areas where groundwater resources should be protected for future development. SAEFL (2000) introduced a method to delineate groundwater protection zones in karst areas based on groundwater vulnerability maps. SUNNA & MARGANE, 2002, proposed the use of this so-called EPIK method for defining groundwater protection zones in the technical cooperation project 'Groundwater Resources Management'. This project is being carried out in conjunction with the Jordanian Ministry of Water and Irrigation.

In order to facilitate the preparation of groundwater vulnerability maps in the Arab region, BGR and ACSAD cooperated in the preparation of a guideline for groundwater vulnerability mapping and risk assessment (MARGANE, 2003). In this guideline the most important mapping methods are presented and discussed. Examples of groundwater vulnerability mapping in the Arab region are presented below. For reasons of comparison an example of a map based on the DRASTC method (ALLER et al., 1985) has also been included. The groundwater vulnerability map of the western Nile Delta was prepared by specialists of the Research Institute for Groundwater in Egypt (KHATER, pers. comm.).

2.2.2 Groundwater Vulnerability Mapping of the Irbid Area, Jordan using the GLA-Method

The Groundwater Vulnerability Map of the Irbid Area was prepared by a team of the Jordanian-German Technical Cooperation Project 'Advisory Services to the Water Authority of Jordan – Groundwater Resources of Northern Jordan' (MARGANE et al., 1997, MARGANE et al., 1999, MARGANE et al., 2002).

The map covers an area of approximately 1,500 km^2. Sequences of limestone units interbedded with marl dominate the geological set-up of the area. In most of the aquifers groundwater levels are very deep and in some cases exceed 300 m. In the mountainous regions rainfall is high and in some areas exceeds 500 mm/a. The climate becomes increasingly arid in the eastern part of the area where rainfall is mostly below 150 mm/a. Therefore, the amount of groundwater recharge varies highly throughout the area.

Groundwater vulnerability is high or very high in much of the outcrop areas of the limestone units (B4 aquifer and A7/B2 aquifer) as seen in Figure 2.2.1. The high vulnerability in these areas correlates with the high level of bacteriological contamination (especially in the B4 aquifer) and high nitrate contents. The Wadi al Arab well field, located approx. 15 km west of Irbid, near the Jordan Valley, is very important for the current water supply of the city. This well field is at risk and could become polluted by sewage water from the city of Irbid and surrounding communities. Therefore, the establishment of groundwater protection zones and recommendations for restricted land use are considered to be top priorities of a new German – Jordanian Technical Cooperation Project 'Groundwater Resources Management' (MARGANE & SUNNA, 2002). These recommendations are based on the map of hazards to groundwater (Figure 2.2.2), which uses a legend proposed by VRBA & ZAPOROZEC (1994), and available piezometric maps of all relevant aquifers.

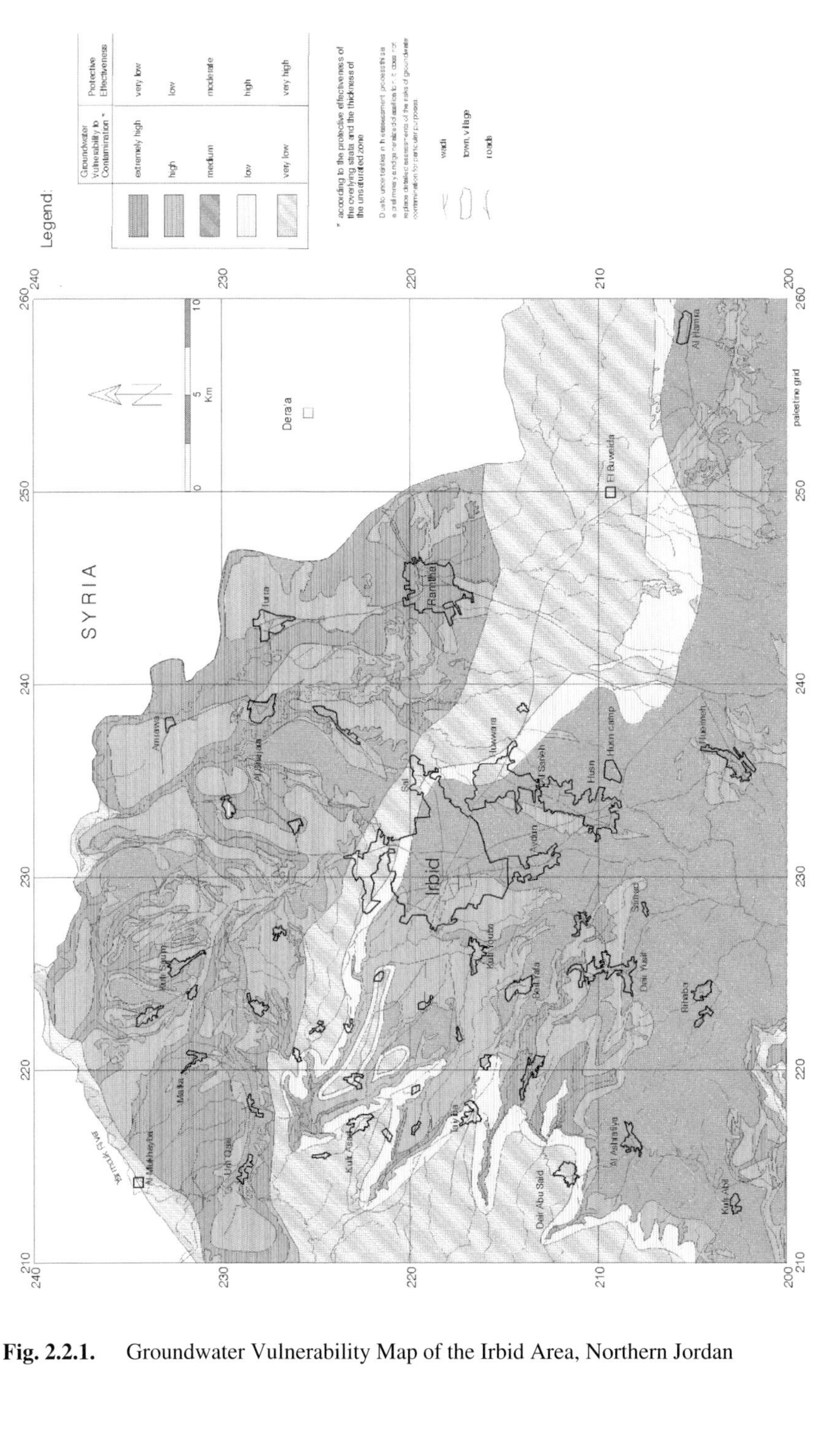

Fig. 2.2.1. Groundwater Vulnerability Map of the Irbid Area, Northern Jordan

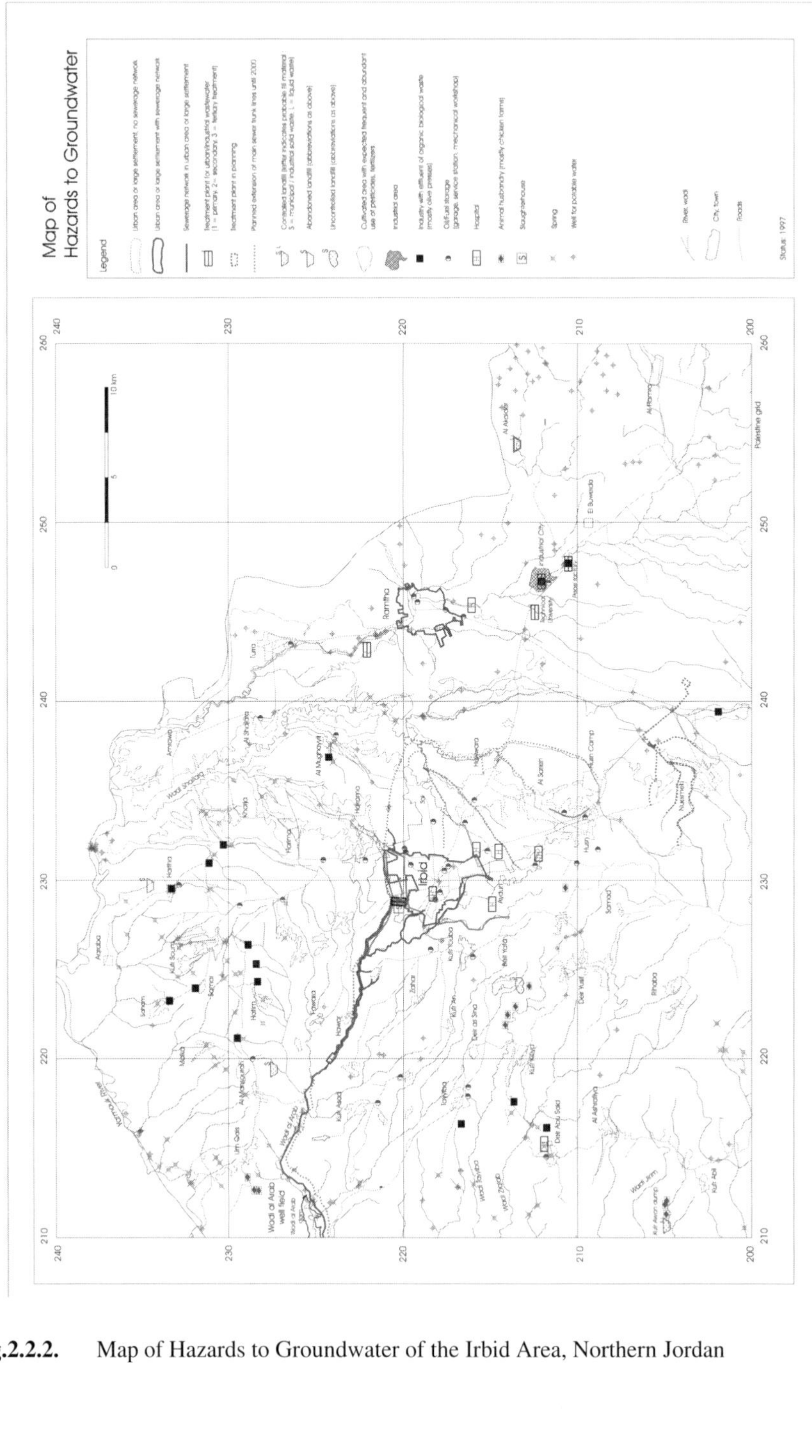

Fig.2.2.2. Map of Hazards to Groundwater of the Irbid Area, Northern Jordan

2.2.3 Groundwater Vulnerability Mapping of the Damascus Ghouta Plain using the GLA-Method

The groundwater vulnerability map of the Ghouta Plain was compiled by a team from the Technical Cooperation Project 'Management, Protection and Sustainable Use of Groundwater and Soil Resources in the Arab Region' (HOBLER & RAJAB, 2002).

Groundwater withdrawal in Damascus and the Ghouta plain is very important for the water supply of the city and the surrounding areas. However, the rapid increase in the demand for water, the decrease in groundwater recharge and successive years of draught led to severe over-pumping of aquifers. Springs and shallow wells dried up in many areas and the productivity of wells has decreased. Lack of water now seriously affects agricultural production, particularly in the southern and eastern parts of the Ghouta. Furthermore, most of the surface water is heavily polluted. Practices, such as the disposal of untreated industrial wastewater into rivers and channels, the use of polluted or insufficiently treated water for irrigation, uncontrolled garbage disposal and intensive use of fertilizers and pesticides endanger the quality of the groundwater.

Today, most of the domestic wastewater of Damascus is pumped to a treatment plant in Adra, about 15 - 20 km northeast of the city. From there, the treated wastewater is pumped back to the northern and central part of the Ghouta and used for irrigation. However, this does not alleviate the problem that groundwater withdrawal far exceeds the average annual recharge in large parts of the Ghouta. A sustainable use of groundwater seems impossible without a substantial change in the management and protection of resources.

Due to the high protective effectiveness of the soils, the levels of heavy metals and organic compounds in the groundwater still appear to be relatively low in most parts of the Ghouta. However pollution poses an ever increasing risk. Some of the scarce water resources could be lost for domestic and even agricultural use. In many locations, Nitrate contents in the groundwater already exceed the permissible value for drinking water of 40 mg/l.

Based on the results of vulnerability mapping (Figure 2.2.3), recommendations for reducing the risk of groundwater contamination have been listed. The implementation of these suggestions would increase the prospects for an improvement in the sustainable use of water resources. Waste disposal sites and uncontrolled handling of hazardous substances (Figure 2.2.4) should by all means be avoided in areas of high groundwater vulnerability. Areas less susceptible to pollution could be taken into consideration as 'search areas' for more suitable locations.

Limestones and conglomerates in the mountain ranges north of the city form productive aquifers. Their outcrops should be considered as high-risk areas. The use of abandoned limestone quarries for garbage disposal poses a very serious threat to the groundwater and should therefore be stopped immediately (e.g. El Tal waste disposal site).

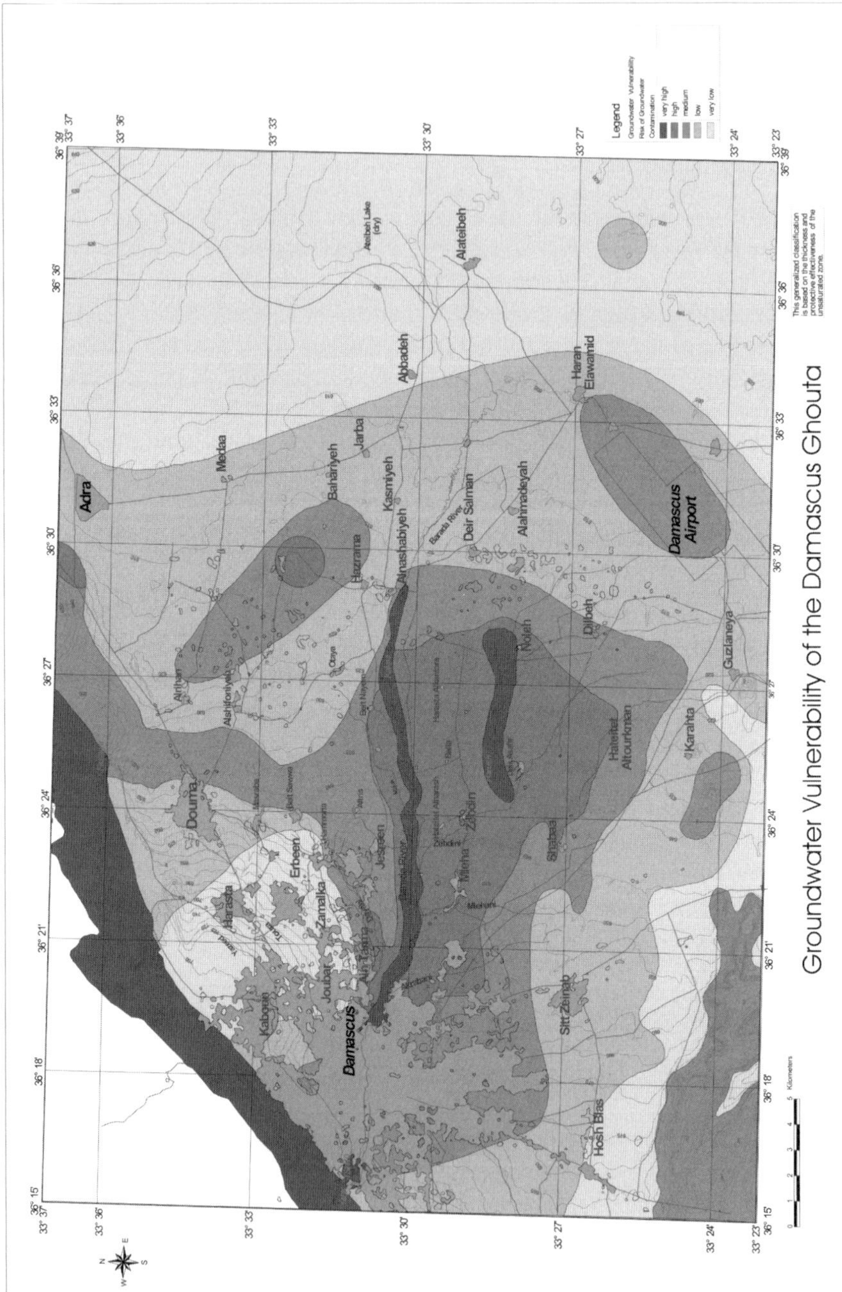

Fig. 2.2.3. Groundwater Vulnerability Map of the Ghouta Plain, Central Syria

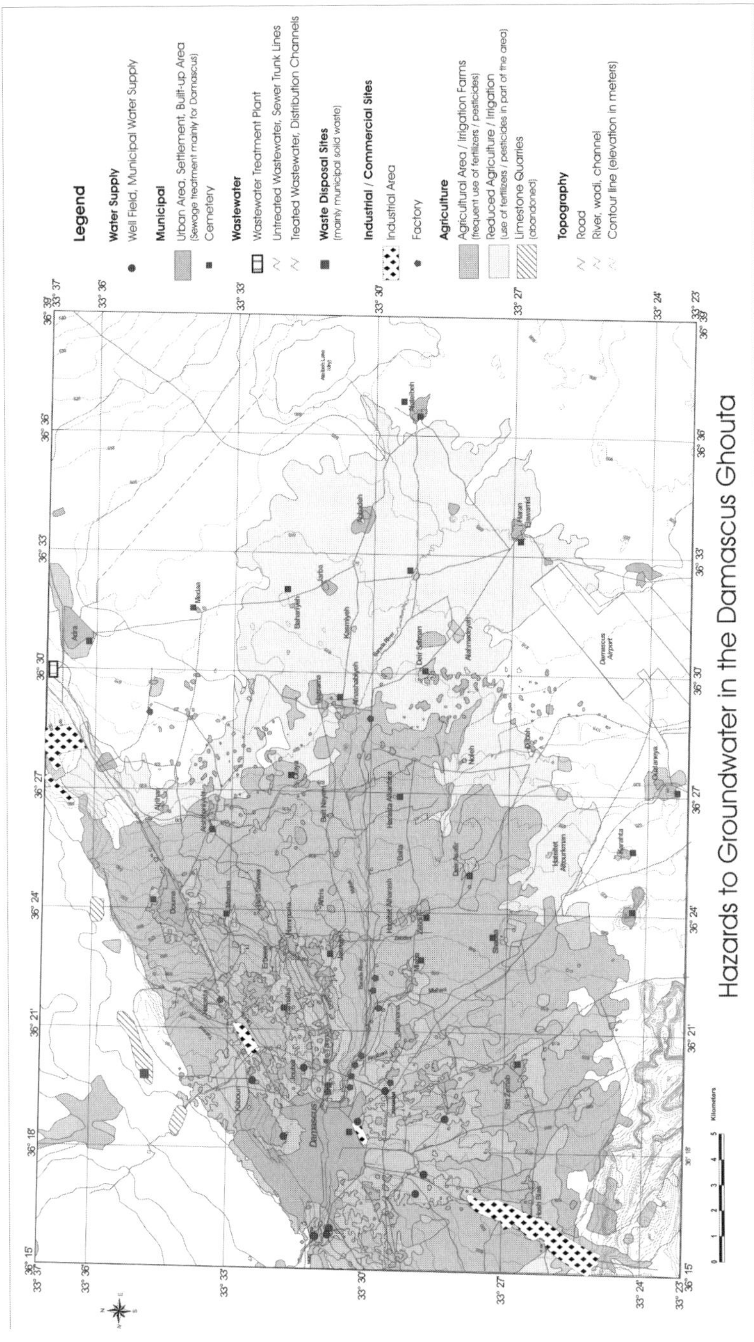

Fig. 2.2.4. Map of Hazards to Groundwater of the Ghouta Plain, Central Syria

During the ACSAD – BGR Cooperation Project the evaluation of the groundwater vulnerability has been combined with studies of soil vulnerability and the present degree of soil contamination. On the whole, the vulnerability of the Ghouta soils can be classified as relatively low. Areas of comparatively high soil vulnerability have been defined in the central part of the pilot area. They coincide with areas, where the protective effectiveness of the unsaturated rock formations above the aquifer is low. In some of these areas, e.g. the tannery area on the eastern rim of the city, the potential for groundwater pollution is high.

Water pollution and environmental problems are increasing concerns to the population and members of governmental institutions. However, efforts to protect valuable groundwater resources are still insufficient.

2.2.4 Groundwater Vulnerability Mapping in the Western Nile Delta, Egypt using DRASTIC

Members of the Research Institute for Groundwater (RIGW), Egypt, prepared a groundwater vulnerability map of the western Nile Delta based on the DRASTIC method (KHATER, pers. comm.). This method was modified to achieve a result that illustrates the observed groundwater quality impacts.

The hydrogeological conditions are such that some of the DRASTIC parameters do not really have a significant influence on the groundwater vulnerability. Examples are the topography (generally flat) and the natural recharge which is evenly distributed throughout the area. Of primary importance to the evaluation of groundwater vulnerability in the Nile Delta are the following parameters: depth to groundwater, aquifer media, soil media, clay thickness (representative of the impact of the vadose zone) and hydraulic conductivity. In order to obtain a vulnerability map that accurately reflects the observed impacts on groundwater quality, locations where groundwater pollution has been recorded (one or more parameters exceed the drinking water standards) were also taken into consideration. Both maps, the one using the 'classic' and the one using the 'modified' approach, are shown in Figures 2.2.5 and 2.2.6.

2.2.5 Conclusion

Groundwater vulnerability maps are a valuable tool for groundwater protection and water quality conservation, particularly in regions where water is scarce, as in the Arab region. The preparation of such maps has been promoted by the BGR within the framework of technical cooperation projects. In Jordan, groundwater vulnerability maps have become a standard tool for the planning and water management authorities.

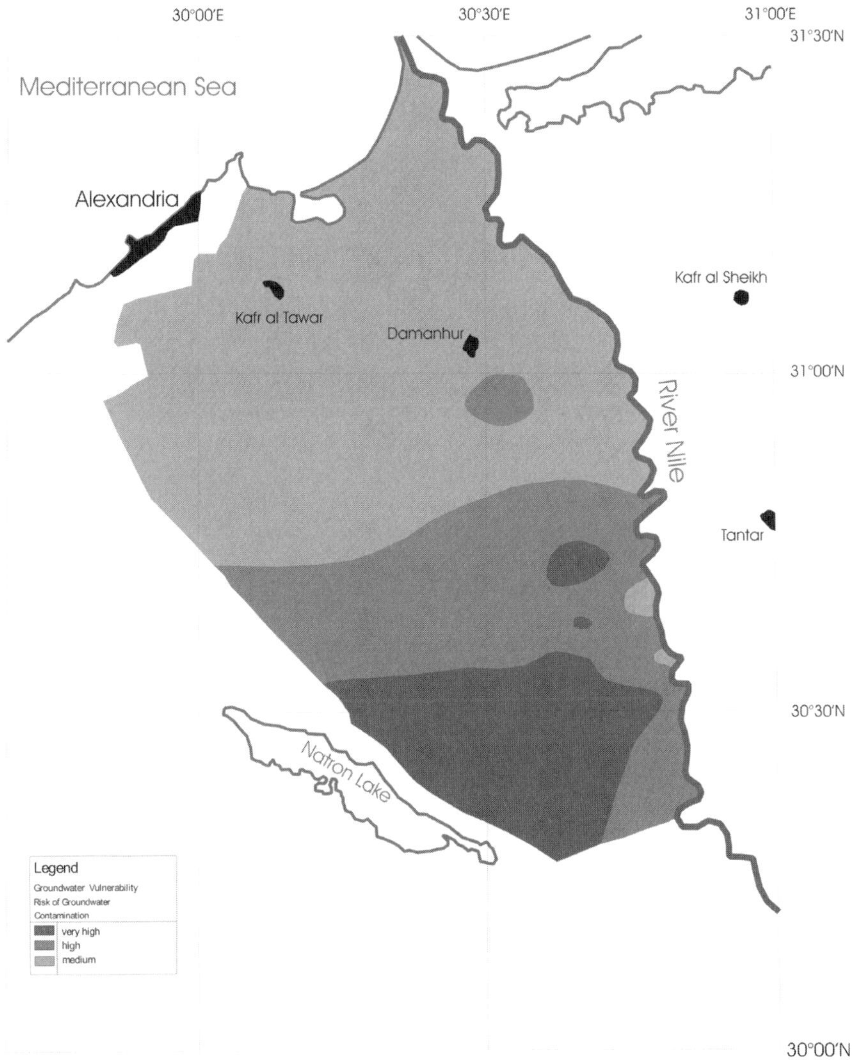

Fig. 2.2.5. Groundwater Vulnerability Map of the Western Nile Delta, Northern Egypt using the Classical DRASTIC Approach

It became evident that close cooperation with national planning authorities is essential for the acceptance of these maps as instruments for land use planning, water quality management and the implementation of any recommendations. For this reason groundwater vulnerability maps should be accessible and understandable to those not familiar with geology. Descriptions of methods, procedures and the degree of accuracy and conclusions based on a hydrogeological point of view should be available as additional information.

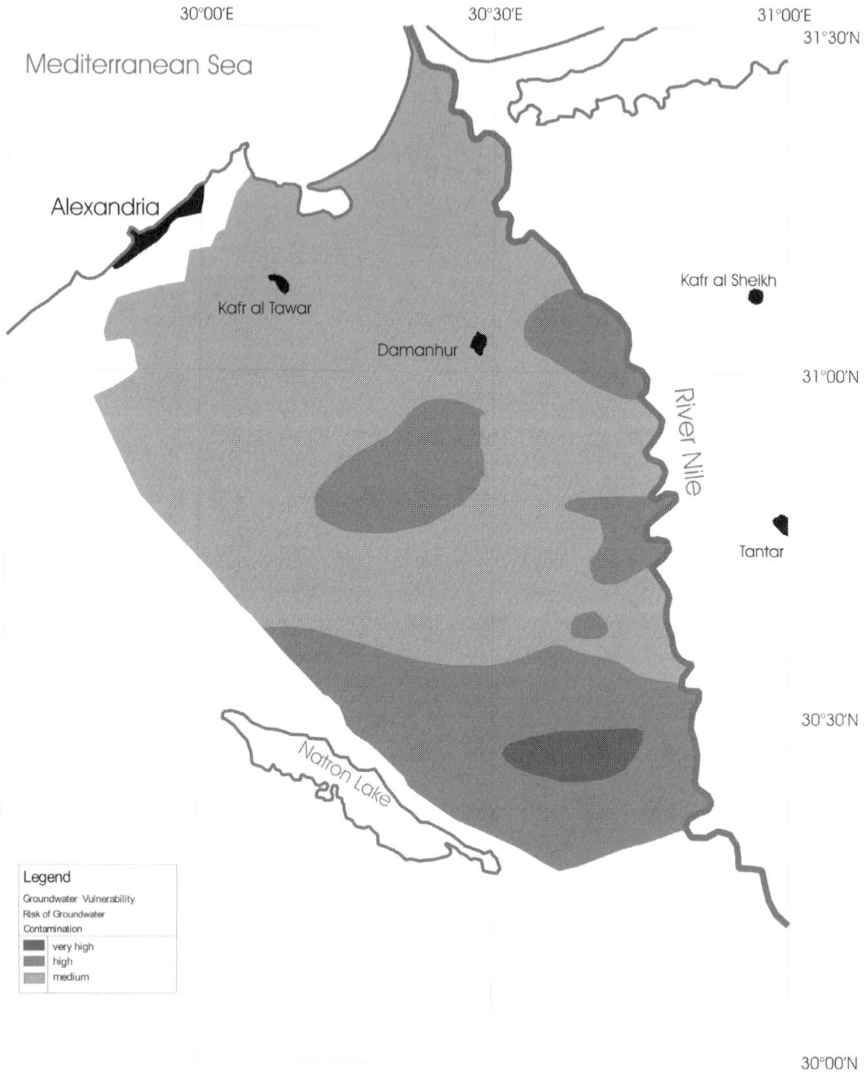

Fig. 2.2.6. Groundwater Vulnerability Map of the Western Nile Delta, Northern Egypt using a Modified DRASTIC Approach

References

Aller L, Bennett T, Lehr JH, Petty RJ (1985) DRASTIC: A Standardized System for Evaluating Groundwater Pollution Using Hydrogeologic Settings. - US EPA/Robert S. Kerr Environmental Research Laboratory, EPA/600/2-85/018, 163 p; Ada (Oklahoma).

Goldscheider N (2002) Hydrogeology and Vulnerability of Karst Systems – Examples from the Northern Alps and Swabian Alp. – PhD thesis, University of Karlsruhe, 229 p.; Karlsruhe.

Hijazi H, Hobler M, Rayyan M & Subah A (1999) Mapping of Groundwater Vulnerability and Hazards to Groundwater in the area south of Amman. - Report prepared by WAJ & BGR, Technical Cooperation Project 'Advisory Services to the Water Authority of Jordan', Special Report No. *4*; Amman.

Hobler M & Rajab R (2002) Groundwater Vulnerability and Hazards to Groundwater in the Damascus Ghouta. - Report prepared by ACSAD & BGR, Technical Cooperation Project 'Management, Protection and Sustainable Use of Groundwater and Soil Resources', Vol. *2.1*, p.55, ; Damascus.

Hobler M & Rajab R (in prep.) Groundwater Vulnerability and Hazards to Groundwater in the Beka'a Valley. - Report prepared by ACSAD & BGR, Technical Cooperation Project 'Management, Protection and Sustainable Use of Groundwater and Soil Resources', Vol. 2.2, p., ; Damascus.

Hoelting B, Haertlé T, Hohberger KH, Nachtigall KH, Villinger E, Weinzierl W & Wrobel JP (1995) Konzept zur Ermittlung der Schutzfunktion der Grundwasserueberdeckung. – Geol. Jb., C, *63*, pp 5–24; Hannover. – [translated into English by the Federal Institute for Geosciences and Natural Resources: Concept for the Determination of the Protective Effectiveness of the Cover above the Groundwater against Pollution.– Ad-hoc Working Group on Hydrogeology, 28 p.; Hannover]

Margane A (2003) Guideline for Groundwater Vulnerability Mapping and Risk Assessment for the Susceptibility of Groundwater Resources to Contamination. – Unpublished report prepared by ACSAD & BGR, Technical Cooperation Project 'Management Protection and Sustainable Use of Groundwater and Soil Resources in the Arab Region, 177 p, Damascus.

Margane A (2002) Mapping of Groundwater Vulnerability and Hazards to Groundwater in the Surat Thani Greater City Area. – Report prepared by Dep.Min.Res. & BGR, Technical Cooperation Project 'Environmental Geology for Regional Planning', Technical Report No. *34*, 36 p, 13 annexes, 2 maps; Bangkok.

Margane A, Hobler M & Subah A (1997) Mapping of Groundwater Vulnerability and Hazards to Groundwater in the Irbid Area, Northern Jordan.– Report prepared by WAJ & BGR, Technical Cooperation Project 'Advisory Services to the Water Authority of Jordan', Special Report No. *3*, 50 p; Amman.

Margane A, Hobler M & Subah A (1999) Mapping of Vulnerability and Hazards to Groundwater in the Irbid Area, N-Jordan. – Zeitschrift Angewandte Geologie, *45*, 4: 175-187.

Margane A & Sunna N (2002) Proposal for a National Guideline for the Delineation of Groundwater Protection Zones. - Report prepared by WAJ & BGR, Technical Cooperation Project 'Groundwater Resources Management', Technical Report No. *1*, 161 p; Amman.

Saefl (2000) Practical Guide Groundwater Vulnerability Mapping in Karst Regions (EPIK). – Report, 57 p.; Bern/CH.

Vrba J & Zaporozec A [eds] (1994) Guidebook on Mapping Groundwater Vulnerability. - IAH International Contributions to Hydrogeology, Vol. *16*, 131 p.; Hannover/FRG (Heise Publ.).

2.3 Intermittent Water Supply and Domestic Water Quality in the Middle East

Fuad Hashwa & Sima Tokajian

Microbiology & Biotechnology Lebanese American University, Byblos – Lebanon

Summary

Many countries in the Middle East suffer from chronic water shortage and as a result supplies generally operate on an intermittent basis, leaving the distribution system un-pressurized for long periods of time. Operating a distribution system in a non-continuous manner leads to conditions that favor the deterioration of the water quality, due mainly to infiltration, regrowth within pipes and the detachment of the bacterial biofilm following variations in pressure and velocity. During no-flow periods individual households store water in roof top tanks for several days, thus providing further opportunity for microbial regrowth and deterioration of the water quality, contributing eventually to incidences with opportunistic pathogens and thus increased public health related problems.

2.3.1 Introduction

Life cannot exist without the most important natural resource, which is water. A cholera outbreak in 1854 in London caused 10.000 deaths, and was linked to bacterial contamination of drinking water by sewage pollution [3]. Ever since, adequate water and sewage purification and hygiene standards have played the most significant part in the eradication of diseases and public health improvements. The problem of water scarcity is a growing worldwide phenomenon. Renewable water resources have declined dramatically. Around 70% of the water that is withdrawn from rivers or from underground sources is used for irrigation. Twenty percent is used by the industry and 10% for residential purposes [4]. It is expected that by the year 2025 the average net water resources in the Middle East will be less than 700 cubic meters per person per year, half of what they are today [27]. Many countries in this area suffer from chronic shortage of both water and capital to invest in water supply.As a result supplies generally operate on an intermittent basis, leaving the distribution system unpressurized for long periods of time. Operating a distribution system in a non-continuous manner leads to conditions that favor the deterioration of the water quality, due mainly to infiltration, regrowth within pipes and the detachment of the bacterial biofilm following in tanks on roof-tops variations in pressure and velocity. Consumers are forced to store the water for several days providing further opportunity for water quality deterioration [18, 38].

Water supply distribution systems are designed to deliver to consumer's safe drinking water that is also adequate in quantity and acceptable in terms of taste, odor and appearance [15]. The change in the quality of drinking water during distribution is of major concern. Knowledge of the changes in the quality of drinking water in distribution systems and the cause of such changes is necessary for good management of the entire water works [14]. Water being transported through distribution systems is subject to both chemical and microbial quality changes. Bacteria are able to grow in most bodies of water, regardless of the variations observed in the chemical or physical conditions [17]. Related to microbial activities during distribution are the development of taste and odors, increased turbidity and regrowth of bacteria surviving treatment [2]. The term regrowth is used in the United States to describe the occurrence of coliforms in water distribution systems [22], while in Europe it refers to the multiplication of heterotrophic bacteria [39].

Intermittent water supply is not widely explored [18]. The literature practically lacks any direct study considering this important aspect and its impact on the quality of water delivered to consumers. All studies deal with systems basically characterized by the continuous mode of flow. However, situations like water storage, retention of water at reservoirs or dead ends, flow interruptions, leakage and pipe breaks, which characterize the intermittent mode of supply, are all examples of problems occasionally facing systems employing the continuous mode of flow. Therefore, although studies in the literature are not directly addressing the intermittent mode many of the studied aspects could be very useful in understanding factors causing quality deterioration problems in these systems [18].

In this chapter we deal with factors causing the deterioration of the water quality in distribution systems and the discussion will specifically highlight the problems associated with the intermittent mode of supply.

2.3.2 Water quality deterioration in systems with intermittent water supply

Biofilm

Biofilms are organized multicellular systems with specific structural and functional properties that influence the metabolic processes, response to nutrients and resistance to antimicrobial agents [21]. In drinking water distribution systems the occurrence of microbial biofilm can be associated with technical and hygienic problems. Microbially contaminated water by its slow flow rate and stagnation (common conditions in systems with the intermittent mode of supply) colonizes the interior surface areas of distribution pipes with extracellular material and embedded cells. The increased physical thickness of this material as a result of continuous layering leads to biofilm formation [40]. However, problems associated with biofilm formation in distribution systems employing intermittent mode of supply could be even more aggravated by the sloughing off the biofilm population and thus the deterioration of the treated quality of water as a result of flow interruptions. Thorough investigations revealed that the unexplained occurrences of coliform bacteria from finished drinking water were attributed to their association and detachment from the surface of pipe walls [6]. Wende et al [42] showed that biofilm detachment accounted for the significant increase observed in the number of planktonic cells present in the bulk water in a pilot reactor system. Tokajian [35] reported the isolation of acid-fast bacilli from a drinking water distribution system in Lebanon employing the intermittent mode of supply. Acid-fast bacilli and coliforms were isolated from drinking water biofilms [10, 28]. Several factors have been implicated in waterborne dissemination of these organisms including prolonged time in the distribution system, optimal growth temperatures, nutrient availability and stagnant or interrupted flow [26].

Water retention

Bacterial regrowth in distribution systems is a common phenomenon but it is a source of concern since it could lead to non-compliance with water quality regulations, taste and odor problems, and may be associated with an increased risk of gastrointestinal illnesses [30]. Levels of nutrients entering the pipe network, pipe characteristics and conditions, water stagnation and the level of residual chlorine are all important factors controlling bacterial regrowth [13, 41]. Intermittent supply of water leads to prolonged water retention in the distribution system at times of no flow [18]. Maul et al. [24] attributed the detection of high bacterial counts in parts of the distribution system of the city of Metz in France, to prolonged retention time of water in the network.

Water storage in household tanks

Storage of treated drinking water in household storage tanks is essential, though such systems cannot be excluded as being sources of post-treatment contamination, introducing coliform bacteria and possible pathogens into the water supply [15, 18, 37]. Treated water in an intermitted supply system does not maintain a constant level of hygienic quality under storage conditions [5, 18, 37, 38].

Geldreich et al. [16] detected a significant decline in emergency water quality upon storage, while water retention in reservoirs led to bacterial regrowth with the opportunistic pathogens comprising a considerable segment of the population [9]. Simmons et al. [34] also showed that roof-collected rainwater systems provided a supply with relatively poor physicochemical and microbiological quality, and Carter et al. [7] attributed the detection of high bacterial counts at a dead end region in Milford distribution system to low flow conditions.

Some coating materials used in lining the interior of a potable water reservoir could have a bacterial growth support potential [11]. The accumulation of sediments in drinking water reservoirs also provides the needed environment for the establishment of a stable association between varieties of different organisms [33].

The quality of treated water is related to the disinfectant effectiveness, the concentration of the disinfectant residual, the age of water in the system and the origin of the intake water [1, 25]. Decay in the disinfectant residual, which is associated with long hydraulic retention times and warm water conditions [5, 38], leads to bacterial regrowth [25, 37]. Water temperature is also an important rate-controlling factor in bacterial regrowth [23, 37].

Storing the distributed water in household tanks (Figure 2.3.1) for periods as long as four to seven days, especially in the summer season intensifies the problem of bacterial regrowth in systems employing the intermittent mode of supply [37, 38]. The crucial role that household storage tanks play in determining the bacteriological quality of drinking water in intermittent systems was previously studied [12, 37, 38]. During a 7-days period of storage the increase in heterotrophic plate count were 1-4 orders of magnitude (Figure 2.3.2). The water quality in the storage tanks deteriorates to such an extent that improvements obtained during treatment are largely negated, and water samples collected from the distribution system for quality assessment bear little relationship to the quality being consumed [18, 37, 38].

The characteristics of bacteria found in household storage tanks supplied with conventionally treated surface water were studied in Jordan [12, 38]. The study revealed that the Gram-positive group constituted the largest percentage of the bacterial isolates of the household tank water with the dominance of *Actinomycetes* species. However, other isolates such as *Pseudomonas vesicularis*, *Pseudomonas / Alcaligenes* and *Moraxella* were also detected in the tanks. Contrary to those results, the identification of the heterotrophic bacteria from household tanks in Lebanon based on the 16S rDNA sequence analysis, revealed the dominance of Gram-negative bacteria [35]. The source water in Lebanon was underground treated with rapid sand filtration and UV. Additionally, the majority of the colonies recovered from the household tanks in Lebanon were pigmented [36].

Long retention times and warm temperatures favor pigmented subpopulations [7, 19, 31]. This however, indicates that the water was not polluted. Guthrie and Cherry [17] suggested that in nonpotable water, pigmented bacterial levels could be used as an indicator of the cleanliness of the water, with the percentage of pigmented forms present being inversely proportional to the pollution level of the source water. Pigmented bacteria are considered to be potentially useful markers that can help in interpreting changes in the microbiological quality of finished water [32]. On the other hand, high levels of pigmented bacteria could pose an increased health risk to immunologically compromised individuals. The bacterial quality of drinking water should be controlled to prevent the development of high concentrations of heterotrophic bacteria, including the pigmented forms [31].

Colonies recovered from storage tanks in Lebanon [Hashwa & Tokajian, in press] were found to be members of the α, β, γ subclasses of *Proteobacteria*. The β-subclass was the dominant type detected (Figure 2.3.3).

Fig.2.3.1. Household water storage tanks. Galvanized cast-iron tanks (left) and the polyethylene (right), each tank has a capacity of 0.5 m^3.

Overall, *Novophingobium rosa* and the potentially opportunistic pathogen *Aeromonas* sp. CDC 715-84 were the dominant types observed (Figure 2.3.4).

The presence of sphingomonads in drinking water distribution systems is not desirable, especially that some strains are considered as potential pathogens [29]. Sphingomonads are relatively slow growing and can be easily overgrown by other bacteria and so they are likely to become overlooked [20]. The presence of sphingomonads in drinking water environment may be much more common than has been reported so far and accordingly further studies are merited. The detection and regrowth of opportunistic pathogens in the household tanks, which are commonly used in systems employing the intermittent mode of supply, highlights the potential of such systems in disseminating infections especially in immunocompromised hosts [Tokajian & Hashwa, in press].

Infiltration

Infiltration is another important problem in systems employing the intermittent mode of supply [36]. When water pumping starts after a period of no flow, poor water quality occurs due to the possible infiltration of contaminated water.

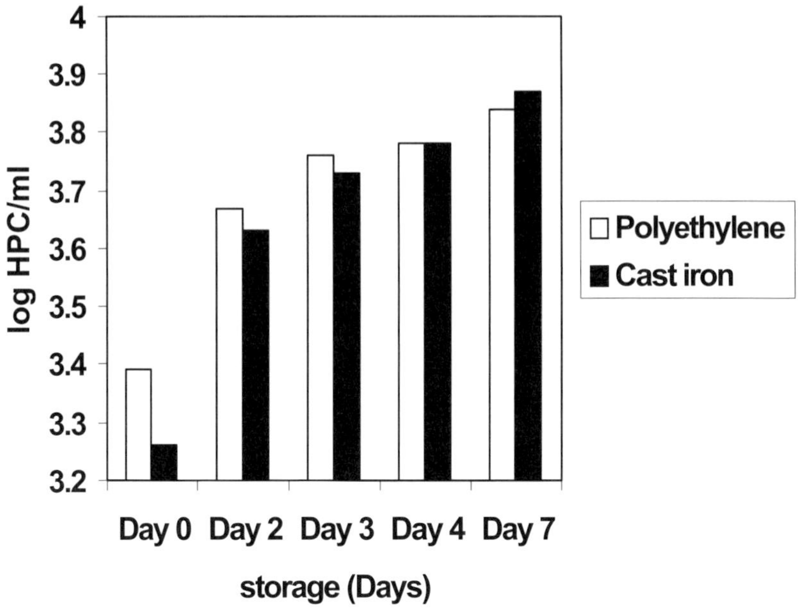

Fig. 2.3.2. Regrowth in the household water storage tanks

The influence of the intermittent mode of supply and/or infiltration when the flow restarts was studied in Palestine [8] and Lebanon (Figure 2.3.5) [36]. Infiltration and /or biofilm detachment due to intermittent supply resulted in deterioration of the water quality in both networks. The first few minutes after reflow was most critical [37].

In an intermittent mode of water supply pressure loss may allow ingression of contaminated water. Water stagnation may promote microbial growth and pressure build up and re-flow may lead to biofilm detachment (Figure 2.3.5).

2.3.3 Conclusion

In a system with an intermittent mode of supply pressure changes, stagnation of water in pipes at times of no flow, regrowth, infiltration and biofilm detachment are all factors causing the deterioration of the water quality. Furthermore, the use of storage tanks by householders to overcome water scarcity problems and storing water up to four to seven days, especially during the warm summer season intensifies the problem of bacterial regrowth and potential risks with opportunistic pathogens.

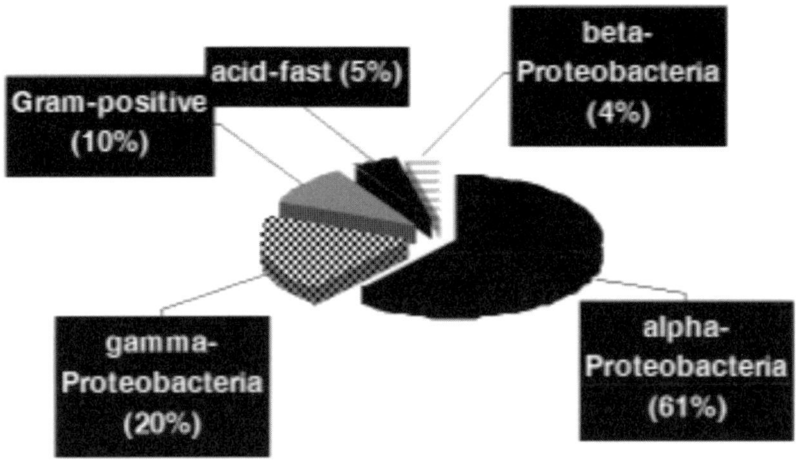

Fig. 2.3.3. Percent occurrence of microbial populations in the household tanks in Lebanon

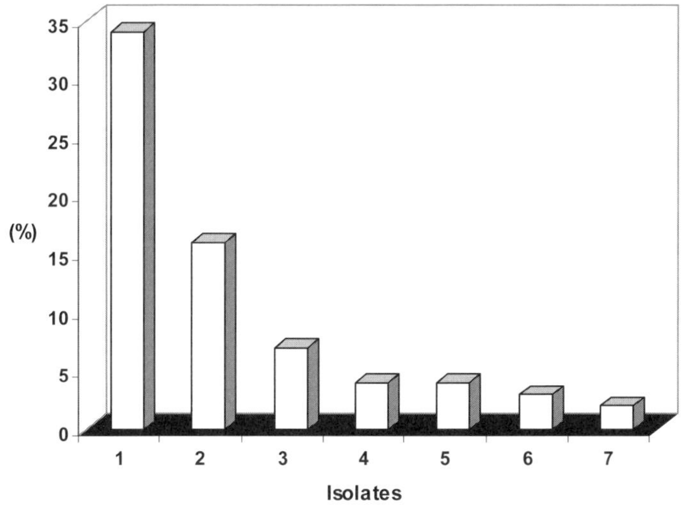

Fig. 2.3.4. Organisms representing the bacterial population re- covered from the household tanks. (1) *Novosphingobium rosa*; (2) *Aeromonas* sp. CDC 715- 84; (3) *Bosea thiooxidans*; (4) *Aeromicrobium erythreum*; (5) *Mycobacterium* sp. N110; (6) *Sphingobium yanoikuyae*; (7) *Brevundimonas vesicularis*.

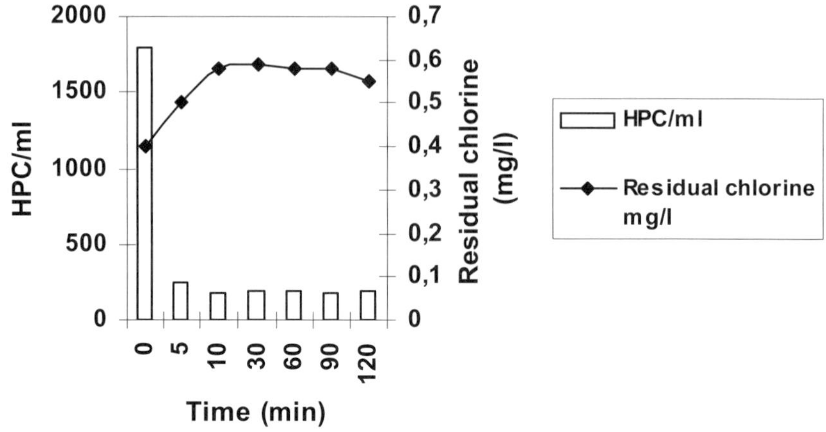

(a)

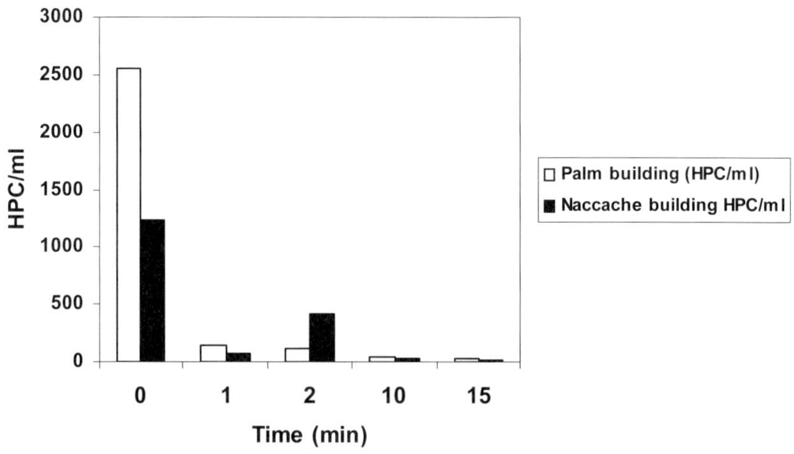

(b)

Fig. 2.3.5. Bacterial infiltration studies in distribution systems of Palestine (a) & Lebanon (b). HPC refers to heterothrophic plate count

References

[1] Allen L, Darby J (1994) Quality control of bottled water and vended water in California: a review and comparison to tap water. J Environ Health *56*: pp 17-22

[2] Allen M, Taylor R, Geldreich E (1980) The occurrence of microorganims in water main encrustations. J Am Water Works Ass *72*: pp 614-625

[3] Bates A (2000) Water as consumed and its impact on the consumer-do we understand the variables? Food Chem Toxicol *38*: pp 29-36

[4] Brown L (2001) How water scarcity will shape the new century. Water Sci Technol *43*: pp 17-22

[5] Burlingame G, Brock G (1985) Water quality deterioration In treated water storage tanks. In: Water Annual Conference Proceedings, American Water Works Association, Denver

[6] Camper A, Jones W, Hayes J (1996) Effect of growth conditions and substratum composition on the persistence of coliforms in mixed-population biofilms. Appl Environ Microbiol *62*: pp 4014-4018

[7] Carter J, Rice E, Buchberger S, Lee Y (2000) Relationships between the levels of heterotrophic bacteria and water quality parameters in a drinking water distribution system. Water Res *34*: pp 1495-1502

[8] Coelho S, James S, Sunna N, Abu Jaish A, Chatila J (2003) Controlling water quality in intermittent supply systems. Water Sci Technol 3: pp 119-125

[9] Committee Report (1983) Deterioration of water quality in large distribution reservoirs (open reservoirs). J Am Water Works Ass *75*: pp 313-318

[10] Covert T, Rodgers M, Reyes A, Stelma G (1999) Occurrence of nontuberculous mycobacteria in environmental samples. Appl Environ Microbiol *65*: pp 2492-2496

[11] Ellgas W, Lee R (1980) Reservoir coatings can support bacterial growth. J Am Water Works Ass *72*: pp 693-695

[12] Evison L, Sunna N (2001) Microbial re-growth in household water storage tanks. J Am Water Works Ass *93*: pp 85-94

[13] Franco E (1997) Defining safe drinking water. Epidemiology *8*: pp 607-609

[14] Gambissini L, Sacco C, Lanciotti E, Burrini D, Griffini O (1990) Microbial quality of water in the distribution system of Florence. J Water Supply Res T *39*: pp 258-264

[15] Geldreich E (1996) Microbial quality of water supply in distribution systems. Lewis Publisher, New York

[16] Geldreich E, Nash H, Reasoner D, Taylor R (1975) The necessity of controlling bacterial populations in potable waters-bottled water and emergency water supplies. J Am Water Works Ass *67*: pp 117-124

[17] Guthrie R, Cherry D (1974) A comparison of thermal loading effects on bacterial populations in polluted and non-polluted aquatic systems. Water Res *8*: pp 143-148

[18] Hashwa, F, Tokajian S (2000) Change of water quality in distribution system of a residential area in Beirut. Proceedings of IV International Conference "Water supply and Water Quality", pp 1125-1136, Krakow, Poland

[19] Herman L (1976) Sources of the slow-growing pigmented water bacteria. Health Lab Sci *23*: pp 5-10

[20] Koskinen R, Ali-Vehmas T, Kampfer P, Laurikkala M, Tsitko I, Kostyal E, Atroshi F, Salkinoja-Salonen M (2000) Characterization of *Sphingomonas* isolates from Finnish Swedish drinking water distribution systems. J Appl Bacteriol *89*: pp 687-696

[21] Lawrence J, Korber D, Hoyle B, Costerton J, Caldwell D (1991) Optical sectioning of microbial biofilms. J bacterial *173*: pp 6558-6567

[22] LeChevallier M (1990) Coliforms re-growth in drinking water: a review. J Am Water Works Ass *82*: pp 74-86

[23] LeChevallier M, Schulz M, Lee R (1991) Bacterial nutrients in drinking water. J Am Water Works Ass *57*: pp 857-862

[24] Maul A, El-Shaarawi A, Block J (1985) Heterotrophic bacteria in water distribution systems. Spatial and temporal variation. Sci Total Environ *44*: pp 201-214

[25] Momba M, Cloete T, Venter S, Kfir R (2000) Influence of disinfection processes on the microbial quality of potable groundwater in a laboratory-scale model. J Water Supply Res T *49*: pp 23-34

[26] Moulin G, Stottmeier K (1986) waterborne mycobacteria: an increasing threat to health. ASM News *52*: pp 525-529

[27] Nachmani A (1994) A community in scarcity: the politics of water in the Middle East. Jerusalem Leet 18 Adar/ 5754

[28] Norton C, LeChevallier M (2000) A pilot study of bacteriological population changes through potable water treatment and distribution. Appl Environ Microbiol 66: pp 268-276

[29] Perola O, Nousiainen T, Suomalainen S, Aukee S, Karkkainen U, Kauppinen J, Ojanen T, Katila M (2002) Recurrent *Sphingomonas paucimobilis*-bacteremia associated with a multibacterial water-borne epidemic among neutropenic patients. J Hosp Infect *50*: pp 196-201

[30] Pruvost O, Couteau A, Perrier X, Luisetti J (1998) Phenotypic diversity of *Xanthomonas* sp. *mangiferaeindicae*. J Appl Microbiol *84*: pp 115-124

[31] Reasoner D, Blannon J, Geldreich E, Barnick J (1989) Nonphotosynthetic pigmented bacteria in potable water treatment and distribution system. Appl Environ Microbiol *55*: pp 912-921

[32] Rusin P, Rose J, Haas C, Gerba C (1997) Risk assessment of opportunistic bacterial pathogens in drinking water. Rev Environ Contam T *152*: pp 57-83

[33] Schreiber H, Schoenen D (1994) Chemical, bacteriological and biological examination and evaluation of sediments from drinking water reservoirs-results from first sampling phase. Zbl Hyg *196*: pp 153-169

[34] Simmons G, Hope V, Lewis G, Whitmore J, Gao W (2001) Contamination of potable roof-collected rainwater in Auckland, New Zealand. Water Res *35*: pp 1518-1524

[35] Tokajian S (2003) Regrowth of bacteria and their genotypic identification in drinking water in Lebanon. Ph.D. thesis, University of Newcastle upon-Tyne, UK

[36] Tokajian S, Hashwa F (2002) Water quality problems associated with intermittent water supply. 11th International Symposium on Health-Related Microbiology, Melbourne, Australia

[37] Tokajian S, Hashwa F (2003) Water quality problems associated with intermittent water supply. Water Sci Technol *47*: pp 229-234

[38] Tokajian S, Sunna N, Evison L, Hashwa F (2000) Microbial regrowth in household storage tanks. 1st World Congress of International Water Association, Paris, France

[39] Van der Kooij D (1992) Assimilable organic carbon as an indicator of bacterial re-growth. J Am Water Works Ass *84:* pp 57-65

[40] Vess R, Anderson R, Carr J, Bond J, Favero M (1993) The colonization of solid PVC surfaces and the acquisition of resistance to germicides by water-organisms. J Appl Bacteriol *74*: pp 215-221

[41] Volk C, LeChevallier M (1999) Impacts of the reduction of nutrient levels on bacterial quality in distribution systems. Appl Environ Microbiol *65*: pp 4957-4966

[42] Wende E, Characklis W, Smith D (1989) Biofilms and bacterial drinking water quality. Water Res *23*: pp 1313-1322

2.4 Sewage Water Treatments and Reuse in Israel

M. Ben-Hur

Institute of Soil, Water and Environmental Science, The Volcani Center, POB 6, Bet Dagan, 50250 Israel

2.4.1 Introduction

Many places, such as Israel, in arid and semiarid regions are characterized by long, dry summers and short, wet winters. Therefore, crop production in these regions relies mainly on irrigation. In 1990, for example, the total cultivated land in Israel was 437,100 ha, of which 205,700 ha was under irrigation. However, in recent decades, although the total cultivated land area has not been changed significantly, the area of irrigated land has been reduced, in accordance with the amount of water available for agriculture.

Conventional sources of good quality water (fresh water) in arid and semiarid regions are scarce. Therefore, in these regions, in order to maintain sustainable agriculture to meet the increasing demands for food, and to combat desertification, unconventional water resources need to be used for agriculture. Additional water for irrigation can be derived from treated sewage water (effluent).

Table 2.4.1. Consumption of fresh water and effluent in agriculture in Israel

Year	Agricultural consumption [million m^3/year]	
	Effluent	Fresh water
1980	52.3	1307.5
1994	254.2	1204.7
2000	290	580
2010	490	500

In 1980, the agriculture consumed 52.3 and 1307.5 million m^3 per year of effluent and fresh water (FW), respectively (Table 2.4.1). In contrast, in 2000, the agricultural consumption of effluent increased to 290 million m^3 per year, and that of FW decreased to 580 million m^3 per year. It is expected that this trend will intensify in the future (Table 2.4.1). This decrease in the consumption of FW for agriculture (Table 2.4.1) was mainly a result of increased demands for FW in the domestic and the industrial sectors. This chapter reviews the types and the quality of the effluents in Israel, the effects of effluent irrigation on the soil, and the accompanying hazard of contamination of the water resources.

2.4.2 Sewage water treatments and quality

Sewage effluent comprises 99.9% water and 0.1% organic and inorganic solids in suspended and soluble forms (Feigin et al. 1991). Raw sewage water contains microorganisms that may be pathogenic (bacteria, viruses, and parasitic protozoa), and parasitic worms. Moreover, some of the chemical materials in the sewage water are toxic. Therefore, in order to protect the public both from consuming contaminated crops and from direct exposure to the effluent applied to them, and to prevent nuisance conditions, operational problems in the irrigation system and adverse effects on soil, crops, and water resources, sewage water intended for irrigation should be treated first.

All the effluent that has been and is being used for irrigation in Israel has undergone at least secondary (biological) treatment. The two most commonly used secondary treatments in Israel are oxidation ponds and activated sludge. In the former, which is the simplest secondary treatment, oxygen is used by bacteria that decompose organic matter (OM), and the nutrients released are consumed by algae. This type of treatment is suitable for rural communities and small or medium-sized urban municipalities, in regions of hot to moderate climate where there is no land shortage. Activated sludge treatment, which is a continuous process based on recycling of the biological sludge, is applied in large communities. The treatment capacity of this method is high, but it needs a careful operational control. Another secondary treatment that is used in Israel, but to a lesser extent, is the trickling filter treatment, in which the wastewater percolates through coarse media covered with a 0.1 to 2 mm layer of biological slime that comprises aerobic and anaerobic sublayers.

The quality of raw sewage waters and of their effluents from some sewage treatment plants in Israel, after various treatments are presented in Fig. 2.4.1. These treatment plants collected sewage water from: two big cities (Tel-Aviv and Haifa), a medium-size city (Netanya) and a small town in a rural region (Migdal-Haemek). The quality parameters in Fig. 2.4.1 are average values for water samples that were taken in May, July and October 1996 and March 1997. The raw wastewater was sampled after undergoing mechanical treatment, and the effluent was taken from the outlet pipe of the treatment plants on the same day.

The electrolytic conductivity (EC) and the sodium adsorption ratio (SAR), which represent the salinity and the sodicity of the water, respectively, differed among the various treatment plants: the highest EC and SAR values were found in the sewage water of Haifa – 2.4 dS m^{-1} and 8.5, respectively; and the lowest in the sewage water of Netanya – 1.6 dS m^{-1} and 5.2, respectively. These differences in the quality of the various sewage waters arise from the differences in the quality of the FW supplied to the several authorities, and from the differences in the substances added to the FW during its consumption and use. No significant differences in EC and SAR were found between the raw sewage waters and the effluents from any of the secondary treatments and treatment plants (Fig. 2.4.1). These results indicate that the secondary treatment does not decrease the contents of the major salts in the sewage water. In contrast, the secondary treatment significantly decreased the contents of OM and total suspended solids (TSS) in the effluent (Fig. 2.4.1). However, the decrease in suspended OM contributed more to the OM decrease than that in soluble OM (Fig. 2.4.1). In general, for OM and TSS reduction in the water, activated sludge treatment was the most efficient method and the oxidation pond was the least efficient one (Fig. 2.4.1).

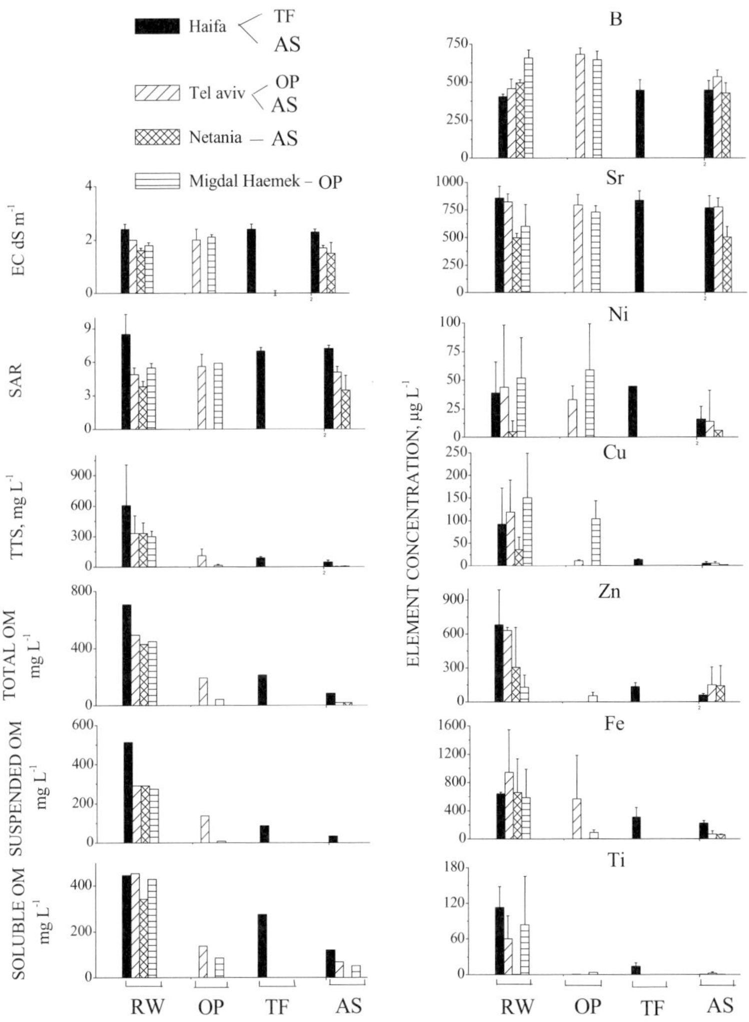

Fig. 2.4.1. Quality of raw sewage (RW) and their effluents after oxidation ponds (OP), trickling filters (TF) and activated sludge (AS) treatments from various sewage treatment plants. Vertical bars indicate standard deviation (from Ben-Hur et al. 1999)

The trace elements in the sewage water and in the effluent can be divided into three major groups with regard to the effects of the secondary treatments on their concentration in the water (Fig. 2.4.1):

1. The elements, B and Sr, whose concentrations in the raw sewage and in the effluent were similar, which indicates that the secondary treatments do not reduce the concentrations of these elements in the effluent.
2. The elements, Cu, Zn, Fe, and Ti, whose concentrations in the effluent were lower than those in the raw sewage, indicating that the secondary treatments are effective in removing these elements from the sewage water. In this case, the activated sludge was, in general, the most effective method.
3. The element Ni, on which the effect of the secondary treatments in reducing its concentration was inconsistent.

2.4.3 Soil salinity

The soil solution in irrigated fields is frequently more saline than the irrigation water, mainly because of evapotranspiration, which leaves the dissolved salts from the water in the soil (Rhoades et al. 1973). Consequently, irrigation with effluent could enhance the salinization of the irrigated lands, which could eventually reduce plant growth (Maas and Hoffman 1977). The EC in saturated paste of soil samples that were taken from plots irrigated with secondary effluent or FW in two sites – the Yizre'el Valley in the north of Israel and the Coastal Plain in the center of Israel – are presented in Fig. 2.4.2, as functions of soil depth. After the irrigation season (fall), the EC values of the effluent-irrigated soils were higher than in the FW-irrigated soil down to >1.2 m depth in the sandy clay loam, and down to 1.0 m depth in the clayey soil (Fig. 2.4.2).

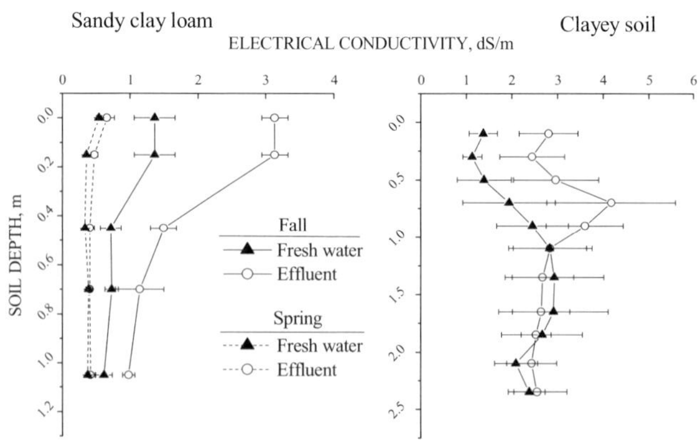

Fig. 2.4.2. Electrical conductivity in saturated paste of soil samples from plots irrigated with water of two qualities, in the Coastal Plain where there is sandy clay loam and the Yizre'el Valley where there is clayey soil. The horizontal bars indicate standard deviation (unpublished data)

To avoid salt accumulation in soil under effluent irrigation, salts should be leached from the root zone. The two major methods of salt leaching are: addition of a leaching fraction (excess water that exits as drainage water) to the irrigation water to hold the salt concentration in the soil below a specific value (Rhoades et al. 1973); and (ii) leaching of salt by rainwater during the winter. The latter method loads less salt on the field and saves water. No long-term accumulation of salt in the soil profile down to 1.5 m was found in a field with clayey soil irrigated with secondary effluent (Fig. 2.4.3).

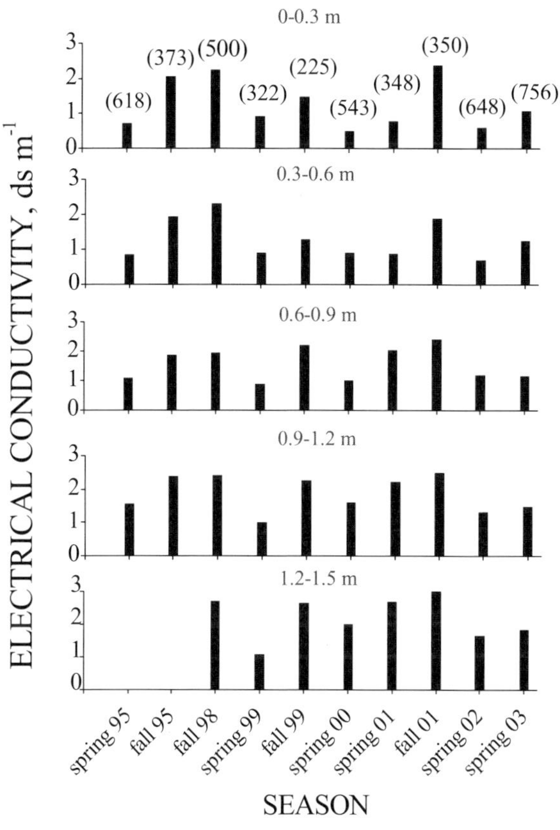

Fig. 2.4.3. Electrical conductivity in various soil layers after the rain (spring) and irrigation (fall) seasons in several years. Values in parentheses indicate the annual rainfall in mm (from Ben-Hur and Ravina 2002)

The average annual rainfall in the region in which this field lies is 500 mm, concentrated mainly into 4 months in the winter. In the fall after the irrigation season, the EC in the saturated soil paste increased to average values of 2.0 and 3.0 dS m^{-1} in the 0 to 0.3 m and 1.2 to 1.5 m layers, respectively (Fig. 2.4.3). These are EC values in which moderately salt-sensitive salt-tolerant plants could be grown without suffering any damage (Maas and Hoffman 1977). In the spring, after the rainy season, the EC decreased to average values of 0.8 and 1.9 dS m^{-1} in the 0 to 0.3 m and 1.2 to 1.5 m layers, respectively (Fig. 2.4.3). These reductions in the EC resulted from the salt leaching by the rainfall. Similar results were obtained in the sandy clay loam from the Coastal Plain (Fig. 2.4.2), and these findings indicate that in regions with similar rainfall characteristics (Fig. 2.4.3), the rainfall is sufficient to prevent long-term salt accumulation, and the consequent reduction in plant growth, in soil irrigated with effluents. However, in regions where the rainfall is too low to leach down the salt, a leaching fraction should be added to the irrigation water.

2.4.4 Soil sodicity and hydraulic properties

In Israel, the SAR of effluents may reach levels of ~7, compared with ~2 in the FW (Fig. 2.4.1). Irrigation with water of such a SAR could increase the SAR of the irrigated soil (Fig. 2.4.4), which would, in turn, impair the hydraulic properties of the soil when it was subsequently exposed to rainfall (Shainberg and Letey 1984).

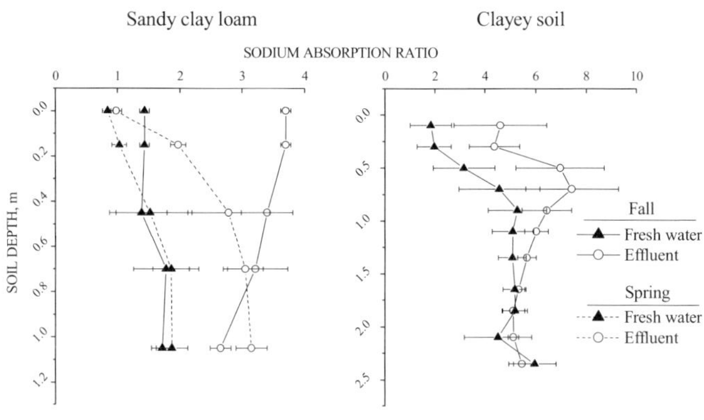

Fig. 2.4.4. Sodium adsorption ratio in saturated paste of soil samples from plots irrigated with water of two qualities in the Coastal Plain where there is sandy clay loam and in the Yizre'el Valley where there is clayey soil. The horizontal bars indicate standard deviation (unpublished data)

The hydraulic properties of soils having such SARs are not likely to be affected during the irrigation season, because of the relatively high electrolyte concentration of the irrigation water, but they could deteriorate when these soils are leached with rainwater (distilled water). Surface seal formation is a common occurrence in many cultivated soils in arid and semi-arid regions (Kemper and Miller 1974; Ben-Hur et al. 1985). Important effects of a soil seal include: reduction of infiltration rate (IR), enhancement of runoff, alteration of erosion, and interference with seed germination (Wakindiki and Ben-Hur 2002). Low electrolyte concentration in the soil solution and high exchangeable sodium percentage (ESP) enhance clay swelling and dispersion, leading to easier breakdown of the surface aggregates and formation of a less permeable seal (Kazman et al. 1983; Ben-Hur et al. 1998).

The effect of irrigation with effluent on seal formation and IR was studied by Mamedov et al. (2000) in four disturbed soils from various locations in Israel. The FW- and effluent-irrigated soils were subjected to simulated rainfall (distilled water) until the final IR was reached. The ratio of the mean final IR of each soil under effluent irrigation to that of the same soil under FW irrigation (FIR_e/FIR_f) was calculated and the results are presented in Fig. 2.4.5 as functions of the rain kinetic energy. For all the studied soils, the final IR values under effluent irrigation were lower than those under FW irrigation, and this was attributed to the relatively high ESPs of these soils, that enhanced the breakdown of aggregates and the consequent seal formation at the soil surface.

The data presented in Fig. 2.4.5 indicate that the loess and clayey soil with 40% clay were affected by the use of effluents to a greater extent than the clayey soils with 48 and 51% clay.

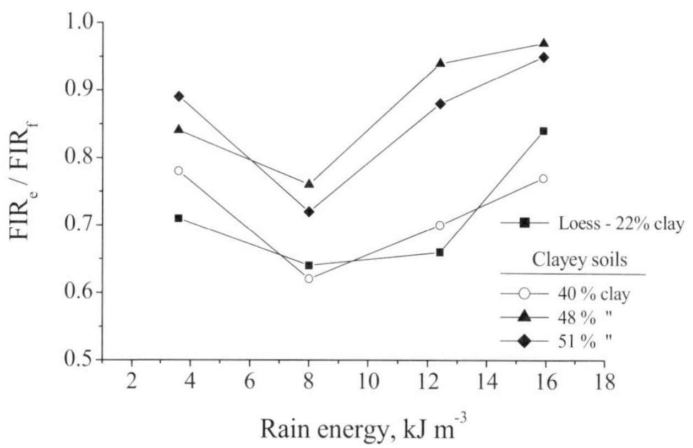

Fig. 2.4.5. Ratio between the mean final infiltration rate of the effluent- and fresh water-irrigated soils (FIR_e/FIR_f) as a function of raindrop energy) from Mamedov et al. 2000).

The higher susceptibility of the former two soils to effluent irrigation was attributed to their lower clay content. Kemper and Koch (1966) suggested that clay acts as a cementing agent that stabilizes soil aggregates: the higher the clay content, the more stable the aggregates, and therefore the higher the resistance of the soil to seal formation. Similar findings were made by Ben-Hur et al. (1985), who found that soils with ~20% clay were the most susceptible to seal formation: those with clay content >40% had stable aggregates and showed less tendency to form seals (Ben-Hur et al. 1985).

The amounts of runoff and of soil loss from field plots that were irrigated in previous summers with FW or effluent are presented in Fig. 2.4.6, as associated with the consecutive rainstorms. The field plots were located in the Coastal Plain with sandy clay loam soil. No significant differences were found in the runoff and soil loss amounts between the FW- and effluent-irrigated plots, indicating that the effluent irrigation had no effect on seal formation, IR and erosion, despite the fact that the SAR values of the effluent-irrigated soil after the irrigation season were higher than those of the FW-irrigated soil (Fig. 2.4.4). This lack of effect of the effluent irrigation on seal formation could be a result of self reclamation of the soil during the winter, since that would reduce the SAR at the soil surface to a value similar to that of the FW-irrigated soil (Fig. 2.4.4).

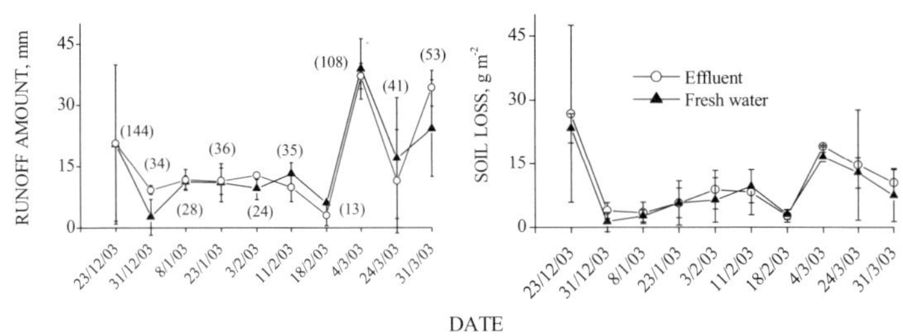

Fig. 2.4.6. Average amounts of runoff and soil loss from the effluent- and fresh water-irrigated soils, as associated with various rainstorms. The vertical bars indicate standard deviation. Values in parentheses indicate the rainfall (in mm) in each storm (unpublished data)

2.4.5 Environmental aspects

Pollution of groundwater and surface water resources is a major concern under effluent irrigation conditions. Water flow below the root zone could carry various contaminants to the vadose zone and eventually further down, to aquifers. Likewise, chemicals and other substances applied to the soil with the effluent irrigation can be transported with the surface runoff and the eroded soil, and could reach surface water reservoirs and contaminate the water.

The concentrations of trace elements in the Israeli effluents are higher than those in the FW (Fig. 2.4.1), therefore, irrigation with effluents could contaminate the irrigated soils and the water resources. Trace elements can be divided into two main groups. (i) Elements, such as B, Zn and Cu, that are essential, as trace elements, for plant growth, but which become toxic to organisms when their concentration exceed certain threshold levels. The margin between recommended and toxic concentrations is often quite narrow. (ii) Trace elements, such as Cd, Hg, Pb, that are not essential to plants and animals.

Concentrations of some trace elements in saturated paste of soil samples that were taken from plots irrigated with secondary effluent or FW in two sites are presented in Fig. 2.4.7 as functions of soil depth. The effects of the effluent irrigation on the accumulation and movement of the trace elements in the soil differed between the two sites (Fig. 2.4.7). In the clayey soil, the concentrations of B and Sr were higher in the effluent-irrigated soil than in the FW-irrigated soil down to depths of 1.2 m (Fig. 2.4.7), but no significant differences were found between soils irrigated with the two waters, in their concentrations of the other studied elements. In contrast, in the sandy clay loam, the concentrations of all the studied elements except Ti were higher in the effluent-irrigated soil than in the FW-irrigated soil (Fig. 2.4.7). These differences between the two soils probably arose from differences in clay content and in dissolved OM (DOM) concentration (Fig. 2.4.8). An increase in the clay content of the soil could increase the adsorption of trace elements and thereby decrease their downward movement in the profile (McBride 1989). Likewise, formation of complexes between DOM and trace elements could increase the solubility of the trace elements in the soil solution, and thereby increase their mobility in the soil (McBride 1989). The fact that the clay content in the sandy clay loam (30%) was less than that in the clayey soil (65%), and the finding that the differences in DOM concentration between the effluent- and FW-irrigated soils were larger in the sandy clay loam than in the clayey soil (Fig. 2.4.8.), probably account for the finding that the effect of the effluent irrigation on the concentration and the movement of the trace elements was greater in the former soil than in the latter one (Fig. 2.4.7).

Another undesirable phenomenon that may be associated with effluent irrigation is the potential formation of complexes of toxic organic substances present in the soil with effluent-borne mobile colloids, macromolecules, and DOM. This would enhance the transport of these toxic organic substances toward the groundwater.

The effect of effluent irrigation on the transport of atrazine applied to a corn field on silt clay loam, located in the Hefer Valley in the center of Israel, was studied by Graber et al. (1995). In this study, cores to 4 m depth were taken from 10 effluent-irrigated and 10 FW-irrigated plots after two growing seasons and two winter rainy seasons (Fig. 2.4.9). In eight of the 10 FW-irrigated profiles, no movement below 1.45 m was noted, but high atrazine concentrations were apparent down to 4 m in two of them (Fig. 2.4.9). The deep penetration of atrazine in these two profiles was attributed to preferential flow through cracks. In contrast, nine of the 10 effluent-irrigated profiles contained atrazine below 1.45 m, and in most of them, atrazine was present at depths >3 m (Fig. 2.4.9). Graber et al. (1995) suggested that formation of complexes of the atrazine molecules with the effluent-borne mobile OM accelerated the transport of atrazine out of the biologically active root zone, which, in turn decreased the atrazine loss in the soil.

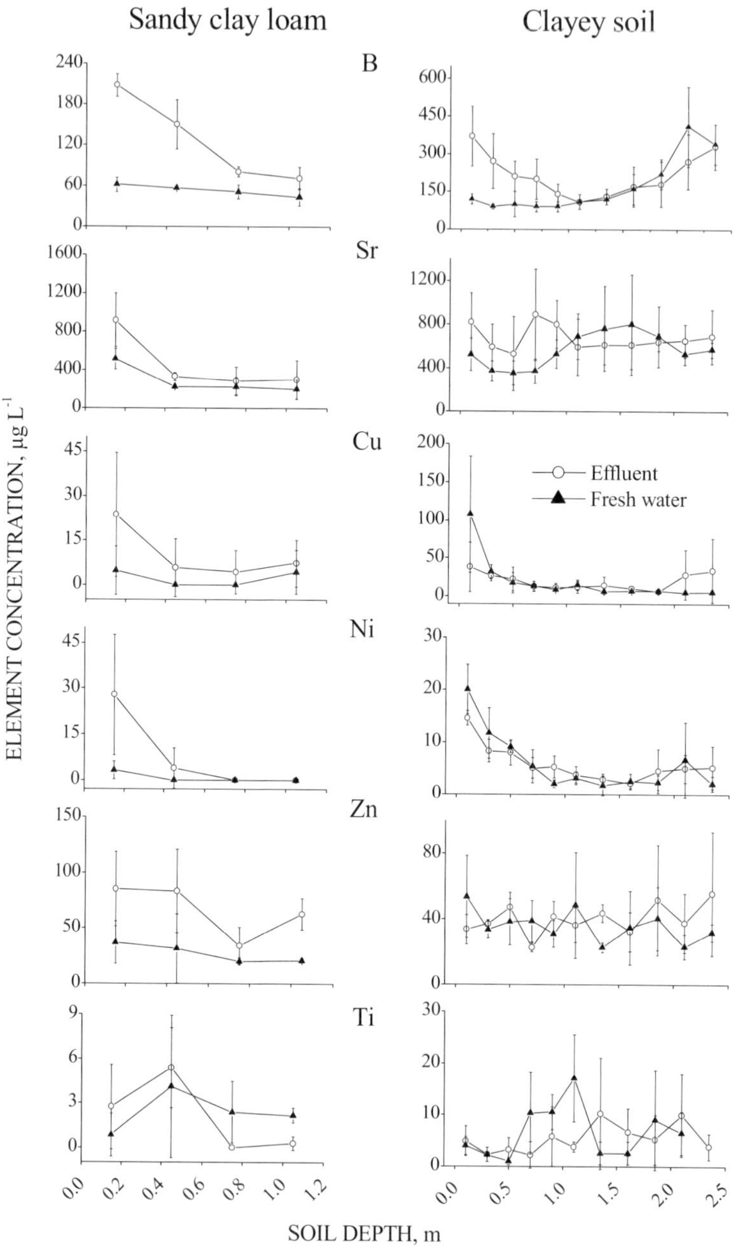

Fig. 2.4.7. Trace element concentration in saturated paste of soil samples from plots irrigated with water of two qualities. The plots were on sandy clay loam in the Coastal Plain and on clayey soil in the Yizre'el Valley. The vertical bars indicate standard deviation (unpublished data)

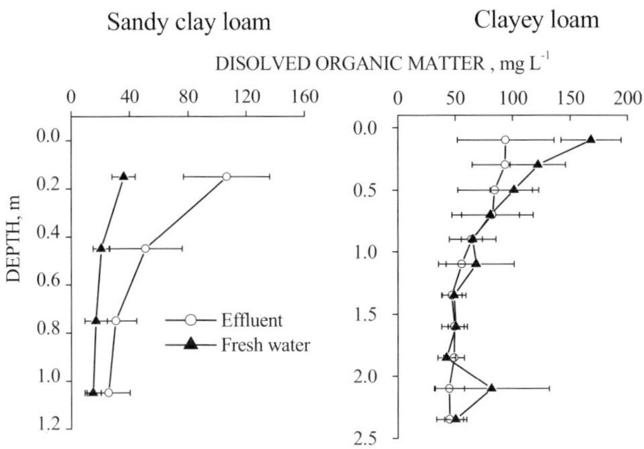

Fig. 2.4.8. Dissolved organic matter of soil samples from plots irrigated with water of two qualities. The plots were on sandy clay loam in the Coastal Plain and on clayey soil in the Yizre'el Valley. The horizontal bars indicate standard deviation (unpublished data)

The main aquifer in Israel is located in the Coastal Plain, and the soils above it have moderate to light textures. Enhancement of the mobility of contaminants in these soils under effluent irrigation could cause serious, long-term problems, if these contaminants reach the groundwater. Therefore, more research should be conducted on the effects of effluent irrigation on the transport of contaminants in the soil, and much more caution should be exercised in irrigating with effluents in the Coastal Plain region.

2.4.6 Nutrients

Nitrogen and phosphorus are essential nutrients for living organisms, and are usually added to agricultural lands in large quantities, as fertilizers to ensure high crop yields. Secondary effluents contain relatively high concentration of these nutrients, and this could increase the nutrient concentrations in effluent-irrigated soils.

The concentrations of nitrogen as NH_4 and NO_3, and of phosphorus as PO_4, in soil samples that were taken after the irrigation season, from plots in two sites that were irrigated with secondary effluent or with FW are presented in Fig. 2.4.10, as functions of soil depth. The NO_3 and PO_4 concentrations in the effluent-irrigated soils were higher than those in the FW-irrigated soil in the upper soil layer; in the clayey soil the higher NO_3 concentration in the effluent irrigation soil was found down to >2.5 m (Fig. 2.4.10). The predominant N com pound in effluent is NH_4, which is a product of the OM decompose during mineralization (Feigin et al. 1991). However, since soil NH_4 is susceptible to oxidation, NO_3 is eventually formed in the soil, and this is probably the reason for the low NH_4 concentrations in the effluent-irrigated soils (Fig. 2.4.10).

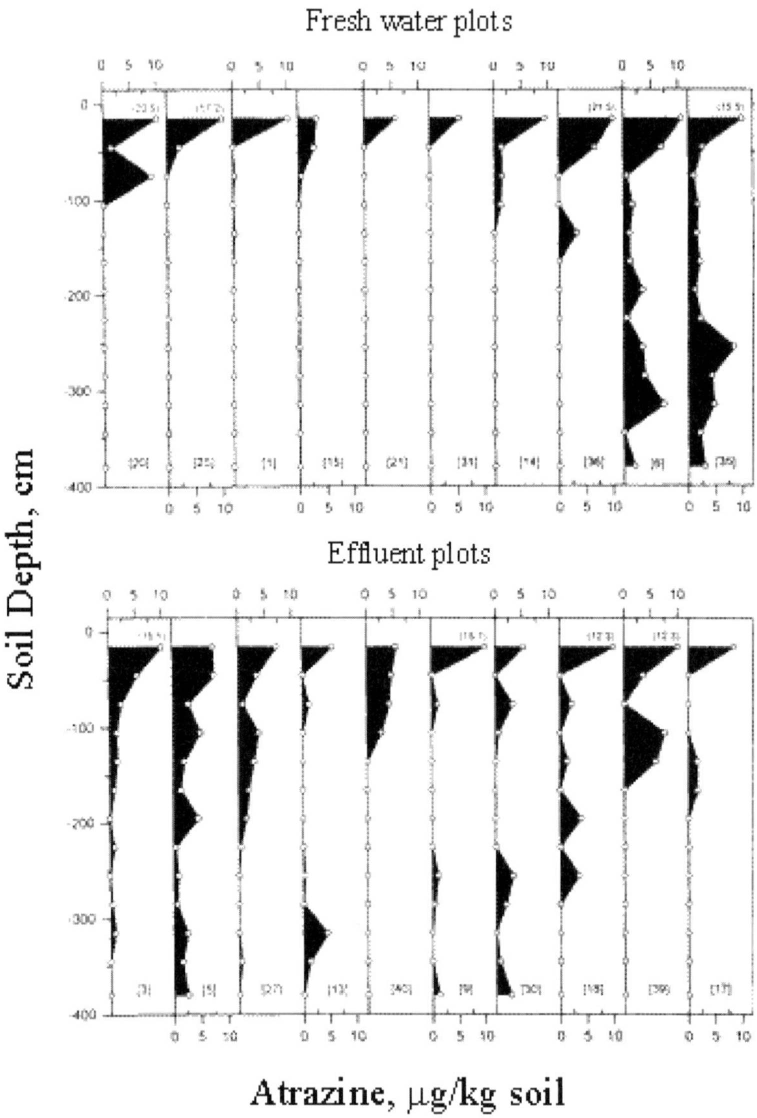

Fig. 2.4.9. Atrazine distribution in 10 fresh water- and 10 effluent-irrigated profiles. Numbers in parentheses are off-scale concentrations; numbers in brackets are profile designations. Data points are marked at mid-depths of segments (from Graber et al. 1995)

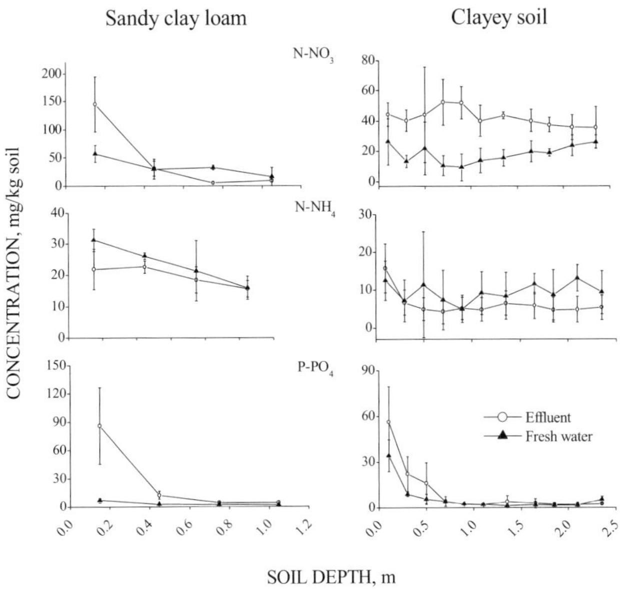

Fig. 2.4.10. NH_4, NO_3 and PO_4 of soil samples from plots irrigated with two water qualities in the Coastal Plain with sandy clay loam and Yizre'el Valley with clayey soil. The vertical bars indicate standard deviation (unpublished data)

The results in Fig. 2.4.10 indicate that irrigation with effluents could affect the uptake of nutrients by crops, because of their escape in the form of gases and leaching below the root zone. Thus, effluent irrigation has potential agronomic impacts, through the addition of nutrients for growing crops, and potential environmental impacts, through the possible leaching of nutrients to the groundwater. However, appropriate fertilizer management in effluent-irrigated soils could, on the one hand, yield benefits from the presence of nutrient in the effluent, and, on the other hand, prevent their excessive application and the consequential environmental problems

References

1. Ben-Hur M, Ravina I (2002) Effluent irrigation: Agronomic and environmental aspects (in Hebrew).Water Engineering, Fluids and Irrigation 20: 13-22
2. Ben-Hur M, Shainberg I, Bakker D, Keren R (1985) Effect of soil texture and $CaCO_3$ content on water infiltration in crusted soils as related to water salinity. Irrig. Sci. 6: 281-284
3. Ben-Hur M, Keren R, Ovaday H (1999) Effects of water treatments on effluent quality in Israel (in Hebrew). Water Water Eng. 38: 34-39.
4. Feigin A, Ravina I, Shalhevet J (1991) Irrigation with treated sewage effluent. Springer-Verlag, New York

5. Graber ER, Gerstl Z, Fischer E, Mingelgrin U (1995) Enhanced transport of atrazine under irrigation with effluent. Soil Sci. Soc. Am. J. 59: 1513-1519.
6. Kazman Z, Shainberg I, Gal M (1983) Effect of low levels of exchangeable Na and phosphorgypsum on the infiltration rate of various soils. Soil Sci. 135: 184-192.
7. Kemper WD, Koch EK (1966) Aggregate stability of soils from western United States and Canada. Colorado Agric. Exp. Stn. Bull. No. 1355: 1-52.
8. Kemper WD, Miller DE (1974) Management of crusting soils: Some practical possibilities. pp. 1-6. In: Carry JW, Evans DD (eds). Soil crusts. Technical Bull. 214. Agric. Exp. Stn. University of Arizona.
9. Mamedov AI, Shainberg I, Levy GJ (2000) Irrigation with effluent water: Effect of rainfall energy on soil infiltration. Soil Sci. Soc. Am. J. 64: 732-737.
10. Maas EV, Hoffman GJ (1977) Crop salt tolerance – current assessment. J. Irrig. Drain. Div. Am. Soc. Civ. Eng. 103: 115-134.
11. McBride MB (1989) Reaction controlling heavy metal solubility in soils. Adv. Soil Sci. 10: 1-56.
12. Rhoades JD, Ingvalson RD, Tucker JM, Clark M (1973) Salts in irrigation waters: I. Effects of irrigation water composition, leaching fraction, and time of year on salt compositions of irrigation drainage waters. Soil Sci. Soc. Am. J. 37: 770-774.
13. Shainberg I, Letey J (1984) Response of soils to sodic and saline conditions. Hilgardia 52: 1-57.
14. Wakindiki IIC, Ben-Hur M (2002) Soil mineralogy and texture effects on crust micromorphology, infiltration, and erosion. Soil Sci. Soc. Am. J. 66: 897-905.

2.5 Integrated Wastewater Concepts for Olive Mills in the West Bank, Palestine

Silke Drescher
EAWAG/SANDEC, Switzerland

2.5.1 Introduction

Like in many countries of the Mediterranean Region, Olives and products made of olives also play an important role for the economic development and income generation of the West Bank. The climatic conditions are favourable for the cultivation of olive trees and it has been a traditional business of the local people since centuries. More than 90 % of the olives are being processed locally to produce olive oil; therefore olive mills are located in almost every village of the region [8].

However, there are also drawbacks of olive oil production, since it requires considerable amounts of process water and therefore generates large quantities of waste water – locally called Zibar. Due to its pollution potential Zibar causes severe effects on surface water bodies during the olive harvest season and also threatens the sensitive groundwater resources. Currently there are no treatment technologies for Zibar available in the West Bank and for recently developed sanitation concepts the Zibar still seems to be an unsolved challenge [2, 8]. Several post treatment solutions are either not suitable for the West Bank, still in the pilot stage or highly sophisticated and uneconomical [1, 5, 13].

A feasibility study conducted by the German Environmental Consultant (DAR) in 2000 searched for an appropriate waste water concept for olive mill effluents in the catchment area of Wadi Zeimar which is located between Nablus and Tulkarem in the West Bank. The study focused on avoiding olive mill effluents via integrated solutions strategies instead of additive end-of-pipe options. An integrated approach requires an overall concept which considers also economic, institutional and social aspects besides the issues of technological change. This article summarises the findings of that study. It briefly explains the characteristics of the current olive oil production as it is the basis for an integrated change. On basis of the case study of Tulkarem Region, it introduces an integrated technical concept and discusses additional measures which can help to accelerate a change of technology in the olive oil sector in the West Bank.

2.5.2 Olives - An Economic Factor of the West Bank

Olive oil production has been a traditional business since centuries and still is an important source of income for the Palestinian people. In 2000 the value of olive oil production was US$ 46 Mio. - 5.7 % of the agricultural production. Nablus, Ramallah, Jenin and Tulkarem are the districts with the largest Olive groves in the

West Bank. Approximately 75 % of the existing trees in the Tulkarem Region are olive trees [9]. Two thirds of the agricultural area of the Tulkarem District is covered with olive groves. Here, the production of olive oil contributes approx. 20 % of total income from agriculture in the area [14]. In the harvest season of 2002, 35 olive presses in the Tulkarem District processed about 15 610 tons of olives with an oil output of 3539 tons with a value of US$ 613 000 (12 % of total West Bank and Gaza) [8].

In contrast to Israeli agriculture, the agriculture in the West Bank is mainly rainwater fed and therefore the harvest yield strongly varies with the annual rainfalls. Due to sufficient rain in winter, the olive yield was exceptionally good in the season of 2002. However, due to the current political situation, the owners of the land have to live with significant losses due to prohibited access to land and a very limited market for olive oil [6, 7].

2.5.3 Organisational Structure of Olive Oil Mills

As mentioned above, olives are processed locally and the olive mills are located in the scattered villages in the West Bank. There is a low degree of joint co-operations among the olive mill owners and the mills are mainly run as private business of individual families [4, 8]. But most of the olive mills are registered in the Palestinian Chambers of Commerce. As the olive mill production is a seasonal event, most operators have an additional business during the remaining months of the year. The interaction between farmers and olive mill owners is restricted to the service of olive pressing during the olive harvest season. There are only very few examples of shared ownership of olive mills between farmers and service providers. Farmers normally choose the olive mill according to the transportation cost, given service and quality. It is common that the farmer receives the olive oil pressed from the own olives. Olive mill owners gain revenues from cash fees and the sale of surplus olive oil, since some farmers pay the pressing fee with a share of oil. According to the *Palestinian Central Bureau of Statistics* more than 50 % of the oil is sold in the local market, due to export constraints. [8]. Due to a fluctuation of yield and market price the olive mill owner has to carry high income risks.

2.5.4 Legislative Framework

In the Palestinian Administration ecological issues are seen as a cross disciplinary issue. With regard to waste water regulations, the *Ministry of Environment* cooperates with several institutions like the *Palestinian Water Authority* and the *Palestine Standard Institute (PSI)* which are responsible for wastewater treatment standards and regulations. They developed regulations for the discharge of effluents from industries, municipalities and settlements. Due to lack of enforcement the regulations and standards are not applied until now. Additionally, the water related issues are still strongly influenced by the Israeli government and the Palestinian institutions need approval for most water related decision.

2.5.5 Technological Aspects of Olive Oil Production

In the West Bank three different processes and technologies are used for the production of olive oil: Traditional presses, half-automatic presses and automatic decanters. The number of modern automatic decanters is steadily increasing because of superior production efficiency. Compared to olive mills in other Mediterranean countries these olive mills are relatively small scale. The average capacity of one automatic mill amounts to 45 tons of olives per day and an output of approx. eight tons olive oil and approximately 40 tons of Zibar per day.

The role of process water and Zibar

All processes have in common that additional process water is needed. The water is mainly added for grinding and mixing the olives and supports the separation process later on. The Zibar has a low pH value and carries significant amounts of organic matter and hazardous substances like polyphenols, a natural by-product of olives. The waste water characteristics are summarised in Table 2.5.1.

Table 2.5.1. Chemical Characteristics of Zibar

Parameter	Value [g/l]]	Average
pH	4,1 – 5,9	5,0
COD	40 – 200	120
BOD_5	20 – 90	40
Total Solids	5 – 30	15
$Phenol_{tot}$	2 – 7	3,5
N_{tot}	0,1 – 1	0,3
P_{tot}	0,1 – 0,5	0,2

Almost all values, especially the values of organic load and polyphenols, exceed the current limits set by the *PSI* by far [11]. Due to the high organic load of the generated wastewater, local water bodies frequently drop to anaerobic conditions. Zibar also contains polyphenols which have herbicidal effects on perennial plants and inhibit the activity of micro-organisms. Therefore, Zibar neither can be used for irrigation of plants nor can be efficiently treated in biological treatment plant for effluents with high organic loads (e.g. anaerobic digestion). Hence avoidance strategies seem to be more suitable than end-of-pipe treatment solutions and it is worth having a closer look at the decanter process itself.

Types of automatic decanters

Automatic olive presses have a modular design. The process steps like washing, grinding, mixing and separation are done in mainly independent devices but are connected to allow continuous operation. Therefore the optimisation of single process steps is possible and allows integrated changes. The central unit of an automatic olive press is a horizontal centrifuge, called decanter, which separates the liquid and solid fractions of the olive mash. According to the separation process conventional 3-phase decanters and modern 2-phase decanters can be differentiated. Conventional 3-phase decanters separate three phases from the olive mesh, namely

- Oil,
- Solids (Djefet),
- Waste water (Zibar).

Since the introduction of 3-phase decanters the average water consumption per litre of oil has significantly increased as the separation process needs more water than traditional presses and therefore generates more waste water. Modern 2-phase decanters need less water and separate only two phases, namely oil and solids. The output of Zibar from the 2-phase-decanter process is reduced by 2/3 and the core amount of all above mentioned substances remains in the solid fraction in a concentrated form. One can argue that the pollution problem is shifted from the liquid to the solid fraction. However, the solid waste is easier to handle as the liquid fraction. Traditionally the Djefet is returned to the olive farmers and used as fuel or for stock feeding. The Djefet from 2-phase decanters still can be dried and used as fuel or further be processed in olive oil extraction facilities. Another option is composting and reuse as fertiliser; however these options still need further investigations like done in [12]. Figure 2.5.1 shows the material balance of both decanter types as well as the composition of the different solid and liquid fractions.

In the West Bank mostly 3-phase decanters are in operation as most of the equipment acquired is second hand and older than 10 years. Since olive presses have a modular design, the technical adaptation of the existing systems is relatively simple, as only the decanter unit needs to be exchanged to significantly reduce the amount of waste water.

2.5.6 The Integrated Approach

For the development of an integrated technical wastewater concept it is necessary to have the whole technological process and its social and natural environment in mind. This approach comprises the local conditions in the villages, existing infrastructure, the current use and treatment of outputs, the technical knowledge of the owners, and geographical distribution of their olive mills.

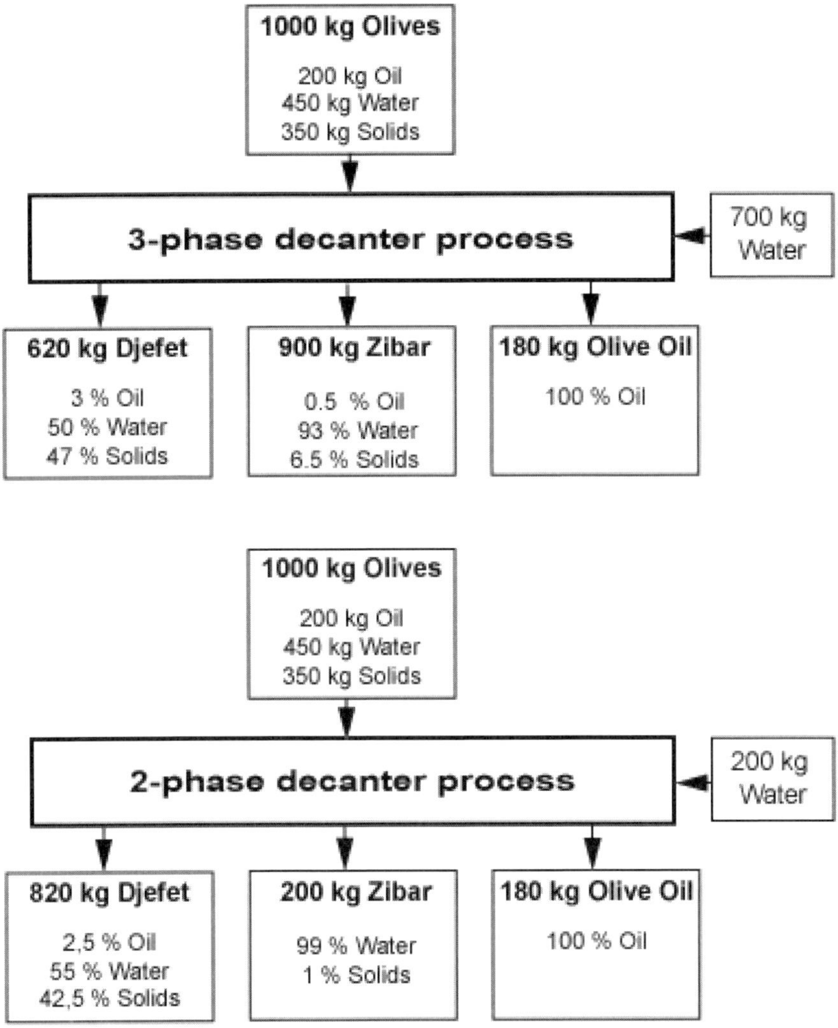

Fig. 2.5.1. Material Balance of different decanter types used for olive oil production [5]

From the technical point of view the only integrated technology currently known is the 2-phase decanter and it suits to the decentralised structure of olive mills in the West Bank. An integrated avoidance strategy is a decentralised approach implying a direct intervention in the production process and individual habits of the olive mill owners. Hence, additional measures are necessary, which are discussed on the case study of Tulkarem District.

2.5.7 The Case Study of Tulkarem District

In the framework of the development of an overall wastewater concept for the lower catchment area of Wadi Zeimar (comprising municipal and industrial wastewater) the effluents from olive oil mills were considered to be an urgent problem. On the one hand, they cause severe ecological and economic damages as the Zibar is discharged without any treatment in the nearby Wadi Zeimar, which already is considered to be a highly polluted water body. Downstream fisheries and water bodies are seriously affected by the pollution during the harvest season. On the other hand, international financing institutions like the *German Kreditanstalt für Wiederaufbau (KfW)* plan to co-finance the construction of new sewerage treatments plants in the Tulkarem Region. Yet, the operation of these new plants would have seriously been affected if the olive mill effluents entered the scheme [2, 3]. As part of the above mentioned overall wastewater concept, 20 olive mills located in the catchment area of Wadi Zeimar were assessed in a feasibility study considering logistics of input and output streams. Seven mills were equipped with automatic decanters while ten were traditional presses or half-automatic presses. As these seven full automatic mills require more water for the process than the traditional mills they generate 70 % of Zibar [3]. Hence, the study focused on appropriate scenarios for these automatic mills without neglecting solutions for the remaining mills.

2.5.8 Technical concept

The exchange of existing 3-phase decanters to 2-phase decanters is an obvious solution, as the reduction of wastewater would be significant. In the long run many of the traditional and half-automatic presses will phase out or be replaced by the new 2-phase decanters. In combination with the avoidance strategy, additional interim measures like covered solar drying beds can treat the remaining Zibar. They enforce evaporation of the water by means of a transparent roof and ventilation systems and need less space than common evaporation ponds. This centralised system is appropriate as it can be reassigned later for the future planned wastewater treatment systems [16]. The Djefet also needs additional treatment. Due to space constraints in the area, a rotating drum oven seems to be applicable. The heating system of the oven is fed by the already dried Djefet hence additional fuel is hardly needed. Figure 2.5.2 gives and overview of the concept with different options.

This concept has to be seen as a medium-term approach considering time for phasing out old equipment and introducing organisational changes. As it will be shown in the next chapter, such a technical concept also needs accompanying measures considering financial, institutional, legislative aspects. It needs to address the olive mill owners on local level as well as institutions on regional and national levels. Such a bundle of measures safeguards a change of technology which is environmentally beneficial and economically and socially acceptable for the olive mill owners.

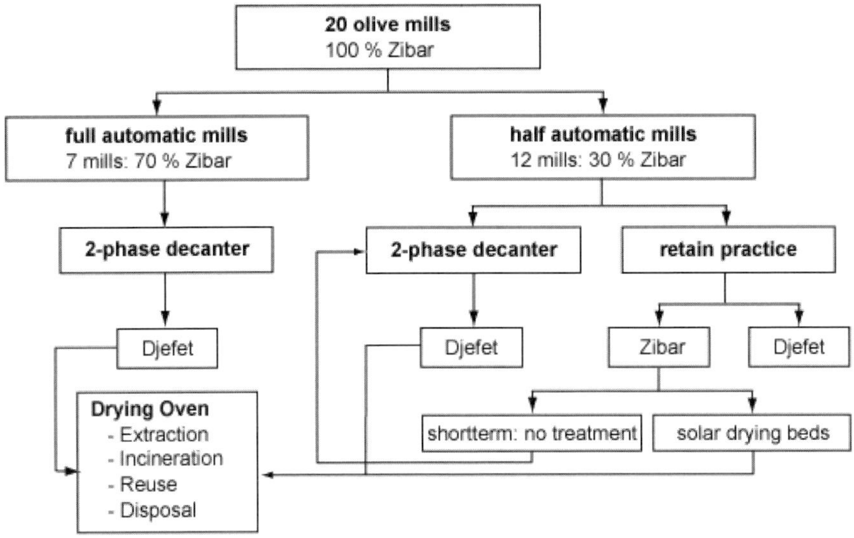

Fig. 2.5.2. Technical waste water concept for olive mill waste water with different options

2.5.9 Financial and organisational considerations

The financial and economic calculation of different technical scenarios revealed that the integrated concept is the least cost solution compared to other centralised end-of-pipe treatment plants [3, 4]. The costs of the individual change of the olive mill process depend on the extent of necessary changes. The exchange of one decanter costs approx. Euro 30 000, a completely new olive mill equipment is approx. Euro 204 000. The investment costs for additional centralised treatment units like solar drying beds, drying ovens and logistics amount to Euro 460 000. Annual costs including depreciation, operation and maintenance cost as well as capital costs are estimated to Euro 160 000 per year. The feasibility study envisaged a co-financing model where both the funding agency and the Palestinian Authority contribute to such an investment. The running costs need to be covered by tax or fees gained from olive mill owners.

Table 2.5.2 summarises the costs of the single components of the integrated concept. The table does not show a total sum of costs as the total investment depends on the strategy chosen. For example, if the decision makers decide for a swift change of technology and concentrate on the introduction of new mills, the solar drying beds are not required, thus the total investment is significantly reduced.

Table 2.5.2. Investment costs of components of an integrated concept

Item	Costs (Euro)
Solar Drying Beds	255 000
Drying Oven	105 000
Transport	98 000
2-Phase-Decanter Module	30 000
Olive Oil Mill (capacity approx. 40 t olives/day)	204 000

A decentralised concept implicates a shift in costs for the changes towards the individual olive mill owner. This is in line with the commonly accepted "polluter pays principle" and need to be enforced by appropriate regulations.

Thus, strong but realistic regulations are the basis for a sustainable introduction of environmental concepts. Existing environmental laws need to be enforced in order to assure a sector wide change and prevent distorted competition among olive mills in different districts. The Spanish environmental legislation addressing olive mills can act as a basic model for the entire Mediterranean region.

Though olive mills are highly profitable for the owner he will realise less revenues when environmental measures like fees or regulations are applied. The government can help those affected by such changes with additional economic and organisational measures. Incentives are necessary encouraging and allowing olive mill owners such high investments for a technical change. Different measures could be applied such as direct subsidies on investments for process changes or environmental revolving funds for credits on assets like olive mill equipment [10, 5]. Such economic instruments accompanied with awareness building and training programmes, ensure the acceptance and implementation of technical changes and protect olive mill owners from economic hardship.

Another issue related to the technical change is the adaptation of organisational structures. Since the technical change of olive mills is accompanied with an increase of investments and production capacity alternative organisational structures need to be considered. As discussed in [4] different organisational models ranging from private companies to co-operatives or management contracts are suitable for both olive mill operation and centralised plants. Private olive mill operators can join efforts for setting up new plants. While a private contractor is regarded as the most suitable option for the operation of a centralised treatment plant. The organisational model depends on several factors like the ability to pay of partners, the willingness of cooperation, competence and the degree of envisaged political regulation. This also means that the conventional financing process for centralised infrastructure projects in the framework of development co-operations is not applicable. International donors mainly focus on governmental institutions and investment in centralised infrastructure systems. In addition to that, initiatives of the local private sector need to be developed and supported as they are a cost effective and transparent alternative to public organisations. One precondition is a stable organisational set-up as well as competency for management and operation.

Several local and regional organisations are able to develop such competency and will be able to manage such centralised plants. For example, an association of individual olive mill owners could act as operators of the drying oven under supervision of the local government. That means that international funding agencies need to consider and develop capacity building programmes as well as environmental education programmes in co-operation with all local stakeholders (e.g. the Chamber of Commerce) ensuring the sustainability of their core infrastructure investments.

2.5.10 Conclusion

The 2-phase-decanter technology is a state of the art technology and significantly reduces the heavily polluted wastewater of the olive oil production. Especially in countries in the Middle East water is a valuable source. Hence, both the reduction of input water and the protection of the existing water resources should have a high priority as it touches the environmental, social and economic situation of Palestinian people.

A change of technology or tasks always goes along with a change of organisational and social structures. This is especially true if integrated technologies are applied. The intervention needs accompanied measures on all levels comprising awareness building, training, financial incentives and finally enforced regulations. An integrated approach does not simplify the tasks and may not lead to an instant single solution to a problem. But it can help to understand the whole picture and to identify constraints beforehand.

References

[1] Benyahia N, Zein K (2003) Analyse des problèmes de l'industrie de l'huile d'olive et solutions récemment développées, Discussion Paper SESEC II, Lausanne

[2] Deutsche Abwasserreinigungs GmbH, Universal Group (1999) Sewerage Project Tulkarem-Region, Site Investigation and Conceptual Planning, Report 2, Mannheim, Ramallah

[3] Deutsche Abwasserreinigungs GmbH, Universal Group (1999) Sewerage Project Tulkarem-Region, Draft Feasibility Study, Report *3*, Mannheim, Ramallah

[4] Drescher S (2001) Entsorgung von Abwasser aus der Olivenölproduktion in der Region Tulkarem, Palästina – Entwicklung eines Abwasserkonzeptes unter technischen, wirtschaftlichen und institutionellen Gesichtspunkten, Final Thesis, Fachhochschule Trier

[5] Gesellschaft für Technische Zusammenarbeit mbH (gtz)/ Ministère de l'Environnement et de Aménagement du territoire (1996) Les experiences méditerranéennes dans le traitement et l'elimination des eaux résiduaires des huileries d'olives, Tunis

[6] Ha'aretz Newspaper, November 15 2002 A8-A9, Jerusalem

[7] Lein Y (2002) The Performance of Law Enforcement Authorities in Responding to Settler Attacks on Olive Harvester, Status Report of B'Tselem, Jerusalem

[8] Palestinian Central Bureau of Statistics (2002): Comprehensive Statistics on Olive Presses, Ramallah, www.pcbs.org

[9] Palestinian Central Bureau of Statistics, (1997) 3rd Agricultural Statistics 1995/96, Ramallah

[10] Palestinian Economic Council for Development and Reconstruction (PECDAR) (1998) Private Sector Credits, Donor Assistance, Ramallah

[11 Palestinian Standards Institution (PSI) (1998) Industry Effluent Wastewater, Draft-Version, Ramallah]

[12] Paredes C et al. (2002) Bio-degradation of olive mill wastewater sludge by its co-composting with agricultural wastes. Bioresource Technology *85*: pp 1-8

[13] Sülzer R, Zimmermann A (1996) Organisieren und Organisationen verstehen: Wege der internationalen Zusammenarbeit, Opladen

[14] Ubay G, Öztürk I (1997) Anaerobic Treatment of Olive Mill Effluents. Water Science and Technology *36*, No. 2-3: pp 287-294

[15] Various (2000) www.palnet.com.

[16] Witte H (1998) Solare Trocknung – Vergangenheit oder Zukunft? Korrespondenz Abwasser *45* No. 6: pp 1139-1146

2.6 Distribution Pattern of Some Inorganic Pollutants in Groundwater and Soil of a Sewage Farm at NE-Cairo/Egypt

Abo el Abas, Yosry M.
Berlin, Germany

Abstract

Overpopulation is still the main problem in all third world developing countries in the 21 century. Egypt for example, planned to reclaim large areas in the Desert with an increasing use of sewage around the Great Cairo. Gabal el Asfar sewage farm is an example, in which more than 80 years treated sewage effluent is applied. This farm covers more than 1,250 ha and located approximately 25 km NE of Cairo, at the fringes of Delta floodplain. To evaluate sewage impact on the groundwater and soil, a well-representative sampling has taken place. 20 groundwater Samples from different water Aquifers were collected. 48 soil samples from upper and lower soil were also selected from many traverses to evaluate the heavy metal distribution horizontally and vertically. The concentrations of heavy metals in the soil were determined by XRF. AAS and ICP Methods have been used to determining the main cations and anions of the groundwater.

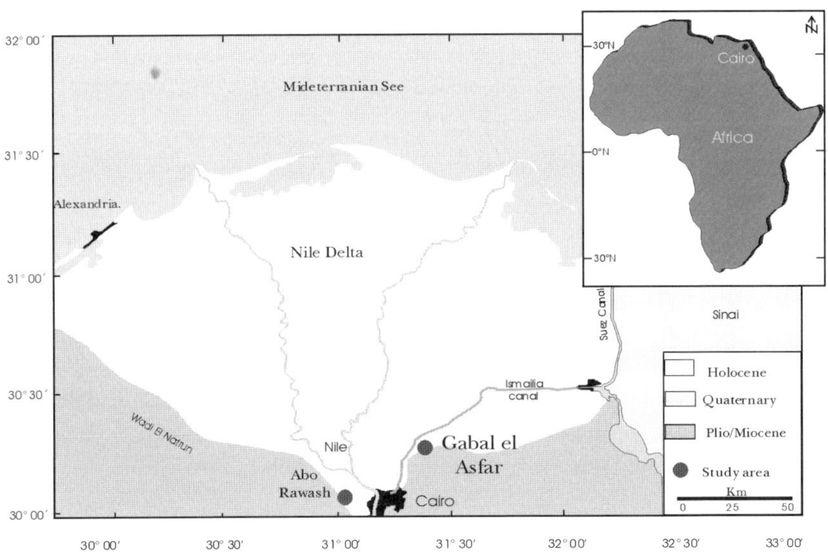

Fig. 2.6.1. Simplified geological and location map of the study area

Only the upper soil in the study area has been contaminated with heavy metals such as Zn, Cr, Cu and Lead in comparison to the reference samples. The horizontal distribution of heavy metals depends mainly on the distance of sewage input on the farm, and vertically on the organic matter, which plays the key role as heavy metal carrier. The groundwater has polluted mainly with nitrite and nitrate. The sulphate and chloride concentrations are also higher in comparison to a non polluted area. Although Fe, Zn and Mn are also detected in the groundwater, but their mean values are lower or equal to the WHO and EC Limits. The elevated level of Zn, Cr and Cu in soil can be referred to industrial wastes originated from the industrial Complex which is very close to the sewage farm. Except of few anomalies of the de termined inorganic pollutants, sewage application leads not only to positive changes in the sandy soil, but also to refreshing the upper groundwater aquifer in comparison to initial groundwater.

2.6.1 Introduction

Using of sewage for agriculture purposes has gained importance particularly in countries of arid and semi-arid regions, where the rainfall and other water resources are depleted (Farid et al., 1993). Sewage is rich with initially nutrients needed for the plants. In other countries like Mexico also sewage is using for irrigation (BGS, 1996). Various countries like Egypt, Jordanian, Mexico, USA and Germany have already gained experience with sewage (Kent et al., 1994, Tröger et al., 1996; Graniel, 1999, Marschner et al. 2000; Böken, 2001, Hoffmann, 2002) Abo el Abas, 2002). After the long term of sewage irrigation, which is rich in organic matter and organic and inorganic contaminants, the objectives of this work is now arising the question: Is there any negative impact of inorganic contaminants on soil and groundwater?. In Egypt, many data in the last 30 years have been published about the influence of sewage applications on plants. Data on water quality and soil are scarcely available.

The reuse of treated and nontreated sewage water has been practised in sandy region around Cairo city on Gabal el Asfar and Abo Rawash since 1915 and 1936 respectively (Fig. 2.6.1). These areas are located in the Eastern and the Western fringes of the Nile Delta floodplain, where sandy soils are initially poor in nutrients and where negative impact of sewage effluent on groundwater quality is largely expected. Sources of the sewage are Cairo effluent treatments states. The Great Cairo Waste Water Organisation established six waste water treatments plants. The total amount of treated sewage is about 5 million m^3/d for Cairo.

Gabal el Asfar sewage farm is the study area with a total area of 1,250 ha (Fig. 2.6.& 2.6.2, RIGW, 1991). The farm receives its main secondary treated sewage water from Birka and gabal el Asfar treatment stations for irrigating citrus trees, maize crops, wheat, alfalfa, eucalyptus, palms and flower plants.

2.6.2 Geological and Hydrogeological setting of the study area

The research area is located 25 km NE of Cairo near the eastern desert area at the fringes of Delta floodplain. The eastern boundary of the floodplain is defined by higher ground, comprising the western edge of the Eastern Desert structural plain. The land surface slopes from about 30 m above mean see level at the eastern border to 15 m above mean see level at western and northern border. The area has an arid climate characterised by warm winters, hot summers, low rainfall and high evaporation rates. The average annual rainfall in Cairo city is approximately 25 mm, increasing to 150 mm in the northern part of Egypt. The average daily temperatures vary from a minimum of around 13°C in January to 28°C in June. Evapo-transpiration rates are high of an annual total of approximately 1,600 mm.

Gabal el Asfar area is located west of the north-south fault, dividing the quaternary sandy deposits on the West side and Miocene/Pliocene deposits east of it (Farid, 1980). The study area is generally underlained by the Pleistocene sands and gravels with the Holocene clay occurring only in the western part of the area (Fig. 2.6.2). To the east, Holocene sand dunes locally cover the Pleistocene deposits before the Quaternary thins out and Miocene sands and limestone of the structural plains are encountered.

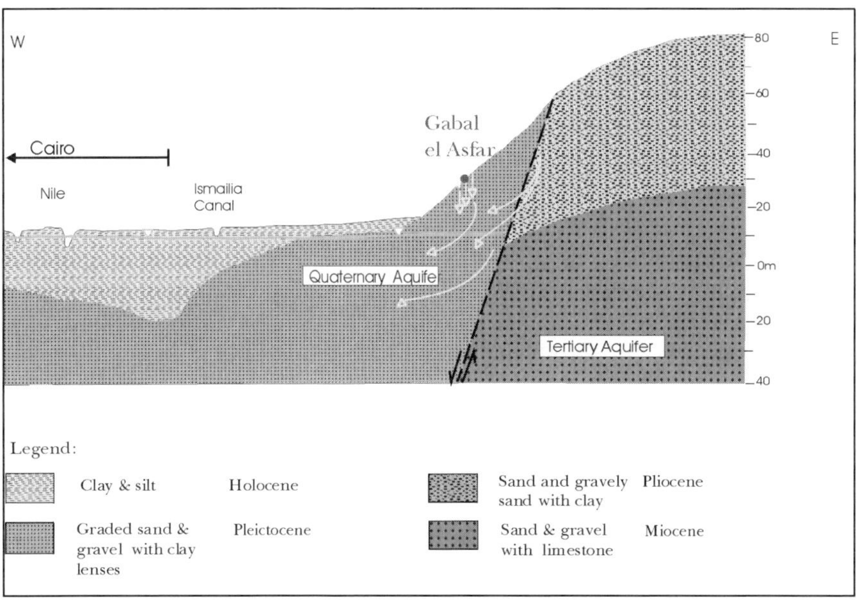

Fig. 2.6.2. Lithohydrogeological section at Gabal el Asfar sewage farm Region

Hydrogeologically, the farm is underlained by the unconfined Pleistocene graded sand and gravel aquifer, which belongs to the Delta aquifer. The thickness of the aquifer is greater than 10 m to SE and is possibly greater than 65 m to the

NW of the farm. Clay intercalation at depths varying from 2 to 10 m in the S and W, respectively, disappearing towards the N (RIGW, 1995). The Pleistocene aquifer is underlained by the Oligocene and Miocene aquifers (El Malky, 1990). The depth to groundwater varies from 1 to 3 m below ground surface. The main direction of groundwater flow is E-W toward the Nile. The aquifer transimissivity is high (5,000 to 6,000 m^2/d) and extended. The main source of groundwater recharge is supplied vertically from irrigation excess water.

2.6.3 Materials and Methods

A comprehensive systematically soil sampling was carried out of sewage pilot with an area of 1 ha. 48 soil samples were collected from 12 soil profiles at different depths (10, 30, 60, and 100 cm depth). Additional 12 control samples were collected from non swage irrigated areas. All samples were dried and sieved at the TU- Berlin (APHA, 1995). Physiochemical parameters like pH, EC, carbonate and organic matter contents were measured. Carbonate and organic matter contents have been determined with loss on ignition method (Schlichting et al., 1995). The soil fraction < 200 μm was chosen to be the optimal for the analysis. The total contents of macro-, and trace elements were determined by XRF analysis.

18 well and hand pump groundwater samples and 2 sewage samples were collected in the same time of soil sampling collection. The water was collected from sewage farm and the surrounding areas. To determine the vertical and the horizontal effect of sewage on the aquifer, samples were collected from the Quaternary aquifer in the farm and from underlined Tertiary aquifer (Fig. 2.6.5). Parameters like pH, EC, Hardness, O_2 content, Cl^-, NH_4^+, PO_4^{3-} were determined in the field. Heavy metals and anion concentrations have analysed in the laboratory of TU-Berlin by AAS, ICP- AES and IC methods, respectively.

2.6.4 Results and Discussion

Soil

In the late 1970s and early 1980s many works have been published about the effect of the sewage application on plants in sludge amended sandy soils. Works about impact of sewage on soil and groundwater are very scarce, very localised and hardly to be available (El Nennah, 1982; El Malky, 1990 and Farid, 1993). Alloway (1991), Matheis (1996), BGS (1996) and Abo el Abas (2002) documented on changing of soil parameters as a result of sewage application. They documented also on metal accumulation particularly in the upper soil. Farid (1993), Asbrand (1997), Graniel and Abo el Abas found that sewage application has, however, a positive effect by freshening the groundwater of upper aquifer.

Table 2.6.1. Physio- chemical parameters of soil and control samples in the study sewage pilot area

Parameter	Depth	pH	EC	$CaCO_3$	Organic matter
	(cm)	($CaCl_2$)	(µS/cm)	(%)	(%)
Upper soil	10	6.5	515	1.1	8.8
	30	7.2	248	1.5	3.3
Lower soil					
	60	7.7	195	2.1	2.1
	100	7.8	142	1.4	1.9
Control Samples					
	10	7.5	227	1.0	3
	30	7.9	166	0.8	2

According to Hölting (1992), the sandy soil contains porosity of 33 % and good hydraulic permeability of 13 m/d. The upper soil has relative lower acidic pH than lower soil and that of control samples (Table 2.6.1). Bacterial activity, raising reduction zone and the newly formed acids might be responsible for the lowering pH of the upper soils. **Carbonate contents** show an increasing trend with depth. Accumulation and precipitation of salts increase by infiltration time with depth. pH and $CaCO_3$- distribution are vertically and horizontally inhomogeneous in the study pilot.

Electric conductivity ranges from 135 and 995 µS/cm with mean values of 215 and 515 µS/cm for upper soil and decreases with depth (Table 2.6.1). EC is the direct indicator for ion and salts concentration in soil and in water. Table 2.6.1 shows a positive correlation between EC and organic matter of soil.

Organic matter contents depend on the enrichment of sewage water as well as decomposition of plant origin wastes. For the upper soil, the lowest value is 6 % and the highest one 12.5 %. The mean values as well as the sample values decrease with depth. Upper soil organic matter values of 8.8 and 3.3 % are relatively high according to Grünewald (1994). The horizontal distribution of organic matter is inhomogeneous and depends on sample distance from sewage input. Organic matter is the main carrier of heavy metals. It seems to play the important role in the distribution and content of heavy metals.

Heavy metals Zn, Mn, Cr, Pb, Cu, Ni and Fe were determined using the XRF method. Only these metals have been detected with higher values than the known detection limits (Table 2.6.2 and 2.6.3). In comparison to the control samples, all heavy metals have specific remarkably higher contents. As ready mentioned, organic matter decreases with depth and heavy metals decrease analogously. Fig. 2.6.3 shows the horizontal distribution of Zn in the upper soil of a selected study pilot. The dark colour shows higher concentration than the lighter colour.

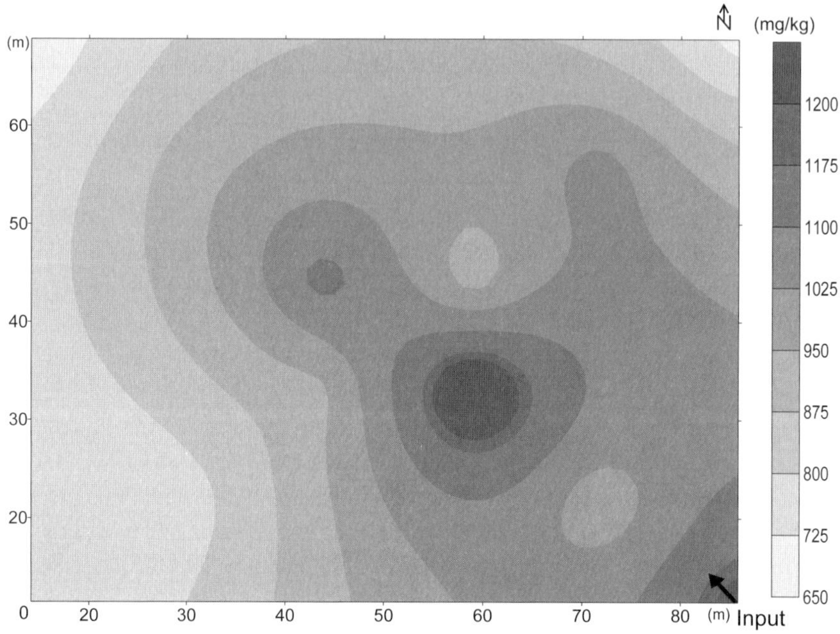

Fig 2.6.3. Horizontal distribution of Zn in soil of the study sewage pilot area

Only the upper soil of sewage pilot is contaminated with heavy metals. In comparison to the EU- limits for agriculture, upper soil has higher Zn, Cr, Pb, and Cu than the permissible values (Table 2.6.3). Zn has the highest concentration in upper soil, after Fe, followed by Mn, Cr, Pb and Cu. The elevated content of Zn, Cr, Pb and Cu are higher than EU- limits. Mn content is higher than control samples, still below the EU- allowed limits. According to Leeper (1972), Sauerbeck (1988), Kent (1994), McGrath (1995), BGS (1996), Cr might be in the non-toxic reduction species (Cr^{III}). Ni has been founded in elevated content in comparison to control samples, but not exceeds the EU- limit.

This implies that, the upper soil of sewage farm Gabal el Asfar is contaminated with heavy metals which not only originated from municipal sources, but should coming from other anthropogenic sources like industry.

Table 2.6.2. Heavy metal concentration of the study area. (Values in mg/Kg)

Upper soil	Ni	Cu	Pb	Cr	Mn	Zn	Fe
(n = 24)							
10 cm	44	274	339	562	734	943	31,004
30 cm	24	78	51	141	414	339	20,277
Lower soil							
(n = 24)							
60	20	50	< 10	84	285	107	17,243
100	19	43	< 10	85	271	70	15,025
Control samples							
10 cm	28	32	16	110	555	115	22,632
30 cm	16	14	10	45	359	51	16,380

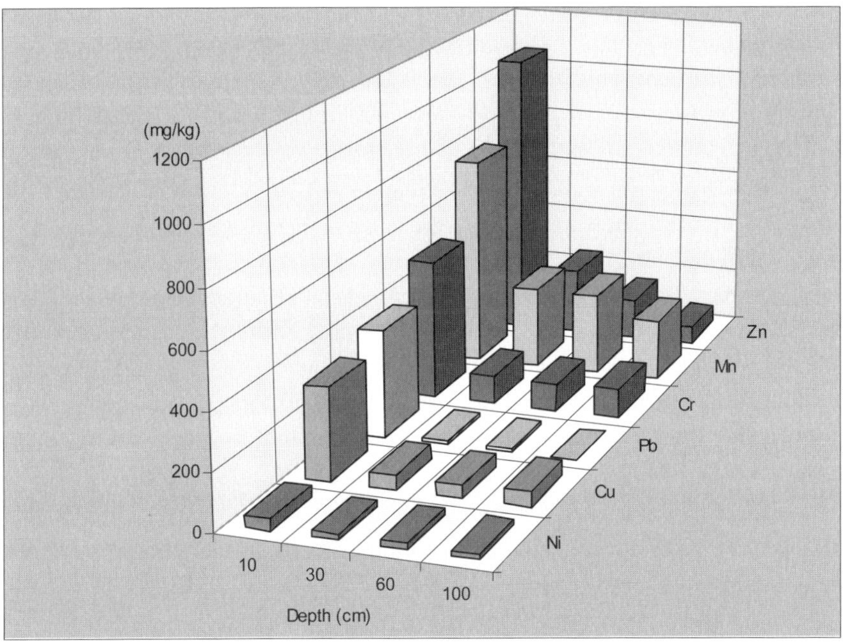

Fig. 2.6.4. Behaviour of metal concentrations with depth in one profile of the study sewage areaThe profile shows the distribution of metals with depth. However, the deeper the soil horizon, the less the content of the heavy metals.

Table 2.6.3. EU- Limits of heavy metal concentration in soil for agriculture purposes (1986)

Metal (mg/kg)	Ni	Cu	Pb	Cr	Mn	Zn
EU- Limit	75	140	300	300	850	300

2.6.5 Groundwater

The groundwater in the study area from the upper aquifer is neutral (pH 7.1) with mean total dissolved solids (TDS) of 1,008 mg/L (Table 2.6.4), much lower than that of lower aquifer (1,677 mg/L; Table 2.6.4). The water is hard and can be as Ca-Na-Bicarbonate typified, with elevated Cl^- and SO_4^{2-} content according to Piper Diagram. **The electric conductivity** (EC, 1,164 µS/cm) is low in comparison to that of lower Aquifer (Tertiary Aquifer) and of drinking water (Table 2.6.4). Also the main ions **Ca^{2+}, Na^+, and Mg^{2+}** contents are low. This implies that dilution and freshening of groundwater of upper aquifer has taken place. The intensive sewage irrigation (10 mm/d) might dissolve the easily solved ions and washed them down. The relative elevated carbonate contents in lower soil further support this interpretation (Table 2.6.1). In contrast, **K^+** is the only cation with higher concentration in the upper aquifer (Table 2.6.4) and can be used as indicator of sewage application on groundwater. The very low mobility of K and its high affinity to clay minerals in both unsaturated and saturated zone are the most likely reasons.

Sewage water application is mainly represented by N- species. Nitrites and **Nitrates** mean values of 9 and 68 mg/L are much higher than the allowed limits for drinking water (Table 2.6.4). Nitrification of N and NH_4^+ might be responsible for the high nitrites. Fig. 2.6.5 illustrates the distribution of nitrate values of the study area. The highest values of nitrate are in the N and NE of the region, where the groundwater table is just below the ground (1 m) and the clay cap of Holocene is very thin. BGS (1996), Asbrand (1997) Graniel (1999), Abo el Abas (2002) have documented from groundwater contamination with nitrate as a result of municipal sewage.

Phosphate concentration of 2,3 mg/L is low, but still higher than that of underlined Tertiary Aquifer. This should be resulting from the availability of strong soil adsorption for PO_4^{3-} ions (Seiler, 1988).

Sulphate mean value of the upper aquifer of (162 mg/L) is relatively high, but still lower than the values of lower aquifer and drink water (Table 2.6.4). Dissolution of gypsum ($CaSO_4$-H_2O) and using of additional fertiliser to sewage water might elevate SO_3^{2-} values. The higher values within the sewage farm belong mainly to the upper Quaternary aquifer.

Table 2.6.4. Water quality data for sewage water, water of upper and lower aquifer in Gabal el Asfar Region. All values in mg/L; EC in µS/cm; n.d. means under detection limit.

Parameter	K^+	Na^+	Ca^{2+}	Mg^{2+}	Zn^{2+}	Mn^{2+}	Fe^{2+}	Cd	Ni	Cu	Pb	Cr
Sewage	12.5	74	46	11	0	0.05	0.2	n.d.	n.d.	n.d.	n.d.	n.d.
Upper Aquifer	23	119	120	21	0.06	0.11	0.41	n.d.	n.d.	n.d.	n.d.	n.d.
Lower Aquifer	12	266	158	36	0.04	0.26	0.78	n.d.	n.d.	n.d.	n.d.	n.d.
EU (1998)	-	200	-	-	0.1-5	0.05	0.2	0.005	0.02	0.05	0.02	0.005

Parameter	T°C	pH	EC	O_2	NH_4^+	NO_2^-	NO_3^-	Cl^-	SO_4^{2-}	PO_4^{3-}	HCO_3^-	TDS
Sewage	28	7.4	708	n.d.	0	38	3	75	74	6.6	261	601
Upper Aquifer	24.8	7.1	1164	0.3	0.06	9	68	125	162	2.3	356	1,008
Lower Aquifer	24.5	7.2	2450	1.1	0.05	5	55	227	397	0.09	415	1,677
EU (1998)	-	6.5 - 9.5	2500	5	0.5	0.5	50	250	250	5	-	1,000

Chloride mean value in the upper aquifer is about 0.4 % of that mean values of the lower one (Table 2.6.4). Chloride considers as conservative elements and one of the first indicators of sewage impact on groundwater (Driscoll, 1989). The higher values of water constituents as Na^+, Ca^{2+}, Mg^{2+}, Cl^-, SO_3^{2-} and EC can be interpreted as upcoming and recharge of salty water from the lower/neighboured aquifer (Fig. 2.6.2).

A similar $Cl^-/NO_{3\ -}$ ratio of about 2:1 as indicator for sewage application on groundwater has been observed beneath several developing cities (BGS, 1994 and 1996; Sahgal, 1989, and Abo el Abas 20002)

Heavy metals concentration in sewage effluent is negligible. The majority of metal concentrations are under the detected limits (Table 2.6.4). In groundwater of the Quaternary aquifer, Fe^{2+} and Mn^{2+} have been detected with values relatively higher than that of Tertiary aquifer. This indicates the reduction condition of the aquifer. This can be forced by the low O_2- values of 1.1 mg/L. Both Fe^{2+} and Mn^{2+} values of Quaternary aquifer is slightly higher than drinking water. Although Zn^{2+} proves to be the most mobile element, the mean value is much lower than EU-limit.

Other heavy metals like Cd, Ni, Cu, Pb, and Cr were not detected. Although Cr value is high in soil, it seems to be completely absorbed in the aquifer. Ni and Cd contents were very low in soil. The mean values of Pb and Cu are relatively high in mean values in upper soil.

Clay lenses and clay intercalation and the neutral pH condition of the saturated zones are probably responsible for absorption and/adsorption of heavy metals. Water table is about 1 m under the ground and the discharge after irrigation through drainage is important way to minimise the Contaminants in both soil and groundwater.

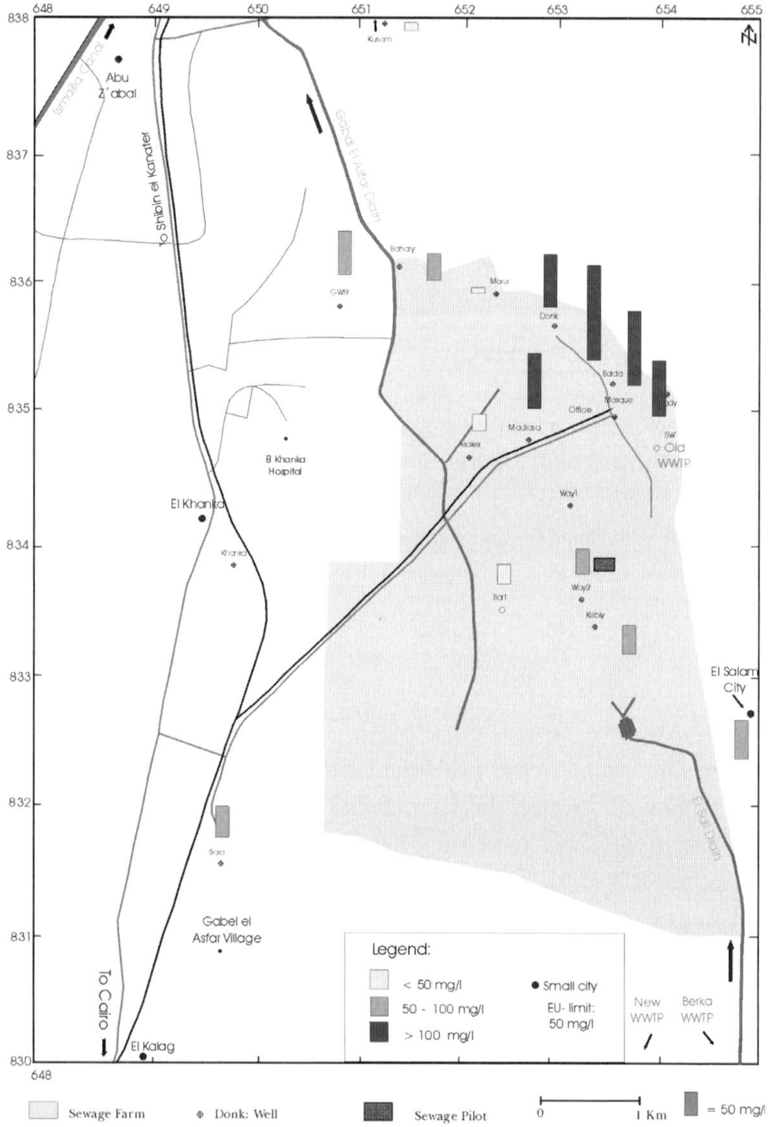

Fig. 2.6.5. Distribution of nitrate concentration in the groundwater of sewage farm area

2.6.6 Conclusion

Using of sewage for sandy soil cultivation leads to positive changing in physiochemical parameters of soil. After more than 85 years of sewage application in sewage farm Gabal el Asfar, the soil is accumulated with heavy metals Fe, Zn, Mn, Cr, Pb and Cu. Except few anomalies of Zn and Cr, the soil is not potentially endangered as one thought.

The intensive sewage irrigation leads to flushing and freshening the upper aquifer and the ground water still chemically good. The positive impact is achieved by decreasing the TDS and EC of groundwater in comparison to the lower Aquifer and the surrounding groundwater. High concentration of nitrite and nitrate could be the most negative impact of sewage irrigation. The detected heavy metals Fe, Mn and Zn show no potential danger in the Region.

The elevated inorganic contaminants in soil and in groundwater could be minimised if severe authoritarian restrictions concerning the industrial sewage and controlling the human anthropogenical activities were applied.

References

Abo el abas YM (2001) Water Quality and inorganic Pollutants in Sewage Farm at Cairo/Egypt.- Proceeding of the 7th Jordan Geological Conference, 2-4 April 2001, 10 p Abstract Vol. p 132, Amman.

Abo el abas YM (2002) Urbanization Effects on Groundwater Resources at west Cairo/Egypt.- The Geology of Arab World Conference (GAW-6), Poster Abstract, vol. 11-14 Feb. 2002, Cairo University, Cairo.

Alloway BJ, & Jackson A (1991) The behaviour of heavy metals in sewage sludge amended soils.-The Sci. of total Envi.,100: pp 151-176, Elsevier, Amsterdam.

American Public Health Association (APHA), American water Association (AWWA), and Water Environmental Federation (1995) Standard methods for the examination of water and waste water.- American Public Health Association, Washington.

Asbrand M (1997) Dreidimentional-regionalhydrodynamisches Verfahren zur Bewertung einer großräumigen Grundwasserbelastung- Rieselbezirke Sputendorf und Großbeeren südlich Berlins.- Dissertation TU- Berlin.

BGS, CnA & UACh (1994) Effects of wastewater reuse on groundwater resources, Leon, Mexico.- Phase 1 Report- July 1994, BGS, Technical Report WD/94/25.

BGS, CNA, SAPAL & UACH (1996) Commission of the European Communities, Overseas Development Administration: Effects of Wastewater Reuse on Urban Groundwater Resources of Leon, Mexico.- Final Report, British Geological Survey.

Böken H (2001) Beeinflussung des Schwermetalltransfers Boden-Pflanze durch meleorative Bodenverbesserung auf flachgründig kontaminierten Standorten. Mitt. Dtsch. Bodenkdl. Ges. 96(2): pp 403-404

Commission of the European Communities (1986) Council Directive on the protection of the environment, and in particular of the soil, when sewage sludge is used in agriculture.- Official Journal of the European Communities No. L181 / pp 6-12.

Driscoll FG (1989) Groundwater and wells, 2nd edition , Johnson filtration System, Inc., New York.

El Malky MG & Atta SA (1990) Groundwater quality at El Gabal el Asfar, north-east of Cairo, Egypt.- Proceeding of the 2nd National conference on environmental studies and research, Institute of environmental studies & research, Ain Shams University, Cairo.

El-Nennah M, El-Kobia T, Shehata A & El-Gamal I (1982) effect of irrigation loamy sand soil by sewage effluents on its content of some nutrients and heavy metals.- Plant Soil, 65: pp 289-298.

European Commission (EU) (1998) Richtlinie 98/83/EG des RATES von 3 Nov. 1998 über die Qualität von Wasser für menschlichen Gebrauch.

Farid S (1980) Nile delta groundwater study.- M.S. Thesis, Cairo Universuity, Egypt.

Farid MS, Atta S, Rashid M, Mounnik JO & Plattenburg R (1993) Impact of the reuse of domestic waste water for irrigation on groundwater quality.- Wat. Sci. Tech. 27(9): pp 147-157, Great Britain.

Graniel CE, Morris LB & Carrilo-Rivera JJ (1999) Effects of urbanisation on groundwater resources of Merida, Yucatan, Mexico.-Environmental Geology 34(4), Springer.

Grunewald K (1994) Grossräumige Bodenuntersuchungen auf Rieselfeldern südlich Berlin.- Z. Pflanzenernähr. Bodenk., 157: pp 125-130.

Hoffmann C (2002) Schwermetallmobilität und Risikopotentiale der Rieselfeldböden Berlin-Buch.- Bodenökologie und Bodengenese 35: pp 1-226, Dissertation, TU-Berlin

Hölting B (1992) Hydrogeologie.- Enke Verlag, Stuttgart.

Kennt DB, Davis JA, Anderson LCD, Rena BA and Waite TD (1994) Transport of Chromium and selenium in the suboxic zone of shallow aquifer: influence of redox and adsorption reactions.- Water Resources Research, 30: pp 1099-1114.

Leeper GW (1972) Reaction of heavy metals with soils with special regard to their application in sewage wastes. Report to Department of the Army.- Corps of Engineers; C.F.J. of environmental quality; 10 (4), 1991, Washington, USA.

Marschner B & Hoffmann C (2000) Mobilisation of heavy metals in soils on a former sewage treatment farm. -1 Intern. Conf. on soils, of urban, industrial, traffic and mining areas, 12-18.07.2000, Proc. Vol. II: pp 375 -380 Univ. Essen

Matheis G, Siad A & Aderhold G (1995) Abschlußbericht Rieselfelder Südlich Berlin.-Teilprojekt II: Hydrogeologie/Angewandte Geochemie: Geogene-Anthropogene Zustandsanalyse und Gefährdungsabschätzung potentieller Schwermetall-Mobilisation.

McGrath SP (1987) Long-term studies of metal transfers following Application of sewage sludge.- PJ Coughtrey, MH Martin and MH Unsowrth (eds): Pollutant, Transport and Fate in Ecosystems.- Blackwell Scientific Publications, pp 301-317, Oxford.

McGrath SP (1995) Chromium and Nickel.- Heavy metals in soils (Ed. Alloway, B. J.), pp 152-178, Blackie.

RIGW (1989) Hydrological map of Egypt-Nile Delta, 1: 500 000.-Research Institute for Groundwater, El Kanater, Cairo.

RIGW/IWACO (1991) Development and management of Groundwater resources in the Nile Valley and Delta; Monitoring and control of groundwater pollution; Groundwater quality study Gabal el Asfar.-Research Institute for Groundwater, El Kanater, Cairo.

RIGW (1995) Preliminary assessment of suitable sites for sewage-based irrigation and sludge disposal, National water research centre, Ministry of Public works and Water Researches, Cairo.

Sahgal VK, Sahgal RK & Kakar YP (1989) Nitrate pollution of groundwater in Lucknow area, U.P.- Proceedings of international workshop on Appropriate Methodologies for Development and Management of Groundwater Resources in Developing Countries.

Sauerbeck D (1988) in: Isnebeck M & Mattheß G (1988) Transfer von Schwermetallen aus der Bodenlösung ins Grundwasser.-Beurteilung von Schwermetall-kontaminationen im Boden.

In: Beurteilung von Schwermetall-Kontaminationen im Boden- Dachema-Fachgeschpräche Umweltschutz, 21-22 Jan., Frankfurt/M.

Seiler HG, Sigel H & Sigel A (1988) Handbook of toxicity of inorganic compounds.- Marcel Dekker Inc., New York.

Tröger U, Blumenstein O, Tessmann J & Portmann D (1996) Abschlußbericht Rieselfelder südlich Berlin-Altlast, Grundwasser, Oberflächengewässer. Landesumweltamt Brandenburg (internal report, TU-Berlin).

World Health Organisation WHO (1993) Guide line for drinking water quality, 2nd edition,1, recommendation, WHO, Geneva.

3 Water Management

3.1 Water Utilization in Semi-Arid Zone, The Hula Valley (Israel): Pollutant Removal, Agriculture and Ecotourism Management.

Moshe Gophen
MIGAL-Galilee Technological Center
POB 831, Kiryat Shmone, (11016), Israel

I dedicate this paper to the memory of Dr. Israel ('Srulik') Levin (1914 – 1998), a resident of the Hula region, a guide , friend and unforgettable leading scientist of agricultural research in the Hula peat soil; and David Nachmias, (1927 – 2000), the head of the department of Land Development of the Jewish National Fund, his personal efforts and involvement, made the accomplishment of the Hula Project possible.

3.1.1 Introduction

Historical Background

Until the late 1950's the Hula Valley (altitude: 62.00-70.00 m above MSL) was covered by old Lake Hula (1.5 m mean depth; 13 km^2 water surface), 3500 ha was a swampy area mostly covered by dense vegetation of *Cyperus papyrus* (85%) and partly by *Phragmites australis* (ca 10%) and the rest by different aquatic plants, including cattail *(Typha domingensis).* The swampy area was completely water-covered in winter and partly covered in summer. The bottom sediments underneath the swamps vegetation were comprised of organic peat, nitrogen rich, under reductive conditions covered by water with low of oxygen content and probably anoxic during the night. The most common nitrogen form in Lake Hula waters and in the swamps was ammonia whilst nitrate was mostly below detection limit (Dimentman et al. 1992). Northern to the swamps there was an area (3200 ha) where water table levels were high in winter time, making agricultural development impossible. During summer periods, when underground water level declined these 3200 ha, were successfully (economically) cultivated.

Bottom sediments in old Lake Hula and a several hundred meters northern to its shoreline were chalk-marl. Chalk-marl was also the major soil component in the area northern to the swamps. Between the peaty swamps and the northern area there was a narrow segment (1-2 km) of transient soils comprised of a chalk and peat mixture.

Regional Hydrology

Three major river flows (Hatzbani, Banyas and Dan), flow from the Hermon mountain region, located in the north part of the Kinneret drainage basin. These rivers joint into one river Jordan which crossed the swampy area in two branch (tributaries) inflows into old Lake Hula. From Lake Hula at an altitude of about 61m ASL the one river Jordan flowing down stream into Lake Kinneret to an altitude of 209m below sea level along about 15 km. The Jordan river contribute about 63% of the Kinneret water budget and above 70% of total nutrient inputs. We have very little information on nutrient fluxes through Jordan river into Lake Kinneret before the 1950's (Dimentman et al.1992). Nevertheless it is suggested that inputs of available nitrogen were high prior to the Hula drying since the dominant nitrogen species was ammonia and it is known to be more available than nitrate for phytoplankton. Long term record (1970-2000) documented that nitrate and organic – N are the dominant N-species in the Jordan loads. The water level in old Lake Hula fluctuated annually with an amplitude of 1.0-2.0 m, flooding the swamps in winter when water level was at its maximum.

The „Hula Project"

During the 1950's, as a result of a government decision of the newly established (1948) State of Israel, the Hula wetland was dried and more than 6500 ha of natural wetland area, with a unique natural composition of fauna and flora of exceptional diversity, was turned over to agricultural use to serve as an income source for residents of the northern part of Israel. During 40 years the dried area was successfully cultivated and agricultural products (mostly Cotton, Corn, Alfa - Alfa) were economically produced and nutrient flux into Lake Kinneret did not threaten its water quality. Nevertheless, as a result of inappropriate management, drainage canals were blocked, irrigation methods were not suitable for optimal soil structure protection and water table declined. Consequently, soil structure of upper layers (0-0.5m) was deteriorated and heavy dust storms occurred quite often in fall and spring by regional east wind (Sharkiya) resulted in subsidence of soil surface (averaged as 7-10 cm/year) and blocking of drainage canals was enhanced. Due to the decline of water table level, lowering of soil moisture, longer time duration of uncovered soil by vegetation (agricultural crops or weeds) and inappropriate agricultural management, an outbreaks of underground fire occurred frequently in summer time and a lot of water was needed for extinguishing which enhanced nutrient flux into Lake Kinneret. Rodent population outbreak caused severe damage to agricultural crops and stability of drainage canal banks. Ten % of the total dried area, in the middle part of the valley (at lowest altitude), went through processes causing extreme soil deterioration and subsidence which made it impossible to maintain a beneficial cropping and these 500 ha were neglected and not cultivated and the threat on Kinneret water quality was enhanced.

Therefore, a reclamation project was proposed and implemented (1990 – 1997) which was aimed at the reduction of nutrient fluxes from Hula soil while continuing the economical utilization of the land by partial modification of land utilization from agriculture to ecotourism. The concept of the new land use was based on man made structural changes of the system by incorporation of natural components (fauna and flora). The reclamation project, entitled "Hula Project", included several elements based on the concept of increasing the level of soil moisture by elevating the water table, changing the irrigation system and renewing the hydrological (drainage) system of the valley. This included 90 km of drainage canals, used both for water supply for establishing a higher water table, and draining excess waters; creating the shallow Lake Agmon (0.6 average depth; 110 ha area) to be operated as a drainage basin for the valley and ecotouristic wetland. Water effluents from Lake Agmon are removed by pumping system and free flow in open canals, from the Kinneret catchment. The function of the lake is not only being part of the drainage system but also to serve ecotourism, therefore its water quality, landscape and the composition and density of its flora and fauna affect its quality.

A plastic sheet (4 mm thickness) was placed vertically (0-4.5 m) along 2.8 km, across the valley southern to Lake Agmon to prevent underground infiltrated flows down stream to Lake Kinneret. The ecology of Lake Agmon was studied during 1994-2002. A comprehensive information about Lake Agmon and around it is presented here with an attempt to emphasis the need to combine economical , limnological and ecological aspects in one project of constructed wetlands. The Agmon wetlands ecosystem is an example of management design and constructed infrastructure as a package deal between three sectors of our society: 1) the land owners (farmers) that the Hula Valley area is an income resources for them; 2) Water managers which are responsible for the limnological protection of Lake Kinneret for drinking water supply; and 3) the "greens" - nature ecologists. Protection of the Kinneret water quality is essential because more than 50% of its external load mass (N and P forms, sulfate, ect.) originate from the Hula Valley zone (Geifman et al. 1970-1998; Rom 1999).

The Hula Project plan was formulated by a committee comprised of all relevant organizations : the Lake Kinneret Authority, Kinneret Limnological Laboratory, Kinneret Drainage Authority, Nature Protection Society, Nature Conservation and National Parks Authority, Regional Municipalities of all Hula Valley, Water Planing for Israel Co., Water Commission, Jewish National Fund (JNF). Three major alternatives were considered: 1) doing nothing; 2) agricultural soil improvements: and 3) A combination of drainage-agricultural improvements and ecotourism. Socio-economical, limnological, and ecological aspects were considered. All the meetings were open for the public and concluded with option no. 3 – entitled "The Hula Project". The money allocation and responsibilities for the Hula project implementation were carried out by state authorities and JNF. A Hula Project Board of Directory was established to follow up activities and budgeting. Responsibilities for the operation of the research program was given to MIGAL, Center for Research in the Upper Galilee region, with the follow up of steering committee comprised of representatives from all relevant sectors. The total investment in the constructions of the Hula Project infrastructure was ca 20 x 10^6 US $ including 2 x 10^6 US $ for research . Constructions were started in 1993 and accomplished at the end of 1997. From 1998 and onwards a monitoring program is carried out to follow the system functioning

Research

The Hula Project constructions and the accompanied research is one of the biggest ecological operations ever carried out in Israel since late fifties. Land owners are looking forward to allocate money for the commercialized development under the frame of the confirmed program. Data and long-term analysis of nutrient inputs from the catchment into Lake Kinneret are given in Geifman et al. (1970-1998) and Rom (1999). Nutrient concentration and calculated inventories, inputs and outputs, and Temp., conductivity, pH, DO, phytoplankton and zooplankton, macrophyte biomass and their N and P contents, water inflows, and outflows into Lake Agmon and in the vicinities, and P content in the sediments are routinely monitored from1994 by Migal and research teams in the Tel Hai academic college and the Kinneret Limnological Laboratory. The information on spatial and temporal distribution of macrophyte vegetation together with the limnological data were incorporated at an attempt to formulate an overview of the present ecological, limnological and agricultural status of the Hula wetlands ecosystem . In this paper I will summerised the updated knowledge of the ecosystem with an insight into its proposed functions: ecotourism and conservation of natural elements, protection of Lake Kinneret water quality by nutrient removal and income resource for the land owners.

3.1.2 Results and Discussion

Lake Agmon Wetlands

Classification of the trophic status of freshwater lakes in the world (Wetzel 1983) and Lake Agmon indicates that Lake Agmon can be classified as closer to Eutrophic than to Hypertrophic Status (Gophen 2000). Lake Agmon has three major runoff water sources: 1) River Jordan waters via its reconstructed route; 2) Drainage water from peat soils in the Hula Valley northern to the lake, via Drainage Canal Z and 3) drainage waters from the eastern peat block through "Eastern-Hula Canal". The Jordan waters carry nutrients eroded by runoffs, loaded with nitrogen (mostly nitrates), phosphorus and suspended matter. The drainage waters are highly loaded with peat originated nutrients: mostly ammonia produced in anoxic conditions , sulfate (from gypsum dissolution) and nitrate from oxic conditions. Because the bottom of Lake Agmon is topographically lower than the northern water table there are artesian hydraulic pressures throughout the lake bottom forming advection flows of about 0.5-1.0 mm/day during the winter (i.e. 500-1000 m^3/lake/ day) (Markel et al. 1997).

Table 3.1.1. Hydrological budget (mcm/year) for 2002 of the total area: cultivated land, canals and Lake Agmon bordered between: North-Connection canal; South-Western Canal; East-Eastern canal; West-Western canal; (800 ha.) (Fig. 3.1.1).

Gain	
From reconstructed Jordan	13.5
From Hula East	0.5
Rainfall	5.6
Total gain	19.5
Loss	
Evaporation	16.4
From 404 canal to Western Canal	1.8
From Z canal to Western Canal	3.9
From Agmon through "Zero" canal to Op. Reser.	1.8
Total loss	23.9

Results in Table 3.1.1 indicate that measured loss is greater than measured gain by 4.4 mcm/y (10^6 m^3/y) from unmeasured sources. An input of underground waters through unmeasured pathways is concluded (Fig. 3.1.1).

There are water losses from Agmon through bottom infiltration to the west-southern and probably to east – southern directions of the valley. Water budget for Lake Agmon and the cultivated area around was calculated and two examples are presented here: for the years of 2000 and 2002 (2001 budgets are given in Gophen 2003). The total area of Agmon is 1100x 10^3 m^2 and the volume –0.6x10^6 m^3 ; Evaporation in Lake Agmon was accounted as 6 mm/day; Water diversion at Kfar Blum Dam (northern to Lake Agmon) from Jordan main discharge and from the "connection canal" flowing through the reconstructed Jordan route and through drainage canals for irrigation and for increasing water table, was considered as total water income to the wetlands region which include supply to Lake Agmon and for irrigation of the agricultural area northern to it (ca 400 ha); water level changes in Agmon were combined (ups and downs) and annual change was calculated (increase was considered as outflow); Water loss in Lake Agmon through bottom infiltration was calculated as changes of total inflows minus total outflows (m^3/ 10^3m^2/day). During 10 months of the year of 2000 total inflows minus total outflows in Lake Agmon indicated 1.4 10^6 m^3 (4.2 $m^3/10^3$ m^2/day) which was considered as underground infiltration loss. In 1999 the loss through infiltration was almost two times higher.

Table 3.1.2. Water balance (mcm/y) for Lake Agmon (2002)

Inflows	
From Z canal	7.0
From Hula East	0.5
From reconstructed Jordan	1.9
WL changes	0.1
Rain	0.7
Total inflows	10.2
Outflows	
To Z canal	2.1
To Zero canal	4.0
Evaporation	2.3
Total outflows	8.4

The results in Tab. 3.1.2 indicate that measured inflows were higher than measured outflows by 1.8 mcm/y. Consequently, 1.8 mcm/y (0.45 m^3/ha/day or 4.5L/m^2/day) infiltrated through unmeasured pathways.

Due to a sequence of three years of drought (1999-2002), and taking into account that the Hula Valley and river Jordan are part of water supply in Israel (Jordan – Hula – Kinneret system), the total diversion from the Jordan to the Hula Project via Kfar Blum Dam passing through the "connection canal" was reduced in 2000. Consequently bottom infiltration in 2000 was lower than in 1999. In 2000 the mean Residence Time (RT) of water in Lake Agmon was 36 days, in 1999 – 29 days and in the first 5 months of 2001 - 63 days. The longer (24%) RT in 2000 could be one of the reasons for the increase of P and chlorophyll (phytoplankton) (data are not given here) content in the water. The considered area include agricultural irrigated land, Lake Agmon and the grass covered open meadow (Safari) around it. Results of this total balance for the period of 10 months (March-December) in 2000 indicated total inflow of 14.3 x 10^6 m^3 and outflow – 12.2 x 10^6 m^3. The difference between inflow and outflow , 2 x 10^6 m^3 is too low to cover total approximated evapotranspiration over the whole area (ca 3-5 mm /day) (Gophen et al. 2003).

It can be concluded that the total inflows into the wetlands area are higher than those which were measured and other underground, not measured, sources are probably involved. Similar budgets were calculated for the first 5 months of the year 2001 and results indicated total water inflows lower than measured outflows to the whole area and particularly to Lake Agmon. Consequently, in 2000/01 additional, not measured, underground waters were involved. In 2002 we measured discharges in more stations (canals). Budgets are presented in tables 3.1.1-3.1.4. The hydrological balances of the Hula Valley for the year of 2002 took into account the following parameters: total area-800ha, Agmon area-110 ha, the area northern to the plastic barrier(along 2.8 km southern to Agmon) is 400ha and southern to the barrier-400ha, rainfall-700mm, evaporation-2054 mm/year.

Table 3.1.3. Hydrological balance for the area northern to the plastic barrier (400ha) (2002) (Fig. 3.1.1).

Gain	
From Hula east	0.5
From reconstructed Jordan	13.5
Rainfall	2.8
Total gain	16.7
Loss	
From Agmon to Z canal	2.1
From Agmon to Zero Canal	4.0
Reconstructed Jordan underneath Zero canal	1.8
Evaporation	8.2
Total loss	16.1

The results in Table 3.1.3 show that gain was just slightly higher than loss, (0.6 mcm/y). Consequently, in the northern part of the Valley no intensive underground not-measured inputs are suggested.

In the southern part of the Valley the loss was higher than gain which is due to irrigation waters pumped from eastern and western canals into the area and infiltration from Zero canal.

The outcome from the results presented in Tables 3.1.1-3.1.4 is that there are a preferential water flows (vertical and horizontal) in an underground lines in the Hula Valley and mostly in the southern part of the Valley but their existence in other parts of the valley are not excluded.

Table 3.1.4. Hydrological balance (mcm/y) for the area southern to the plastic barrier that was placed along the southern Agmon shoreline and Zero canal (400 ha) (2002) (Fig. 3.1.1).

Gain	
From Agmon to Z canal	2.1
Through Jordan underneath Zero canal	1.8
Rainfall	2.8
Total gain	6.7
Loss	
From Z canal to Western Canal	3.9
From 404 canal to Western Canal	1.8
Evaporation	8.2
Total loss	13.9

3.1.3 Nutrients

Monthly means (1994 – 2002) of TP concentrations in Lake Agmon fluctuated between 20-40 ppb in Januar-March and a 300-500 ppb in November-December. Concentrations of TP significantly increased after 1999. The major sources of P in summer are external inputs by peat drainage water and internal plant mediated recycled P, dust and birds excretions. Phosphorus is incorporated by benthic algae and submerged macrophytes from bottom sediments and released afterwards from decomposed plant material. The biogeochemical P influx from bottom sediments-pore water is probably low as a result of the slow advection flows (Markel et al. 1996;1997;1998; Nishri and Gophen 2002).

The lake is functioning as a nitrogen sink where N is removed by sedimentation and denitrification. Total nitrogen, nitrate, total dissolved nitrogen, and ammonia in the Agmon outflow were significantly (ANOVA of monthly means: $p < 0.0001 – 0.0006$) lower than in Canal Z whilst those of TP were significantly (ANOVA of monthly means: $p = 0.0061$) higher in the effluent. Difference between sulfate concentration in the effluent and Canal Z were insignificant. Based on monthly mean concentrations in 2000 (inputs and outputs) and discharge measurements, nutrient mass balances for Lake Agmon were calculated. Total income minus total outcome = mass balance, where: positive value = input > output; negative value = output > input for the year of 2000. The annual removal in 2000 through Agmon outflow, of ammonium, nitrate, TN and TP was, 6.9, 14.4, 3.2, and 0.64 g/m^2/year respectively. TN and particularly Nitrate are reduced in the Agmon waters by denitrification and sedimentation; ammonium decline by nitrification; TP mass balance is including inputs by atmospheric dust, aquatic birds and vegetation recycling. Nishri and Gophen (2002) aproximated the following phosphorus inputs during 2002: 0.4 t of TP by atmospheric dust, 0.5-2.0 t from birds exretion, and 0.5t was incorporated by plant from the bottom beds and the same amount degraded and decomposed later.

Because inputs and outputs of TP through water flows were similar, the total balance is (input minus output) positive by about 1.0 t. The total balance of nutrient removal from the Kinneret loads include not only the Agmon outflow but also the sedimented nutrients within the lake. The sedimented particles originates mostly from biotic sources (phytoplankton) and inputs of suspended matter. Phosphorus removal from the Lake Agmon system through water outflows was averaged as 1-1.6 t/year. The major source of sulfate to lake Agmon is dissolved gypsum in the peat soil (Markel et al. ;1998).The sulfate is transported to the lake mostly in peat drainage waters through Canal Z and Hula East, and together with phosphorus from the bottom and through side (shoreline) advective infiltration from the surroundings into the lake (Nishri and Gophen 2002) when its water level is lower than the water table of the peat land around in summer.

The sulfate has probably the main effect on water conductivity. Close relation between conductivity and sulfate values in Agmon outflows and Canal Z were indicated ($R^2 = 0.747$; $p < 0.0001$). The major internal P source is shallow rooted macrophyte . P is taken from the upper (<5 cm) sediments layer by roots and incorporated into plant tissues during the vegetation onset (May – September) and released back into the water column during decomposition of the macrophytes (Gophen 2000; 2003). The P stock in the upper layer of the sediments is renewed by sedimentation of particulate P inputs during winter floods through Canal-Z and Jordan waters.

Plant mediated P has a significant role in P dynamics in Lake Agmon. It is likely that P limitation in the bottom sediments suppressed *Typha* plants (Miano and DeBusk 1999). Particulate P (phytoplankton, and abiotic particles) accumulate on the top of the sediment layer (< 5 cm) and became available and utilized by benthic algae and shallow rooted macrophytes *(Potamogeton spp., Ceratophyllum sp. Najas spp.* etc.). This plant material is decomposed during late summer and fall. Due to proposed water managements, perennial increase of P load is not predicted.

To improve water quality, mechanical removal of macrophytes can be used as a management tool. Total biomass of 270 - 450 tonnes dry weight of macrophytes containing ca. 1000 kg P was measured during vegetation peaks of 1998 - 2000 (Kaplan 1997 - 2002). If the value of macrophyte mediated P release to the water column measured by Smith and Adams (1986), i.e. $2gP/m^2/year$, is applied to the Agmon ecosystem ($1100 \times 10^3 m^2$), the flux of 452 kgP for the whole lake during 5 summer months is predicted.

Table 3.1.5. Multiannual averages (1994 – 2002) (SD) of physico – chemical parameters in Lake Agmon, Canal Z, Hula East (HE) and Jordan water.

Parameter (units)	Lake Agmon	Jordan	Canal Z	HE
DO (ppm)	9.5(2.4)	9.0 (1.8)	6.4(1.4)	7.6(2.0)
Alkalinity (ppm $CaCO_3$)	130(46)	191(27)	243(61)	228(116)
Sulfate (ppm)	330(214)	25(5)	364(343)	739(380)
TDS (ppm)	716(428)	255(51)	941(847)	1482(695)
Turbidity (NTU)	40(20)	19.5(8.3)	12.0(5.7)	33(65)
Ammonia (ppm)	0.7(1.3)	0.05(0.1)	3.3(4.9)	1.2(2.4)
Nitrate (ppm)	2.6(5.0)	1.2(0.3)	13.9(22.7)	3.2(4.2)
Nitrite (ppm)	0.04(0.06)	0.01(0.01)	0.2(0.2)	0.1(0.6)
TN (ppm)	6.0(7.3)	2.0(1.2)	18.8(27.0)	6.4(5.0)
TDN (ppm)	4.6(7.4)	1.8(1.2)	16.9(24.8)	5.6(5.0)
TP (ppm)	0.2(0.1)	0.2(0.1)	0.1(0.1)	0.2(0.2)
TDP (ppm)	0.02(0.02)	0.09(0.2)	0.02(0.02)	0.1(0.2)
Conductivity (mS)	0.9(0.5)	0.4(0.1)	1.3(0.9)	1.7(0.7)
pH	8.1(0.5)	7.8(0.2)	7.3(0.2)	7.4(0.6)
Temp. (°C)	19.4(5.5)	17.2(4.6)	18.8(5.3)	20.2(5.4)

Due to higher temperatures and production in Lake Agmon it can be suggested that the plant mediated P release might be even higher. In the chalk-marl sediments in the southern part of the lake P is probably more available than in the peaty-iron-rich-P-binding soil on the bottom of the northern part of the lake and therefore *Typha* did not develop there. The *Typha* root system is mostly developed in the sediment layer deeper than 5 cm. Therefore the high P availability in top sediment layers can not be efficiently utilized by the cattail plants but can be intensively incorporated by filamentous benthic macroalgae and shallow rooted macrophytes like *Potamogeton spp., Najas spp.* and *Ceratophyllum* sp. During the winters of 1998 - 2002, water management in lake Agmon implied low level and the large area of sediments in the eastem-southem part of the lake were uncovered. The exposed organic sediments were oxidized. Phosphorus that was bounded to organic matter was released due to oxidation of this organic matter and/or Iron (Markel et al. 1996; 1997; 1998;) and its availability was probably enhanced. As a result, *Typha* stands were intensively developed on areas where this plant was not grown during 1995-6 (Richardson et al. 2000).

Results in Tab. 3.1.5 prominently show that concentrations of nitrogen forms, sulfate, and TDS, are higher in peat drainage waters (Z canal and Hula East) than in Jordan water and partly in Lake Agmon. Data collected during 1994-2002 show a decline of nitrogen forms and increase of phosphorus concentrations during the last 4 years. Increase of the concentrations of phosphorus forms originating from vegetation decomposition and soil degradation, and decline of Nitrogen forms resulted in by enhanced nitrification/denitrification activities. Turbidity, was increased and TDS, conductivity and sulfate declined. The chemical parameters from lake Agmon outflows were applied to discharge measurements on a monthly basis to calculate nutrient mass balances in order to asses the potential of nutrient removal from Kinneret input loads by the Hula wetlands system. Lake Agmon effluents are diverted out of the Kinneret catchment for irrigation.

Removal of 25% of sulfate loads is probably not critical to Kinneret water quality because Sulfur reductor bacteria in Lake Kinneret reduce SO_4 to sulfide and in this process Carbon is utilized. Carbon consumption might have a positive effect on water quality. On the other hand removal of 11% of TP and 16% of ammonia can improve Kinneret water quality. Removal of TP and ammonia by the diversion of Agmon effluents is efficient because TP and ammonia concentrations in the late summer outflow are high. The lower effect of nitrate removal by the Agmon effluents diversion is due to: 1) high contribution from sources other than the peat soil in the Kinneret catchment and 2) there is a nitrate removal within Lake Agmon waters by denitrification and sedimentation and therefore its concentration in the outflow is low. The sulfate originate from gypsum dissolution in the peat soil and transported by underground water infiltration. In winter, the soil moisture and water flows enhance sulfate dissolution and flux into Agmon water. The nitrogen forms (ammonium, nitrate and TN) in winter season also represents high concentrations as a result of inputs by drainage Canal Z and Jordan waters as well as low rate of denitrification.

The annual budgets of nutrients in Lake Agmon during 2002 were calculated on a monthly basis and annual balance was computed, the results were (-=input<output; +=input>output): TSS (measured as turbidity, and 1NTU unit=1ppm) –106t, Ammonium +7t, TDN +45t, Nitrate +43t, TN +47t, TP –0.3t sulfate +162t, TDS +1501t. The outflow loads from lake Agmon during 2002 were: Sulfate-1763t, Ammonium-3.3t, Nitrate-13t, TN-31t, TDN-22.4t, TP-1.2t, TDP-0.2t. Most of these nutrients in the Agmon effluents were eliminated from Lake Kinneret input loads by the Hula system.

3.1.4 Macrophytes

The distribution of macrophytes in Lake Agmon was qualitatively studied since 1995 and from 1997 their biomass is routinely estimated (Kaplan et al. 1998; 1997 – 2002). The dominant species are : *Ceratophyllum demersum,*, *Potamogeton nodosus, P. brechtoldii, P. pectinatus, Najas marina, N. minor, Typha domingensis, Phragmites australis altissimus* and *Chara braunii* (Kaplan et al. 1995-2002; 1998).A seasonal pattern of onset (March-July) and offset (August-November) of the submerged vegetation (Kaplan 1997 - 2002; Kaplan et al. 1998) was observed. An onset of *T. domingensis* during late 1994 through mid 1996 and an abrupt decline afterwards was documented. The peak of submerged macrophyte biomass occure in June-July and in November-December all (except *P. australis* and *T. domingensis* when existed) plants are degraded and disappeared. All macrophytes, except stands of *Typha* and *Phragmites*, grow below water surface with canopy of floating leaves of *Potamogeton* sp. The degradation and decomposition of plant material starts in May-June when the benthic algae die-off and continues intensively in summer. In fall, macrophyte vegetation mass is decomposed and particulate and dissolved degradants are fluxed into the water. Particles are partly sedimented and the rest is fluxed out in the outflow.

3.1.5 The Typha collapse and recovery

From late autumn 1994 stands of *Typha domingensis* were developed in Lake Agmon on the chalk-marl bottom sediments in the southern part. This vegetation became very dense during 1995-early 1996 but totally collapsed during the 2nd half of 1996. The *T. domingensis* plants developed long leaves and their upper part was 2-3 m above water surface. A remarkable colony of several thousands of herons (5 species) utilized the top parts of the *Typha* stands as nesting ground. In 1998-1999 a massive renewal of *Typha* stands was observed in the eastern-southern part of the lake after lowering of Agmon water level.

3.1.6 Birds

Observations on birds distribution were carried out by the Nature Conservation and National Parks authority on a monthly basis during 1994-2002 (Oron et al 1995-2001; Kaplan and Niv 2002; Ashkenazi et al. 1998). Two major factors attract birds to the Hula Valley in general and particularly to the Hula Project area: 1) water bodies (Lake Agmon and drainage canals) of aquatic (shallow and deep) and semi-aquatic habitats; and 2) agricultural crops and open fields. Food resources in the aquatic habitats are macrophytes, phytoplankton, macro-algae, aquatic invertebrates and fish. In terrestrial habitats of cultivated and non - cultivated land, diversity of food resources for birds is much wider: agricultural crops (such as Peanuts, Sunflower, Wheat, Corn and others), terrestrial invertebrates (mostly insects) , rodents prey, green and underground vegetation and more.

Birds might cause damage to crops but on the other hand might be very attractive for bird watchers and therefore important for ecotourism. The Hula Project components of wetland, drainage canals and agriculture, generated a very sensitive combination. This system is formulated as a package deal between the land owners (farmers), water managers and naturalists. The optimal existence of birds population in the valley can not be achieved without the close linkage between those two interests. The aim of the new wetland ecosystem managers is to design an ecological maintenance in the valley where agriculture is economically carried out together with the protection of unique bird flocks all year around as an attraction for bird watchers, and water quality protection. Immediately after filling Lake Agmon with Jordan and drainage waters, a remarkable highly diversified bird flocks populated the valley in and around the new lake and drainage canals. The renewal of the avifauna in the Hula Valley was done naturally and very fast. Species diversity (ca 200 species) and number of specimens (30000 – 40000 in winter) were high beyond predictions. Food and suitable habitats attracted winter and summer migratory birds.

As part of the Hula Project research activity a stocking program of lake Agmon by fish was carried out. The stocked fish were fingerlings from adjacent fish ponds. In these ponds an uncontrolled reproduction of unwanted fish species occure. During commercial harvest these unwanted fingerlings were removed to Lake Agmon where they were intensively preyed by Pelicans (*Pelecanus onocrotalus*) and Cormorants (*Phlacrocorax spp.*). It was documented by T. Oron and Y. Vaadya (2000) that Cormorants came partly to Lake Agmon instead of Lake Kinneret during their daily migration for feeding. The Pelicans are temporal visitors in Israel twice a year when they migrate to south during fall-early winter and coming back to the north to Europe in spring. Fish stocking in Lake Agmon attracted pelicans to stay longer time in the Hula area and a flock of about 500 individuals spent all winter month in the Valley.

The most common genera recorded during 2000-2002 were:

Tachybaptus, Phalacrocorax,Pelecanus,Nycticorax,Ardeola, Bubulcus, Egretta, Ardea, Ciconia, Plegadis, Platalea, Tadorna, Anas, Marmaronetta, Aythya, Milvus, Circus, Bueto, Falco, Francolinus, Gallinula, Fulica, Grus, Himantopus, Hoplopterus, Vanellus, Philomachus, Limosa, Tringa, Larus, Actitis, Sterna, Streptopelia, Halcyon, Alcedo, Ceryle, Galerida, Motacilla, Sturnus, Cormorant, and more.

Nutrient contribution by Birds:

Birds monitoring indicated the following numbers of the most common birds which are closely related to aquatic habitats: 20000-Cranes, 6000-Pelicans, 6000-Cormorants and about 20000 smaller aquatic birds (Herons, Ducks, Mallards, Seagulls etc). These are migrating birds and the time they stay in the Hula Valley was considered. Excretion rate and P content was taken from literature: for cranes and Pelicans 5.24 gP/ind. /day; for cormorants-3.5 gP/ind./day (excluding Pellet), and small aquatic birds-1.0 gP/day. The total excretion of those migrators , Cranes during 150 days, Pelicans-30 days, Cormorants-150 days and small aquatic birds during 150 days is about 23 tons of which 0.5-2.0 t probably influxed into Lake Agmon.

3.1.7 The Crane (*Grus grus*) case(Alon and Yom-Tov 1999-2001)

Until early 1990's Cranes did not come to the Hula Valley for wintering. From mid 1990's the Valley is populated annualy from November to March by approximately 20000 noisy cranes. The peanut crops were the attracted element for these birds in the valley. Peanuts were found to be an economical crop suitable to the peat soil of the Hula Valley. The Cranes are coming late November when peanuts are harvested and partly left uncollected on the ground. The Cranes were strongly attracted to this type of food and are able to dig in the ground to find it.

Consequently they landed intensively on peanut fields but not for a long time. After 1-2 months rainfall start and soil moisture increased and the wet peanut seeds are fermented. The Cranes do not like them any more and are looking for another source of food in the Hula Valley. Consequently they are dispersed on fields in the vicinity and cause damage. The Crane is a protected bird by an international law and the farmers became very frustrated. The Crane attraction is significant for tourism in the newly constructed wetlands but with the price of a heavy damage to agriculture. A collaborative solution was implemented by cooperation between Nature authorities, water managers, land owners, regional municipalities and the German Flight Co. "Lufthanza": money was allocated to rent 20 ha area that was dedicated to be a feeding station where costly corn seeds are distributed daily and the cranes were concentrated there (other birds as well) and crops damage declined significantly.

3.1.8 The impact of the Hula Valley on the Water budget of Lake Kinneret (Fig. 3.1.2)

During the last 30 years a long term trend of water input reduction from the drainage basin into Lake Kinneret was documented (Rom 2001; Berger, 2001; Schanzis 2002). On the other hand, according to Berger (2001) water utilization northern to the lake was intensified but practically decreased. Berger (2001) suggested that annual quantity of available waters in the drainage basin was declined by 17% and discharges at Huri Bridge (south end of the Hula Valley) station was decreased by 27% during 1990-2000 (excluding 1991/2 floody season). Rom (2001) documented decline of 35% (from 500 mcm/y to 320 mcm/y) of Lake Kinneret available waters during 1970-2000; Schanzis (2002) documented a decline of 38% of Kinneret available water (300 mcm/y) during 1963-2001. The table of Kinneret water budget (this volume, Kinneret chapter) indicate water deficiency at the Huri Bridge station mostly during droughts and to a lesser extend under normal regime. Immediately after the Hula drying in late 1950's a processes of soil structure destruction took place as a result of decreased moisture and consolidation. The outcome of those processes was intensification of the formation of subsurface cracks, tunnels, open spaces at different capacities which enhanced the preferential water flows in the underground of the peat soils. The control of discharge flows in the drainage canals and regulation of ground water table (GWT) became difficult. These phenomena made water supply to agricultural irrigation very difficult also.

It is suggested that the unmeasured underground flows (Tab. 3.1.1–3.1.4) are partly involved in the process of water deficiency in Lake Kinneret. The long term decline of rainfall together with the decline of the "Runoff Coefficient" (RC) (runoff discharge[mcm/y]/rainfall [mm/y]) explain the total decrease mentioned above. Halpert and Ben-Zvi (2001) documented long term decline of rainfall and increasing level of evaporation together with the high levels of extreme regimes (drought and flood) in the Kinneret drainage basin. Nevertheless water utilization northern to the Huri Bridge station was not intensified . On the contrary, as a result of droughts (1998-2002) water supply for agriculture was reduced and crops with high water demands for (like Avokado, Apple trees, fish-ponds) were restricted.

The total supply of water for irrigation in the Hula Valley declined from 1997 (32.5 mcm/y) to 2002 (22.7 mcm/y) by 30%. Prior to 1998 the total supplied water for agriculture and drinking northern to the Huri Bridge Station was about 130-150 mcm/y and during 1999-2001 it was 100-110 mcm/y. Fishponds total area during the 1970's was 1700 ha and presently-360 ha, Avokado and Apple Tree crops area was dramatically restricted too. Linear regression between "RC" and years indicate 40% of significant (R^2=0.233, p=0.0212) decline: from 1.1 to 0.7 mcm/mm/year. The significance of these data is that more measured inflows from the northern headwaters does not influx into Lake Kinneret. Linear regression between Kinneret available waters in summer (April-September) and the winter (October-March) RC values indicated significant (R^2=0.518, p<0.0001) positive relations: the lower the winter RC is the lower is Kinneret available waters in summer.

Linear regression indicated that since 1980 there is a negative significant relation (R^2=0.239, p=0.0209) between water deficiency in Huri Bridge station and winter RC value: the lower RC is the higher is the deficiency. Most of the hydrological studies in the Hula Valley referred to the upper 8 m. During the last 30 years several deep drills were carried out as well. The monitoring of GWT level in the upper 4 m is carried out routinely on a monthly basis and in recent years continuous monitoring in 6 drills are done as well whilst the deep water layers (>4-8 m) are not done routinely. Preliminary observations (E. Yasur person. Commun.) suggest that in early winter quantities of water required to increase moisture level to "Field Capacity" is increasing. Fig. 3.1.2 indicate that during droughts when water utilization is restricted, the loss of water between the measured station of the northern headwaters and Huri Bridge is increasing and GWT decline (Gophen et al. 2003). In early 1990's professional committee (Ben-David 2001)concluded that the minimum water quantity needed for the peat soil irrigation in order to maintain reasonable income, slow down of soil deterioration process by GWT at 0.5-1.5 m below surface and high level of moisture, is 1000 m^3/0.1 ha/year, excluding rain. Due to the severe drought during 1998-2002 this conclusion was renewed but because of water scarcity it was lowered to 800 m^3/0.1 ha/year. The final implementation was even lower and therefore water loss was enhanced and agricultural crops declined.

3.1.9 Conclusive remarks

The Hula Project as part of the whole valley is an example of constructed wetland based on natural components which were partly introduced by man and partly renewed naturally. The maintenance of this agro-ecological system depends upon cooperation between farming, nature preservation and water quality protection. The resulted dialog between the partners is good and the future prediction is promising if collaboration will be strong. The water supply for irrigation and other uses in this region should be significant both for the appropriate management of the peat soil and for the national water management to save water and protect the quality of Kinneret waters.

Acknowledgements

Data sources for the information presented in this paper are: water chemistry-the analytical Lab. MIGAL- I. Bar-Ilan and G. Melman; Water discharges-Water Management and Drainage Cor., Upper Galilee Municipality-H. Milard and M. Simchai; Macrophytes and Birds monitoring- Nature conservation and National parks, D. Kaplan, T. Oron, T. Niv and Y. Vaadiya; Water table level monitoring-MIGAL, upper Galilee- M. Meron, Y. Tsipris and V. Orlov; I am grateful to collegues and technicians who contributed valuable support in sampling and chemical analysis, especially E. Yasur (MIGAL) and E. Naim (National Jewish Fund). The research project was funded by Water Commission of Israel and Jewish National Fund (Keren Kayemet LeIsrael).

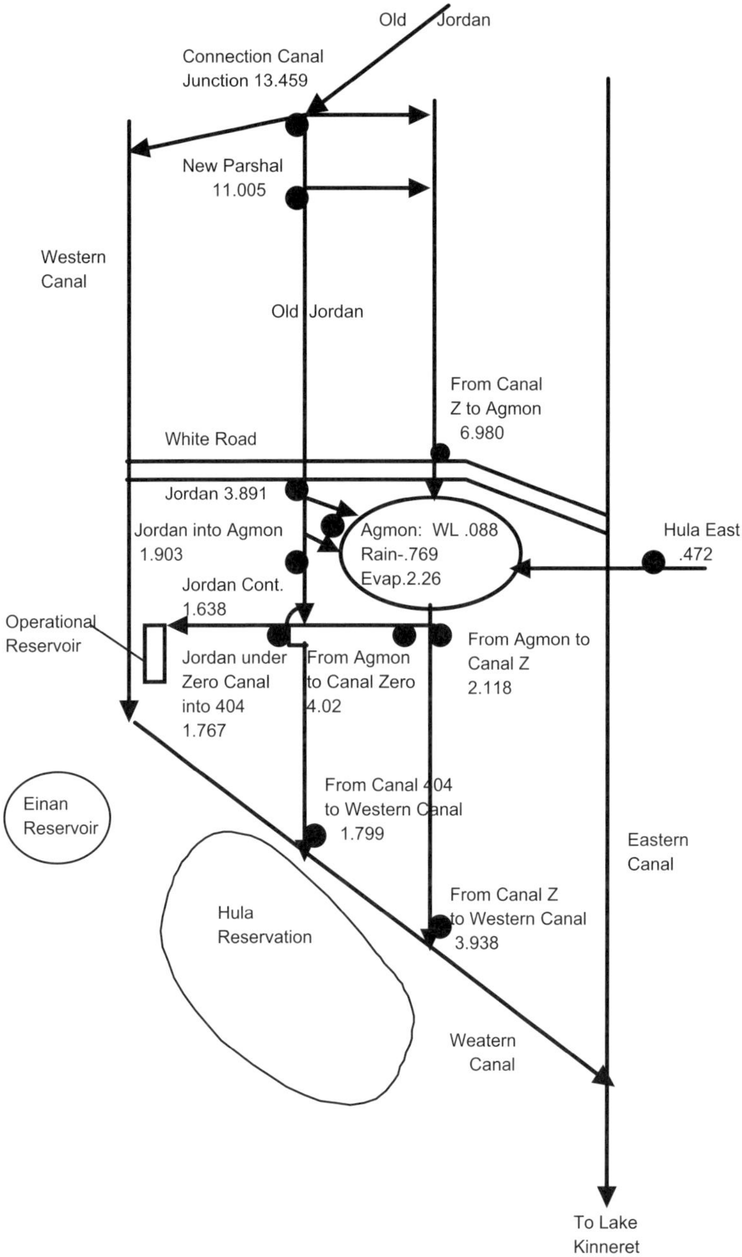

Fig. 3.1.1. Schematic map of Hula Valley with major drainage canals, Jordan canals, reservoirs, Lake Agmon and Hula reservation. The annual discharges in canals and Lake Agmon annual evaporation and rainfall are indicated (mcm/y).

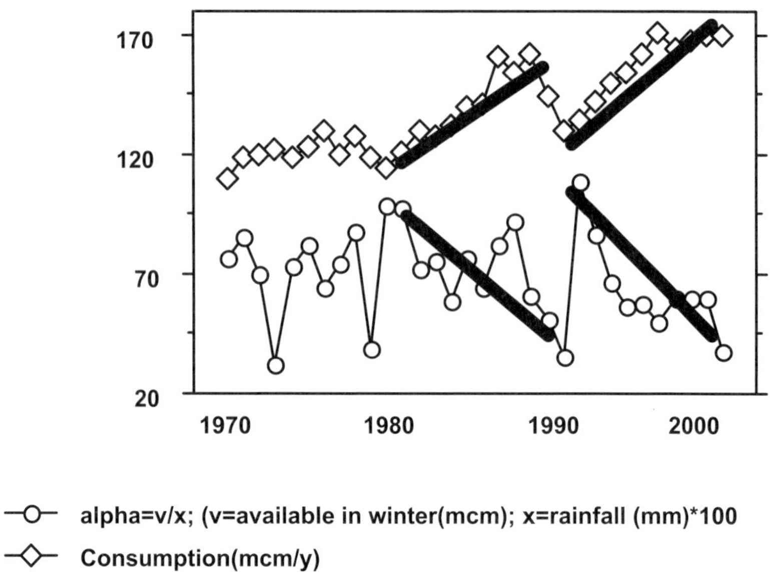

Fig. 3.1.2. Annual (1970-2001) changes of water consumption northern to Huri Bridge (mcm/y) (Berger 2001) (open cycles), and RC coefficients calculated as follows: Kinneret winter balance=inflows(mcm/y) minus evaporation divided by rainfal (mm/y)x100. Trends during 1980-2001 are indicated (solid line).

References

Alon, D. &.Y. Yom – Tov. 1999-2001. Cranes in the Hula Valley. In Hula Project annual report (M. Gophen ed.). pp 72-86.

Alpert, P. & A. Ben-Zvi, 2001. The impact of climatological changes on availability of water resources in Israel.Wate-Water engineering, October, pp.10-15.(in Hebrew).

Ashkenazi, S., Markel D., & Kaplan D. 1999. The catastrophic decline of cattail *(Typha domingensis)* in Lake Agmon: Possible machanisms and remedial measures J. Mediter. Eco. 1: 85-100.

Ben-David,y. 2001. The impact of water price reform on the Upper Galilee region. Memorandum. 8 p.(in Hebrew).

Berger. D. 2001. Assesment of natural discharges in the upper drainage basin. . Mekorot, Water Supply Co,. Drainage basin unit, 14 p. (in Hebrew).

Dimentman, H., Bromley H. J. & Por F. D.. 1992. Lake Hula. Israel Academy of Sciences and Humanities. 170 p.

Geifman, Y., Dexter H., & M. Shaw. 1970-1998. Nutrient fluxes into Lake Kinneret, Annual Reports. Mekorot, Water Supply Co. Nazareth, Jordan District, Catchment area monitoring Unit. (in Hebrew).

Gophen, M., 2000. Nutrient and plant dynamics in Lake Agmon Wetlands (Hula Valley, Israel): a review with emphasis on Typha domingensis (1994-1999). Hydrobiologia 00: 1-12.

Gophen, M., Y. Tsipris, M. Meron & I. Bar-Ilan. The Management of Lake Agmon Wetlands (Hula Valley, Israel. Hydrobiologia, (in press).

Kaplan, D. 1997 - 2002. Submerged Macrophytes in Lake Agmon. In: Hula Project. Annual Report. (Gophen, M. & D. Levanon, eds.). pp. 85-96 (In Hebrew).

Kaplan, D., Oron T., & Gutman M. 1998. Development of macrophytic vegetation in the Agmon wetland of Israel by spontaneous colonization and reintroduction. Wetland ecology and management, 6 (2-3) 143-150.

Markel, D., Sas E., Lazar B. & Bein A., 1996; 1997. Biogeochemistry. In: Hula Project. Annual Reports (Gophen, M. & D. Levanon, eds.). pp. 194-213; 33-37 (In Hebrew).

Markel, D., Sas E., Lazar B. & Bein A. 1998. Biogeochernical evolution of a sulfur-iron rich aquatic system in a reflooded wetland environment (Lake Agmon, northern Israel). Wetland ecology and management, 6 (2-3) 103-120

Miano, S.L., & DeBusk W.F. 1999. Effects of phosphorus enrichment on structure and function of Sawgrass and Cattail communities in the everglades. in: Phosphorus biogeochmistry in subtropical ecosystems (K.R. Reddy, G.A. O'Connor and C.L. Schelske eds.)Lewis Publishers pp.275-299.

Nishri, A. & M. Gophen, 2002. Lake Agmon: phytoremediation and chemical balances. GLOWA- annual report. 7 p.

Oron, T., Y. Va'adya & E. Shay 1996 – 2001. Birds in the Hula Project. In: Hula Project annual reports (M. Gophen & D. Levanon, eds.), pp83-121.(in Hebrew).

Rom, M. 1999. The Kinneret water catchment: trends and chages in nutrient contribution. Mekorot, national Water supply co. Water catchment unit, Jordan District. Special Report, 63 p. (in Hebrew).

Rom, M. 2001. New assessment of the Kinneret recharging and its implications on the water potentialthrough the Ntional Water Carrier. Mekorot, Water Supply Co,. Drainage basin unit, 14 p. (in Hebrew).

Schanzis, B. 2001. Present situation of Kinneret Water availability and consumption in upper drainage basin. Proceedings of the Israeli association of Water resources annual meeting. Kfar Giladi, May 2001. pp 159-164.(in Hebrew).

Smith, C. S. & Adam M. S. 1986. Phosphorus transfer from sediments by *Myriophyllum spicatum*. Limnol. Oceanogr. 31: 1312-1321.

Starmach, K. 1996. P excretion by *Larus ridibundus* and *Mallard platyrinchos*. Ekologia Polska Vol. 44 No.3-4. pp. 289-297.

Wetzel, R. G. 1983. Application of predictive models and data. In: Limnology. Sounders College Publ. Pp. 292-295.

3.2 Regionalization of Hydrogeological Data for Use in Groundwater Modelling: Application to “Insel Hengsen” and at Northern Algeria

Blel Azouzi and Jürgen Schröter

Universität Bremen

3.2.1 Introduction

Groundwater is, particular in the developing countries, one of the most important natural drinking water resources. Therefore good knowledge of the complex hydrodynamic mechanism of the groundwater flow is an important basis for the investigation and exploitation of this resource. The aim of this paper is to present two applications of parameter estimation methods and regionalization procedures for the development of boundary conditions as input for regional hydraulic models.

The first example describes the application of conventional geostatistics for aquifer boundaries as an input for a groundwater model of the “Insel Hengsen” in Schwerte, FRG. The hydraulic properties of the aquifer were estimated from detailed bore log description data, using hydraulic conductivity distributions based on open literature data for the very heterogeneous sediment structures. For the calibration and validation of the model parameters a groundwater model was build, using PMWIN 5.0 / MODFLOW96 (Chiang and Kinzelbach 1998). The model results were compared and calibrated to an existing FE-model of the area (Schöttler and Sommer 1988).

The second example illustrates the application of Fuzzy-Kriging for the enhancement of the standard Kriging error of a Transmissivity map of the Mitidja area (Algeria). The given dataset of measured transmissivities was supplemented by transmissivity range estimates derived from imprecise hydrogeological information. Since in the context of the geostatistic procedures the hydrodynamic parameters are considered to be random variables, mostly very irregularly distributed, they can be regarded as fuzzy data in the sense of the Fuzzy logic, which gives the opportunity to incorporate additional expert knowledge based information to fill in blank areas.

3.2.2 Application 1: Insel Hengsen (Schwerte)

The investigation area "Insel Hengsen" (a drinking water supply station of the Dortmund energy and water supply GmbH (DEW)) is located in Schwerte near Dortmund (FRG) (Fig. 3.2.1).

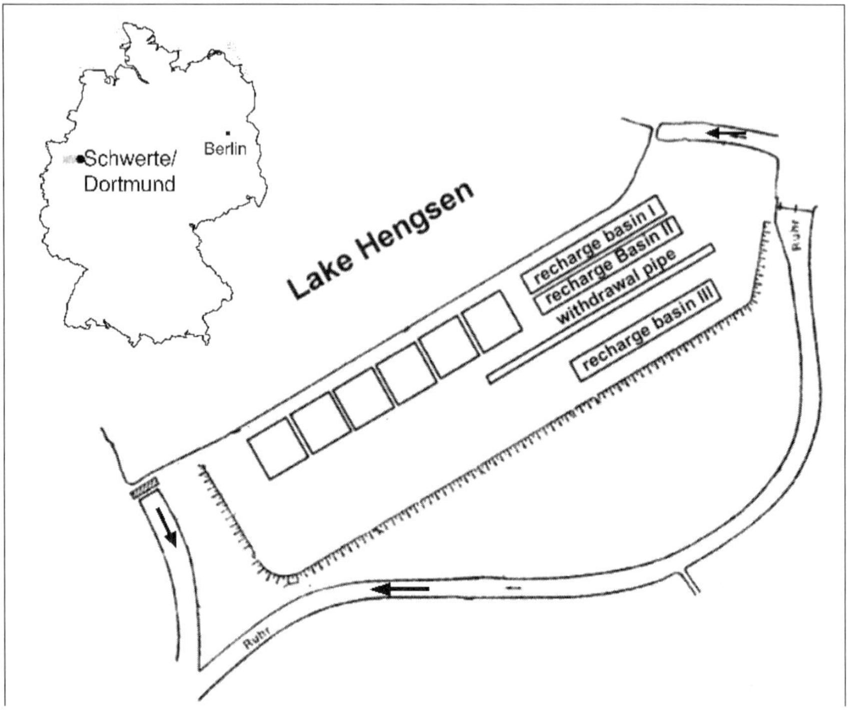

Fig. 3.2.1. Location of the study area at Insel Hengsen (IfW Schwerte, after Schöttler and Sommer, 1988).

Vertical model delimitation

The upper and lower boundary of the aquifer Hengsen were determined from layer listings (drilling data) and regionalized by means of Kriging. A statistical analysis and Variographie of the data were performed with the programs GEOEAS (Englund and Sparks 1991) and VARIOWIN (Pannatier 1996). The experimental Variogram serves the description of the spatial variability of a measured variable. The result of the regionalization of the upper and lower boundaries was used in the groundwater model.

Hydraulic conductivity

The aquifer consists of gravel and sand with fine grained lenses of the river Ruhr showing obvious inhomogeneities in different scales. The hydraulic conductivities were estimated by means of the lithostratigraphy of the monolayers from the sediment description based on literature data. Consequently the transmissivity was determined for the layer model (Daly 1982; Spitz and Moreno 1996; Davis and Dewiest 1966; Morris and Johnson 1967).

The hydraulic conductivity averages at 4.4 E-3 m/s with a standard deviation of 1.7 E-4 m/s. The Variogram of the hydraulic conductivities and their logarithms as well show pure Nugget effects suggesting the spatial correlation of the data below the sampling distance of the drilling grid. Therefore the "Inverse Distance to a Power (=2)" interpolation was used.

Results

Using four different boundary conditions documented for the existing FE-model of the Schwerte Institute for Water Research with different operating quantities (recharge and/or withdrawal) for every recharge basins as well as for the discharge area the hydraulic conductivities were calibrated.

The comparison of the results of the hydraulic heads of these simulations with the results of the Schwerte Institute for Water Research clarifies the quality of the stationary model calibration (Fig. 3.2.2).

The comparison of the relative frequency distributions of the logarithmical K_f-values of the finite elements from the Schwerte model with the relative frequency distribution of the values of the hydraulic model before and after the calibration provided by PMWIN was accomplished with the following result in all four variants: the values of the FE model show a log-normal distribution (average value -2.20 m/s standard deviation of 0.51 $(m/s)^2$ (Fig. 3.2.3 middle). The K_f-values regionalised with the inverse Distance method, used in the model, show an irregular distribution with an average value of -2.30 m/s and a narrow standard deviation of $0.17 (m/s)^2$ before the calibration (Fig. 3.2.3 left) and a shift to an average value of -2.60 m/s with a standard deviation of 0.26 (Fig. 3.2.3 right).

The average k_f-value shifts during calibration from 5.0 E-3 m/s to 2,5 E-3 m/s, i.e. the changes lie within a factor of 2, which holds for the changes of minimum and maximum of the data as well:

K_f before the calibration	1,0 E-3 - 9,3 E-3 m/s
K_f after the calibration	4,6 E-4 - 7,1 E-3 m/s.

Both ranges are substantially smaller than those used in the finite element model, which is 1,5 E-4 - 3,3 E-1 m/s.

The difference of the two models is not directly comparable, since the original model from Schwerte consists of irregular nodes spacing.

The groundwater measuring points show a very irregular distribution, they are grouped in infiltration and withdrawal zones, so that the calibration of the field is limited. The continuous long time abstraction of water might lead to a substantial washout of fine grain and the local existence of very high hydraulic conductance in the abstraction area which is not represented in the bore sample based estimates of the K_f-values.

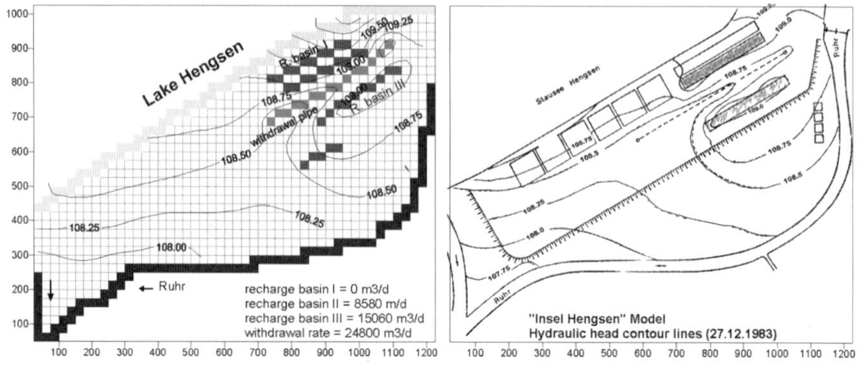

Fig. 3.2.2. Comparison of one result of the four model simulations from the FE-model (IWF Schwerte, SCHÖTTLER and SOMMER, 1988, right) with our MODFLOW simulation (left).

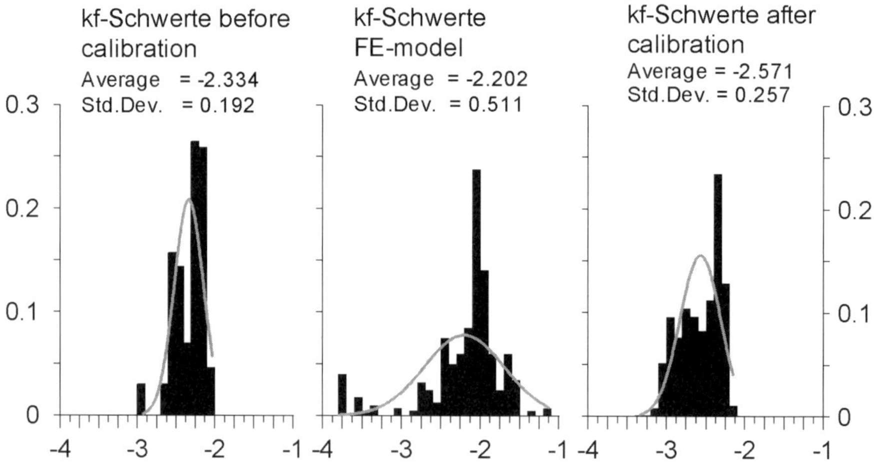

Fig. 3.2.3. Comparison of the frequency distribution of the K_f-values, computed by PMWIN, before (left) and after (right) the calibration with the existing groundwater model of the institute for water research at Schwerte (middle).

The accomplished simulations show, that the estimation from the profile description can give K_f-values which are good start conditions for the calibration, at least in a simple situation as given in this example.

3.2.3 Application 2: Mitidja

The investigation area Mitidja, which is the most expanded, sublitorale, Algerian plain and covers approximately 1300 km^2, stretching 12-18 km southwards to the Atlas, from the river Boudouaou, the eastern borderline, over 100 km westwards to the basin of the river Menaceur over Hadjout, is located in the centre of North Algeria. It is limited by small Sahel from the north and the east (Fig. 3.2.4).

Hydrogeology

The Mitidja is built up of two important aquifer reservoirs, which are separated by the impermeable marl layer of El Harrach on almost the entire territory with an exception in the east:

- the aquifer Astien, consisting of calcareous sand,
- the overlaying quaternary alluvium, the aquifer Mitidja, consisting of loam and gravel deposits of different thickness and expansion.

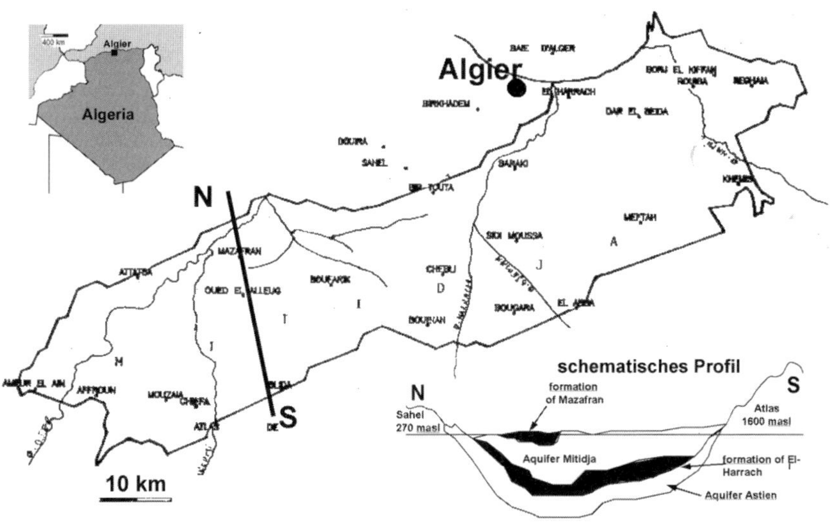

Fig. 3.2.4. Location map of the study area Mitidja (Northern Algeria).

The lower boundary of the aquifer Mitidja is formed by the marl of the formation of El-Harrach (Fig.3.2.4). In some places in the east of the territory these marl are not found, so that both aquifers are in contact. With an exception of the Mazafran zone, where it is confined, most of the aquifer is unconfined. It is limited to the west by the loam formations of Haloula, the eastern border consists partially of the marl of the Piacencium and its southern border is shaped by the layers of the Atlas (Bennie 1983).

For our investigation we have a dataset with 90 transmissivity measuring values, which were assembled by the ANRH (Agence Nationale des Ressources Hydriques) from 1968 to 1978, and which are distributed irregularly in Mitidja aquifer.

The sediments deposited by the large rivers Djer, Bouroumi, Chiffa and Mazafran show generally higher Transmissivities around 23 E-4 m^2/s than those deposited at the Piedmont mountains (T = 33 E-4 - 77 E-4 m^2/s). In the eastern part of the area near Harrach Hamiz and in Rouiba, i.e., in the contact zone of the aquifers of Astien and Mitidja, the transmissivity is increased up to E-2 m^2/s (Azouzi 1993). It is possible that these transmissivity result from the contact of the two aquifers, since the thickness in this zone is large.

Fuzzy geostatistics

The fuzzy-logical principle was introduced in 1965 by Lotfi A. Zadeh (Berkley, USA) as an extension of binary mathematics and gradation of bivalence between 0 and 1 (Kosko 1993). Thus, a fuzzy set or fuzzy number A is defined with the help of a characteristic function F_A between 0 and 1, which is designated as Membership function (Bothe 1993). Then the set A of the pair (x, FA(x)) with

$A = \{(x; F_A(x)) \mid x \in X, F_A(x) \in R\}$ is a fuzzy set on X (a set of elements x) of the membership function $F_A(x)$,

where:

$$F_A(x) = \begin{cases} 1 & x \subset R \\ 0 & sonst \end{cases} \tag{1}$$

The membership function was characterized by three values estimated subjectively by an expert, **a.** the most likely value, **b.** the lowest possible value and **c.** the highest possible value (Fig.3.2.5).

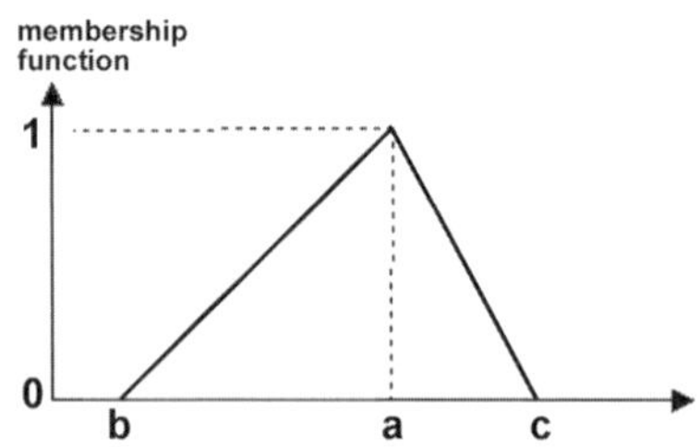

Fig. 3.2.5. Parameter of fuzzy number.

In order to improve the spatial estimation of a variable, experts have the opportunity of adding a number of fuzzy data to real measurement values (crisp data) on the basis of their subjective knowledge by means of this method (Bandemer and Gottwald 1993). The principle of the Fuzzy logic as the means of interpolation of the hydraulic geological parameters was used e.g. by Bardossy et al, 1989 and Piotrowski et al, 1997. Fuzzy Kriging is an extension of the normal Kriging. The computation and the consideration of the spatial structure through the Variogram are exactly the same as for the normal Kriging. Depending on the different types of data input and variograms 3 different Fuzzy Kriging Types defined (Bardossy et al. 1989; Piotrowski et al. 1996):

- fuzzy kriging type 1: with both crisp and fuzzy data and crisp (no fuzzy) variogram,
- fuzzy kriging type 2: with crisp data and fuzzy variogram,
- fuzzy kriging type 3: with both crisp and fuzzy data and fuzzy variogram.

Fuzzy kriging type 1 was selected for the computation logarithmically transformed transmissivities in this study. Since the Fuzzy numbers are defined by the membership function, and the membership function is fixed by α-cuts (a, b and c in Figure 3.2.5, $\alpha \in [0, 1]$), the fuzzy kriging is performed on the basis of the extension principle of the normal kriging via formed intervals ($I_1 \ldots, I_n$) of the fuzzy variable ($Z(x_1), \ldots, Z(x_n)$) between 0 and 1 of the membership function. Thus the Fuzzy variogram (2) and the main fuzzy kriging equation (3) are formed:

$$2\delta(h, I_1, \ldots, I_n) = \{y \in R \forall \exists (x_1, \ldots, x_n) \in (I_1, \ldots, I_n) : y = 1/2n(h) \sum [Z(x) - Z(x + h)]^2\} \quad (2)$$

$$Z^*(x, Z(x_1) \ldots Z(x_n)) = \sum \lambda_i \cdot Z(x_i) \quad (3)$$

Regionalization of the Transmissivity

For the application of the Fuzzy Geostatistics during the regionalization of the transmissivity the program FUZZEKS 1.0 developed by Bartels (1997) was used. First an interpolation of the log-transformed measurements was performed by means of the Kriging procedure on a mesh of 150 x 76 cells of 500 x 500 m cells. The regionalized transmissivity is represented in figure 3.2.6. The experimental Variogram which was computed only with exact measurement data, increases on a range of 10 km and then is stable and constant (Fig. 3.2.6). The Variogram model is spherical with a nugget effect = 0.13 and a sill = 0.29 $(m^2/s)^2$. These Variogram parameters are very similar to the Fuzzy Variogram parameters.

High transmissivity values are located in the northeast (Rouiba) and in the zone Mazafran. The high values in the east are probably caused by hydrodynamic exchange in the contact zone between the alluvium aquifer of Mitidja and that of Astien. Around Mazafran the increase of the Transmissivity can be explained by the higher permeability of the aquifer sediments. Particularly in the central part and in the west of the area strongly varying Transmissivity can be found, which result from irregular data distribution. The lack of observations and data concerning the entire investigation area affects directly the quality of the interpolation. The Kriging variance of the regionalised log-T-values close to the measured values ranges below 0.05 $(m^2/s)^2$ and rises to 0.15 $(m^2/s)^2$ in extrapolated external areas.

In order to improve the regionalization inside the area with lower data density, in points with highest Kriging variance 20 data were added as Fuzzy numbers. For each point 3 parameters a, b and c of a fuzzy number (Fig. 3.2.5) were estimated on the basis of the membership function and geological and hydraulic geological knowledge of the investigation area.

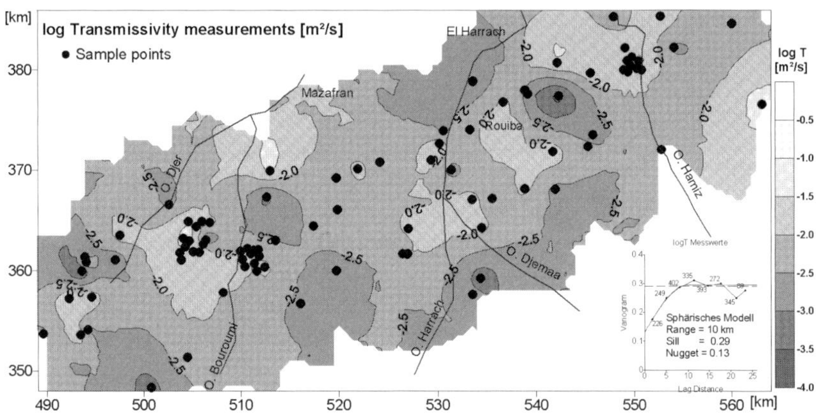

Fig. 3.2.6. Contour lines of Log transmissivity estimated by Kriging

Since a small number of additional fuzzy data points does not affect the spatial structure of a parameter very much, the new fuzzy variogram (Fig. 3.2.7) is very similar to the Variogram of crisp data (same range = 10 km, nugget effect = 0,1157 and sill = 0,269 $(m^2/s)^2$). The new fuzzy contour lines of the transmissivity differ little from the normal Kriging contour lines (using no Fuzzy data) except in the regions where Fuzzy data was used, particularly in the southwest of the investigation area. The Kriging variance is strongly reduced by the introduction of the additional fuzzy data in this area. It decreases in this computation below 0.05 $(m^2/s)^2$.

A characteristic of this method is the estimation of the lowest and highest possible values in form of a contour lines map. In the context of environmental questions it is an important means for the simulation of the different scenarios of the groundwater dynamics and the material transfer in the groundwater (Piotrowski et al. 1997). The probable deviation of the transmissivity in the Mitidja is represented by the difference map of the estimated lowest and highest possible values in Figure 3.2.8. Larger local deviations are especially in the southwest area (Fig 3.2.8).

3.2.4 Summary

Groundwater models are suitable means for the description of the spatial and temporal attitude of the flow of the groundwater with consideration of the water exchange with surface waters, the withdrawal of groundwater and interaction of geological, hydraulic geological and meteorological conditions.

The investigation at "Insel Hengsen" area describes the generation and regionalization of hydraulic conductivities from the hydrogeological interpretation of borehole log description and the regionalization of the upper and lower model boundaries for a numerical hydraulic model. The accomplished simulations indicate that the estimation from the profile description can give K_f-values which are good start conditions for the calibration of hydraulic models, at least in simple situations as given in this example.

The aim of the investigation in Mitidja is the reconstruction of a Transmissivity distribution by integration of fuzzy information about geological structure and description of sediment of an aquifer in the north of Algeria. By addition of the hydrogeological knowledge of the Aquifer, in the eastern part of Mitidja, as well as that of Harrach Hamiz and a part of Mazafran, intervals of Transmissivity were estimated from thickness and sediment descriptions in analogy to existing evaluations of this parameter, in order to reach a quality improvement of the implemented estimation in areas of small data density.

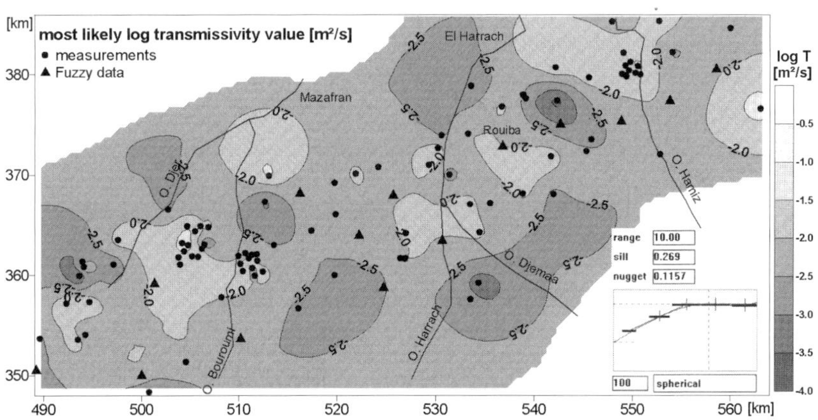

Fig. 3.2.7. Contour lines of the most likely transmissivity values estimated by fuzzy kriging

In applications, the generation and regionalization of hydraulic conductivities and geometrical boundaries for the "Insel Hengsen" model and the second study describing the improvement of Transmissivity distribution calculations by supplementation of Fuzzy numbers, examples are shown for the parameter generation for hydraulic models. Both cases are as well examples for the benefit deriving from the combination of expert knowledge combined with existing mathematical tools for the enhanced interpretation of existing data.

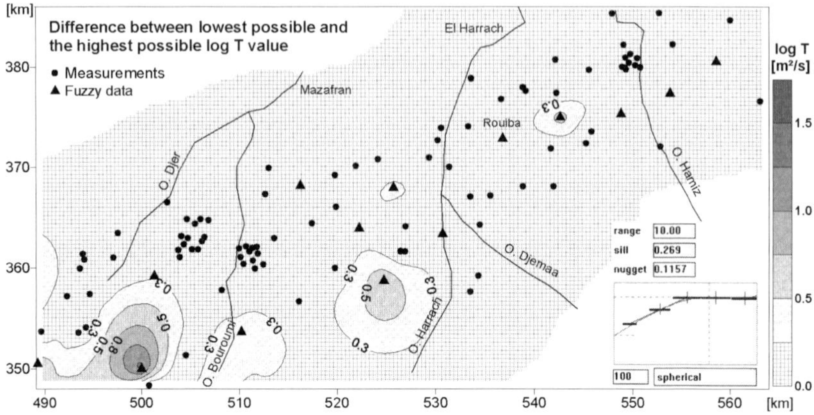

Fig. 3.2.8. Difference between lowest possible and the highest possible value of transmissivity

References

Azouzi B (1993) Contribution à la cartographie automatique des grandeurs hydrochimiques Application à la Mitidja. 163 p. Thèse de Magistère, Ed. INA d`El Harrach, Alger

Bandemer H, Gottwald F (1993) Einführung in Fuzzy-Methoden. 264 p, Akademie-Verlage, Berlin

Bardossy A, Borgardi I, Kelly WE (1989) Geostatitics utilizing imprecise (Fuzzy) information. Fuzzy-Sets and Systems 31: pp 311-327

Bartels F (1997) Ein Fuzzy-Auswertung- und Krigingsystem für raumbezogene Daten. 92 p, Dipl. Arbeit, Christian-Albrechts-Universität Kiel

Bennie P (1983) Schémas d'aménagement des ressources en eau de la région d'Alger et Sebaou. Mission B et F, Vol *4.*, ANRH. Alger

Bothe HH (1993) Fuzzy Logik, Einführung in Theorie und Anwendungen. 226 pp, Springer-Verlag Berlin, Heidelberg

Chiang WH, Kinzelbach (1998) Processing Modflow, A Simulation system for modeling groundwater flow and pollution. 346 p, User's manual.

Daly C (1982) Evaluation of procedures for determining selected aquifer parameters. Prepared for U.S Army Toxic Hazardous Materials Agency. CR REL Report pp 82-41.

Davis S, Dewiest JM (1966) Hydrogeology. 463 p, John Wiley & Sons, Inc., New York, London, Sydney.

Englund E, Sparks A (1991) Geo-Eas 1.2.1 User's Guide. 198 pp, US-EPA Report 600/8-91/008, EPA-EMSL, Las Vegas, Nevada.

Morris DA, Johnson AJ (1967) Summary of hydrological and physical properties of rock and soil materials as analyzed by the hydrologic laboratory of the U.S Geological Survey USGS, Water Supply paper 1839-D.

Pannatier Y (1996) VARIOWIN: Software for spatial data analysis in 2D. In Statistics and Computing. 91 p, Springer, Berlin, Heidelberg, New York, London, Paris.

Kosko B (1993) Fuzzy-Logisch ,eine neue Art Denkens. 352 p, Carlsen-Verlag GmbH, Hamburg

Piotrowski J, Bartels F, Salski A, Schmidt G (1996) Geostatistical regionalization of glacial aquitard thickness in north-western Germany, based on Fuzzy Kriging. Mathematical Geology *28*(4): pp 437-452

Piotrowski J, Bartels F, Salski A, Schmidt G (1997) Regionalisierung der Durchlässigkeitsbeiwerte mit unscharfen (Fuzzy) Zahlen: Der Natur näher? Grundwasser Zeitschr. der Fachsektion Hydrogeologie *2* (1): pp 3-10.

Schöttler U, Sommer H (1988) Systemanalyse quantitativer und qualitativer Grundwasserverhältnisse zur optimalen Bewirtschaftung eines Aquifers. Bohrtechnik-Brunnenbau-Rohrleitungsbau 38: pp 59-66.

Spitz K, Moreno J (1996) A practical guide to Groundwater and solute, Transport Modeling. 461 p, John Wiley & Sons, Inc., New York, Chichester, Brisbane, Toronto, Singapore

Zadeh LA (1965) Fuzzy sets: Information and control. Vol *8*, no. 3, pp 338-353.

3.3 Water Resources Management in Palestine: Political, Technical and Financial Obstacles

Fawzy Naji

DFN-GEOCONSULTING, Braunschweig, Germany

Abstract

The strategy of the water resources management in Palestine aims to satisfy the water requirements of all Palestinian citizens to secure health and economic prosperity.

Political, technical and financial obstacles are facing the effective water resources management in Palestine. The distribution of the limited water resources between Israel and Palestine is unsatisfactory and biased in favor of Israel. Water available to the Palestinians does not meet their requirements.

In regard to the water quality issue, the Palestinians are facing two major problems: the high salinity and the high concentration rate of nitrate. The concentration of chloride is up to 1,763 mg/l in Jordan Valley area, West Bank. The nitrate concentration in the domestic well in Khan Younis Refugee Camp, Gaza Strip is 600 mg/l.

The high nitrate concentration in drinking water affects infants and causes methemoglobinemia. The major causes for salinization are over-pumping of the wells, seawater intrusion and geological factors. The main sources of nitrate pollution are fertilizers, wastewater and cesspits.

To achieve an effective water resources management in Palestine, the following subjects have to be taken into consideration:

- Strengthening the peace process in the Middle East and opening the doors to effective cooperation between the different countries in the region
- Allocation of the water resources between Israel and Palestine on an equal per capita basis
- Expanding the supply and distribution networks
- Loss reduction
- Public awareness
- Storm water harvesting
- Wastewater reuse
- Seawater and brackish ground water desalination
- Modern irrigation technologies

3.3.1 Introduction

Palestine is located in a semi-arid region of the Middle East, with hot, dry summers and mild, wet winters. Rainfall varies between 100 mm/yr at the Dead Sea and 700 mm/yr in the regions of Jerusalem, Ramallah and west of Jenin. The water resources are limited and the population is growing. Therefore an economic utilization of water represents one of the most important key policies in the development strategy of Palestine.

After the Oslo II Agreement of 1995 between the Israelis and Palestinians five major problems have to be discussed within the final negotiations about the establishment of the Palestinian state. These problems are:

1. Jerusalem
2. Refugees
3. Settlements
4. Security and Borders
5. Water

3.3.2 Water Resources

Ground Water

The ground water basins lie partly in Palestine and the rest in Israel. The ground water resources can be subdivided into the following:

The Mountain Aquifer System

It extends from the foot of the Carmel Mountain near Haifa in the north to the area of Bir Alsabe' (Beer Sheva) in the south, and from the Jordan River and Dead Sea in the east to the beginning of the Coastal Aquifer in the West. This aquifer system consists of:

- The North-Eastern Aquifer which is called in Israel Nablus-Gilboa Aquifer
- The Western Aquifer which is known in Israel as Yarqon Taninim Aquifer
- The Eastern Aquifer

Geologically this Aquifer System consists mainly of limestone and dolomite of Cretaceous to Quaternary age.

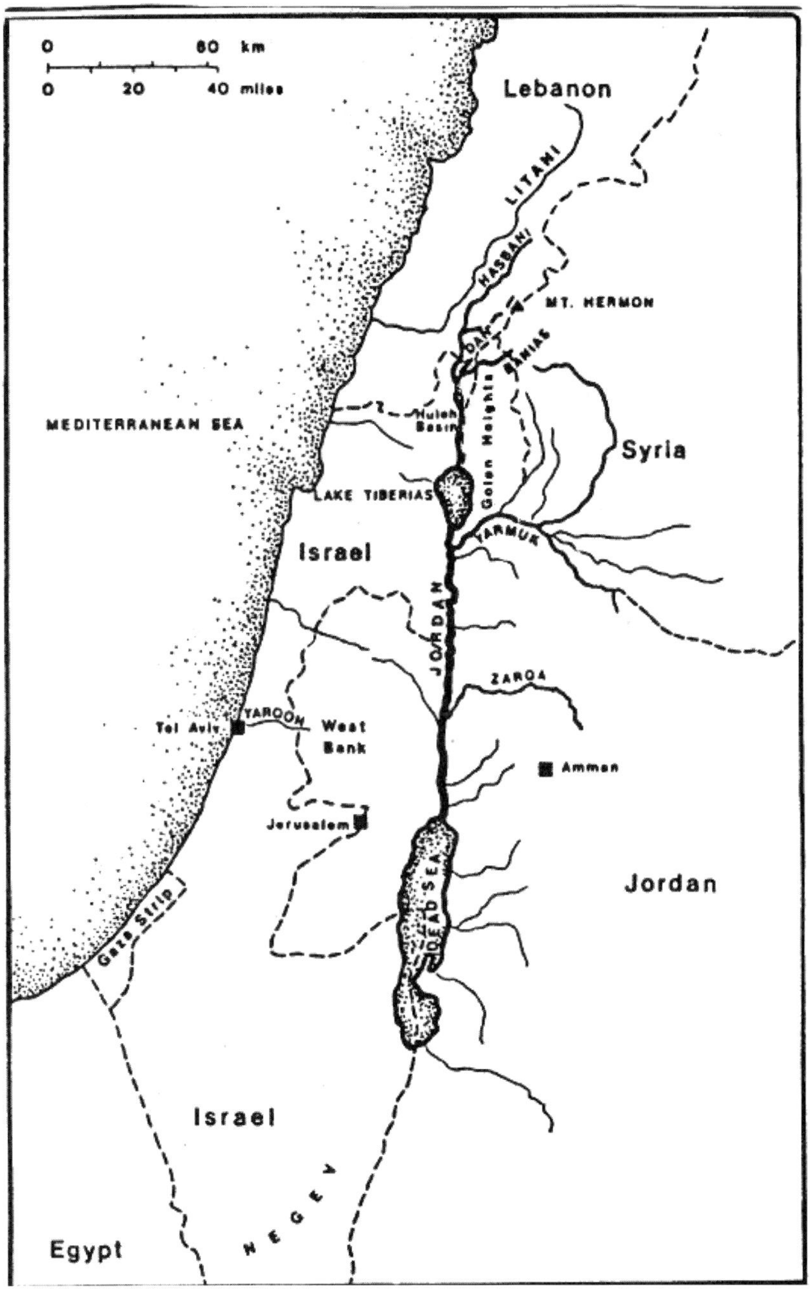

Fig.3.3.1. Location of Palestine (West Bank and Gaza Strip), Israel and Jordan River. (from Naff & Matson, 1984).

The currently annual withdrawal for Palestinian use from the Mountain Aquifer System is 118 MCM (Million Cubic Meters). This means around 17.4 % of the estimated sustainable yield of 679 MCM/year.

The Coastal Aquifer

It stretches from the slopes of the Carmel Mountain in the north to the northern Sinai in the south, and from the foothills of the mountains in the east to the Mediterranean Sea in the west. This aquifer consists mainly of sand, sandstone and pebbles of Pliocene to Pleistocene age. Its safe yield is around 300 MCM/year. Around 70 MCM/year is the total recharge of this aquifer in the Gaza Strip.

The Mountain and Coastal Aquifers lie in Israel and Palestine. The effective management of these aquifers require the cooperation between the Israelis and the Palestinians. Therefore, the recreation of the peace process in the Middle East is crucial.

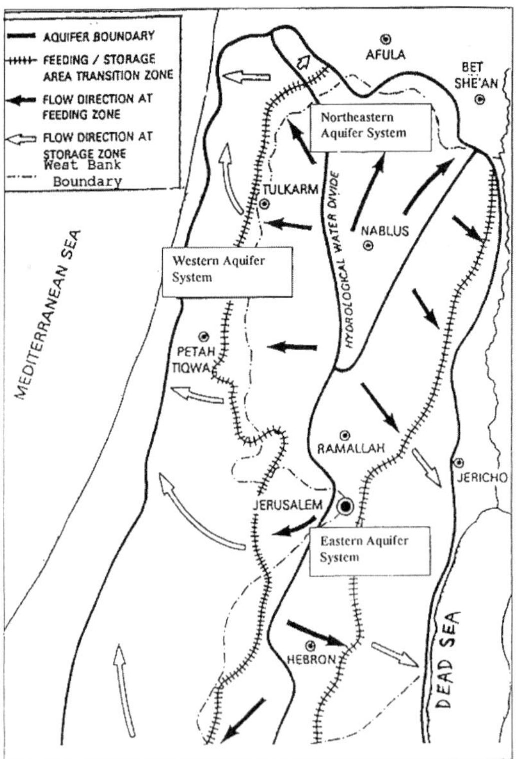

Fig. 3.3.2. The Eastern, Northeastern and Western Aquifers, slightly modified after Gvirtzman, 1994

3.3.3 Surface Water

The Jordan River Basin

It can be subdivided into the Upper Jordan with Lake Tiberias (Kinneret) as its main reservoir, and the Lower Jordan with the Yarmouk River as its main tributary. The annual water flow from the Jordan River Basin is around 1300 MCM.

"The Dan Spring, the largest of the sources of the upper Jordan, lies holly within Israel close to the border with Syria. The spring sources of the Hasbani River lie entirely within modern Lebanon. The spring source of the Banias River is in Syria" (Naff and Matson 1984).

The Jordan River is an international river. Accordingly its water has to be shared between Jordan, Palestine (West Bank and Gaza Strip), Israel, Syria and Lebanon. Since the Israeli occupation of the West Bank 1967, the Palestinians are not allowed to use their share from the water of the Jordan River. After El-Musa (1997) the share of the Palestinians from the Jordan River Basin is 215 MCM/year, i.e. 17% of a total amount of 1287 MCM/year.

Wadis: The major eastern wadis in the West Bank are Fara, Quilt and Auja. Wadi Zeimar and Wadi Qana represent the important wadis in the western part of the West Bank. The main wadi in the Gaza Strip is Wadi Gaza.

The amount of the surface runoff into wadis depends on the intensity and duration of the rainfall. The annual water flow from wadis varies between 70 – 100 MCM.

Dead Sea: It is not allowed to the Palestinians to use the water of the Dead Sea for industrial and tourist activities.

Mediterranean Sea: The Palestinians use of the Mediterranean Sea for fisheries and transport is restricted from the Israelis.

Springs and Seeps

There are 297 springs and seeps in the West Bank, of which 105 springs with a yield of more than 0.1 litre/sec. The average annual flow is around 100 MCM. The fresh water amount is 55 MCM and mainly used for irrigation. The rest is brackish water.

Cisterns

For storm harvesting there are around 80,000 cisterns in Palestine with additional 6-7 MCM/year of water for domestic and livestock purposes.

Non-Conventional Water Resources: Wastewater reuse, seawater and brackish ground water desalination and water importation from water rich countries like Turkey are considered in the Palestinian water strategy. While wastewater reuse and water desalination face technical and financial obstacles, the water importation from Turkey through Syria, Jordan to Palestine, Israel, Saudi Arabia, Arab Emirates etc. can not be implemented due to the current political situation.

3.3.4 Water Quality

In the water quality issue, the Palestinians are facing two major problems, the high salinity and the high concentration rate of nitrate.

Brackish ground water is found to be present in the Jordan River Valley, West Bank and in different areas in the Gaza Strip.

After the recommendation of the World Health Organization (WHO) for drinking water, the chloride concentration has not to exceed 250 mg/l, and the concentration of the total dissolved solids (TDS) has to be less than 500 mg/l.

The sodium adsorption ratio (SAR) of more than 3 would restrict the water use for irrigation.

Chloride concentration is up to 1763 mg/l in the Jordan Valley area. In the Gaza Strip, the TDS reaches values of 3200 mg/l in Khan Younis and 4000 mg/l in Rafah. The SAR is up to 8.8 in Jericho area, West Bank.

The major causes for salinization are over-pumping of the wells, sea water intrusion in the coastal aquifer, the presence of deep saline water and geological factors. WHO recommended that the concentration of nitrate in drinking water has to be less than 50 mg/l. In the northern part of Gaza Strip, the nitrate concentration is up to 150 mg/l. After MOPIC (1996) the nitrate concentration in the domestic well in Khan Younis refugee camp is 600 mg/l.

The main sources of nitrate pollution are excessive use of fertilizers, contamination with wastewater and the presence of cess pits. High nitrate concentration in drinking water affects infants and causes methemoglobinemia.

The majority of the springs of the West Bank are suitable for drinking from the chemical point of view, but all the springs of the West Bank are usually contaminated with coliform bacteria, and are thus unsuitable for drinking unless disinfected properly (Abed Rabbo et al. 1999).

3.3.5 Water consumption

After PECDAR (2001) the present Municipal and Industrial (M&I) supply amounts are around 100 MCM/y.

The sources of supply are:

74% from wells and springs, 24% from Israeli utilities and 2% from cisterns.

Table 3.3.1. Water Consumption in Palestine and Israel 1997 in MCM (from Naji 1999 b)

Water Use	Palestine	Israel**
Domestic		571
Industrial		136
Subtotal	101 (M&I)*	707
Agricultural	177	1252
Total	278	1959

* M&I: Municipal and Industrial; **Source: Statistical Abstract of Israel, 1998

Currently the average consumption is 71 l/c/d. For the short-term horizon in the Palestinian Strategy, the assumed rate will be the recommended WHO minimum of 100 l/c/d. For the long-term horizon, the consumption rate will be the recommended WHO average of 150 l/c/d.

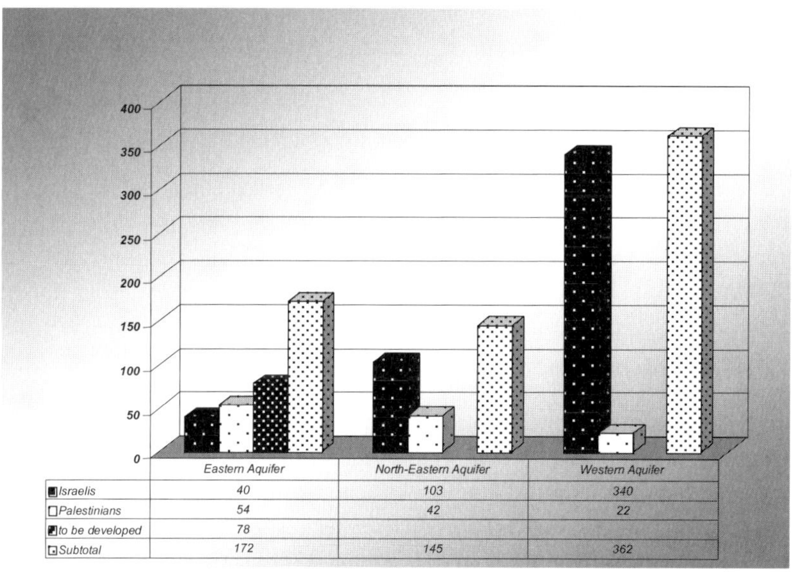

Fig. 3.3.3. Water extraction, utilization and potential of the Mountain Aquifer, after Article 40 of the Oslo II Agreement in MCM

The present agricultural supply amounts are around 170 MCM/y. Sources of irrigation water are: 70% from wells and 30% from springs.

Agriculture is currently the largest water consumer in Palestine, accounting for around 62% of total current available water supply (Naji 1999 a). Traditionally agriculture has been the major productive sector in the Palestinian economy, accounting for about 30% of GDP (Gross Domestic Product) during 1968 – 1992, and about 13% during 1992 – 1998.

Also in Israel agriculture is the largest water consumer, accounting around 64% of total water supply, but representing only around 3% of GDP (Naji in press).

There was a gap between the water available and the water demand in Palestine of around 60 MCM in 2000. This gap utilizing low quality sources will be around 144 MCM in 2020. The gap using developable fresh-water sources will be –286 MCM in 2020 (PECDAR 2001).

The traditional irrigation technologies like open channels are still used. 30 – 50 % of the water amount is lost in areas, where high evaporation rate exists. This is the case in the Jordan Valley area. The adopting of modern irrigation technologies like sprinkler and drip systems leads to:

1. Optimizing the utilization of water resources
2. Reducing the cost of agricultural activities
3. Improving the quality of agricultural products
4. Reducing the potential of pollution of water

3.3.6 Water Management

Municipal and Industrial Water Supply Systems

86% of the population in the West Bank, including Jerusalem, and 98% in the Gaza Strip are served with water supply networks. But not 24 hours per day.

Network losses ranges from 25% - 65% in the West Bank, and about 45% in the Gaza Strip (PECDAR 2001).

Wastewater Management

At present, about 24% of the total population in Palestine is served by a central public urban sewer system. About 73% of the households in the West Bank have cesspit sanitation and almost 3% without any sanitation system.

Wastewater management in Palestine has been a neglected issue over the past years. No comprehensive data on wastewater characteristics and amounts discharged are yet available. The effectiveness of the existing urban sewage collection and treatment facilities is usually constrained by limited capacity, poor maintenance, process malfunction and lack of experienced staff (Al-Saed 2000).

The technologies that are used in the wastewater treatment plants are:

1. Activated sludge in Salfit
2. Extended aeration in Al-Bireh
3. Aerated lagoon in Ramallah and Jenin

Al-Saed (2000) reported that the sewage treatment plants, entailing oxidation ditches and sludge management units are working effectively.

There are four privately owned treatment plants. Two of them are located in the district of Bethlehem, one in Jericho and the other at Birzeit University. A pilot treatment plant has been constructed in Nablus. The technologies that are tested were trickling filters and extended aeration system.

Trickling filters preceded by septic tanks have the best feature for onsite treatment in rural areas. However, utmost care should be taken towards nitrogen pollution control.

The main objective of wastewater strategic plan of PWA is within 2020, 75% of the population in municipalities and 25% of the population in areas under 5000 PE, should be connected to treatment plant.

Also 50% of the population in rural areas must be connected to appropriate treatment within 2020.

The highest priority in the Palestinian Environmental Strategy is setting up an effective wastewater management system in Palestine. This includes maximizations of wastewater collection, upgrading existing collection and sewer systems, rehabilitation or upgrading of existing wastewater treatment plants or construction of new treatment plants (MEnA 1999).

Sbeih (2001) recommended that the location of new wastewater treatment plants should be studied carefully, taking into consideration that these plants have to act as regional plants.

Obstacles

Politically, the main obstacles of the effective water resources management in Palestine are:

- The Israeli occupation of the Palestinian territories
- The Israeli military orders (92 of 15 August 1967; 158 of 19 November, 1967; 291 of 19 December, 1968; 498 of 4 November, 1974), which issued after the Israeli occupation of the West Bank and Gaza Strip 1967 and make the Israeli authority responsible for:

1. Controlling the abstraction of water resources in the Palestinian Territories
2. Setting the prices and quantities allowable for use by Palestinians
3. Changing the ownership of Palestinian water resources from private to public
4. Granting, stopping or adjusting permits
5. Preventing the Palestinians to obtain their share of the water of the Jordan River Basin

- The neglecting of the water infra structure in the West Bank and Gaza Strip during the Israeli occupation
- The establishment of the Israeli settlements in the Palestinian territories and subvention of the water prices for settlers
- The ineffective cooperation between the Palestinians and Israelis in the joint water committee, which was established after the Oslo II Agreement of 1995
- The improper water allocation between the Israelis and Palestinians

The technical obstacles can be summarised as the following:

- Expanding the water supply and distribution networks in Palestine
- Replacing the old sewage systems, which does not meet the current requirements of the growing population in the West Bank and Gaza Strip
- Reducing the losses of water, which varies between 25% - 65%
- Replacing the open channels for irrigation and using modern irrigation technologies
- Using wastewater treatment technologies of low maintenance and operation costs
- Training of the Palestinian scientists and technicians on different technologies of wastewater reuse, water desalination, water monitoring and water conservation

The implementation of the Palestinian water strategy faces the financial obstacle. Shuval et al. (1997) estimated that the cost per cubic meter of treated wastewater is US$ 0.1, to meet WHO Guideline of 1000 faecal coliform per 100 ml, and US$ 0.45 to meet USEPA / USAID Guideline of 0.0 faecal coliform per 100 ml.

The cost of water desalination varies between US$ 0.6 – 1.0 per cubic meter. The cost of water importation from Turkey is cheaper, but without peace in the region no chance exists for implementation.

It is estimated that the total cost in the water sector in Palestine until 2020 is US$ 3.8 billion (PECDAR 2001).

Political, technical and financial contributions of the international community are needed to solve the water and other problems between the Israelis and Palestinians.

3.3.7 Recommendations

To achieve an effective water resources management in Palestine the following is crucial:

- Strengthening the peace process in the Middle East which opens the doors for cooperation between all countries in the region
- Allocation of the water resources between Israel and Palestine on an equal per capita basis
- Expanding the supply and distribution networks
- Protection of the ground water systems
- Rehabilitation of wells and springs
- Drilling of new agricultural wells
- Expansion of irrigated areas
- Replacing of open water channels to closed piped systems
- Selection of crops with high tolerances to salt
- Loss reduction
- Public awareness
- Storm water harvesting

- Wastewater reuse for restricted irrigation
- Replacing of cesspits in new sewage systems
- Sea water and brackish ground water desalination
- Adopting modern irrigation technologies like sprinkler and drip systems

3.3.8 Conclusion

Currently the spirit of cooperation between the Palestinians and Israelis as partners does not exist. Without this cooperation the effective water resources management in Palestine and Israel is not possible. Therefore the recreation and strengthening of the peace process in the Middle East represents the top priority, not only to manage the water resources effectively, but also to bring security and prosperity for all nations in the region.

References

Abed Rabbo A, Scarpa D, Qannam Z, Abdul Jaber Q, Younger P (1999) Springs in the West Bank, Water Quality and Chemistry. Bethlehem.

Al-Saed R (2000) Wastewater Management for Small Communities in Palestine. Amman

El-Musa S (1997) Water in the Palestinian Israeli negotiations. (in Arabic). Beirut

Gvirtzman H (1994) Groundwater allocation in Judea and Samaria. In: Isaac J, Shuval H (eds) Water and Peace in the Middle East, Stud. Environ. Sci. *58*, Elsevier, Amsterdam, pp 205 - 218

MEnA (Ministry of Environmental Affairs) (1999) Palestinian Environmental Strategy(PES). Al-Bireh

MOPIC (Ministry of Planning and International Cooperation) (1996) Coastal Zone Plan – Regional Plan for Gaza Governorates, Ramallah

Naff T, Matson R (1984) Water in the Middle East: Conflict or Cooperation. Boulder

Naji F (1999a) Irrigation Management in Palestine. Proceedings of the Regional Symposium on "Irrigation Management and Saline Conditions". June 21st – 23rd, 1999, p. 440-447, Irbid

Naji F (1999b) Water Crisis in Palestine. Scenarios for Solutions. Palestinian Academic Society for the Study of International Affairs (PASSIA), Jerusalem.

Naji F (in press) Belastung und Behandlung des Wassers und Abwassers am Beispiel Palästina. (Pollution and Treatment of Water and Wastewater in Palestine), (in German), Proceedings of the workshop: Dialogue of cultures, held at the Technical University of Braunschweig, Germany on October 28th to November 1st, 2002. Braunschweiger Beiträge zur Wissenschaftsgeschichte, Braunschweig

PECDAR (Palestinian Economic Council for Development and Reconstruction) (2001) Palestinian Water Strategic Planning Study. Ramallah

Sbeih M (2001) Proper Management, Measures as well as laws and by-laws needed toward Effluent Treated Wastewater Reuse in Palestine. Amman

Shuval H, Lampert Y, Fattal B (1997) Development of a Risk Assessment Aproach for Evaluating Wastewater Reuse Strategy for Agriculture. Water Science and Technology, Vol. *35*, No. 11. London

3.4 Conceptual and Numerical Flow Models of the Western Aquifer - Palestine / Israel

Clemens Messerschmid

SUSMAQ – Sustainable Management of the West Bank and Gaza Aquifers; Centrum für Internationale Migration und Entwicklung – CIM.

Abstract

The flow model is created in 2 steps. First a conceptual model assesses stratigraphy, lithofacies and structure of the basin, the distribution, connections and parameters of aquifers and the cell size, boundaries and geometry of the model. In a second step, when the conceptual model is transformed into a numerical model setup, a steady state run is executed and calibrated for various parameters, most of all the aquifer properties, but also the recharge and water levels. This numerical model uses Geo-Modelling-Software (GMS) and serves as a basis of subsequent transient conditions models.

3.4.1 Introduction

The SUSMAQ project on the Sustainable Management of the West Bank and Gaza Aquifers for the first time assesses this basin. Due to the severe restrictions to Palestinian water use from this basin, the Palestinians have never thoroughly investigated its geology and hydrogeology.

Flow modelling in the Western Aquifer Basin is crucial for future water supply of the Palestinians, since this basin is not only the largest and most productive but also a shared groundwater basin between Palestine and Israel. It therefore will play a key role in future water negotiations. In addition, SUSMAQ evaluates different management options to develop several demand scenarios that are then integrated together with the results from the flow model to create a Sustainable Management Planning platform for the water resources in the Western Aquifer Basin.

3.4.2 Geography and Geology

The Western Aquifer Basin (WAB, or in Israeli terms: Yarkon Taninim Basin) extends over 11000km^2 from Mt. Carmel to the Sinai and from the heights of the West Bank to the Mediterranean coast. The model area ends at the line Gaza-Beer Sheva, where the Afiq channel, a Neogene erosion channel fully or partially dissects the aquifer. The model area covers 9000km^2 (Table 3.4.1).

The basin consists of carbonates from the Upper Albian (Lower Aquifer) and Middle Cenomanian – Turonian (Upper aquifer), separated by Lower Cenomanian clays and marls of tens to hundreds of metres thickness. The oldest formations crop out in the Ramallah and Hebron anticlines along the Eastern border, Cenomanian outcrops are concentrated in the foothills region and parts of the Negev, while in the coastal plain the aquifer system is covered and partly truncated by 100's of metres of Senonian and Eocene chalks and Neogene to Quaternary pelitic clastics (Saqiye group).

Table 3.4.1. Outcrop area of aquifers and aquitards

Aquifer Formation	Outcrop	Aquitard Formation	Outcrop
Upper Albian	292.4 km^2	Lower Cenomanian	165.7 km^2
Middle Cenomanian-Turonian	1366 km^2	Senonian – Quaternary	4210.5 km^2

In more detail, the actual situation is a complicated pattern of different lithofacies types and terminologies that up till now have not been unified completely (Fig.3.4.1).

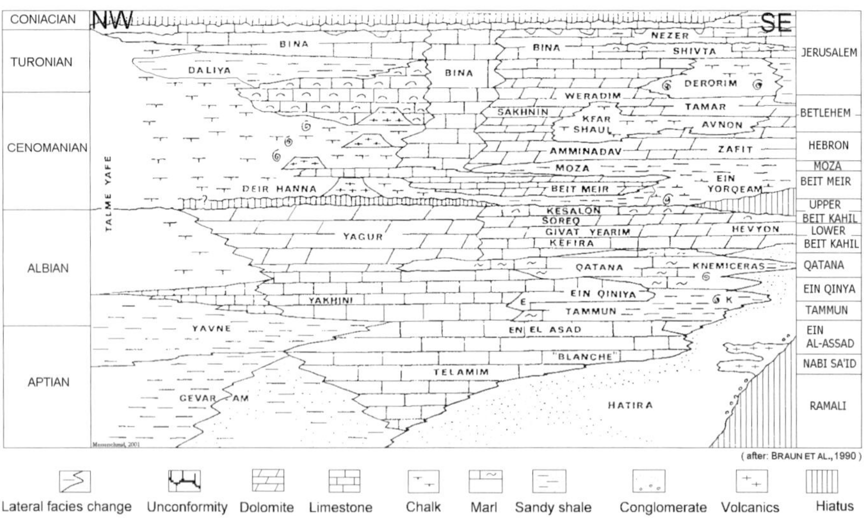

Fig. 3.4.1. Lithostratigraphic scheme of the Western Aquifer Basin

The middle aquiclude Yatta formation (or Beit Meir and Moza formations) crops out in moderate elevations (300-500m) in the West Bank and West of Jerusalem. Not all of Yatta is an aquitard. Therefore, zones of different aquifer potential were differentiated to quantify possible rainfall recharge over this formation.

3.4.3 Basin geometry

Top and Bottom of the aquifers were constructed by structural contour maps. Over 16000 data points were chosen and inserted to give a reliable geometry (Fig.3.4.2).

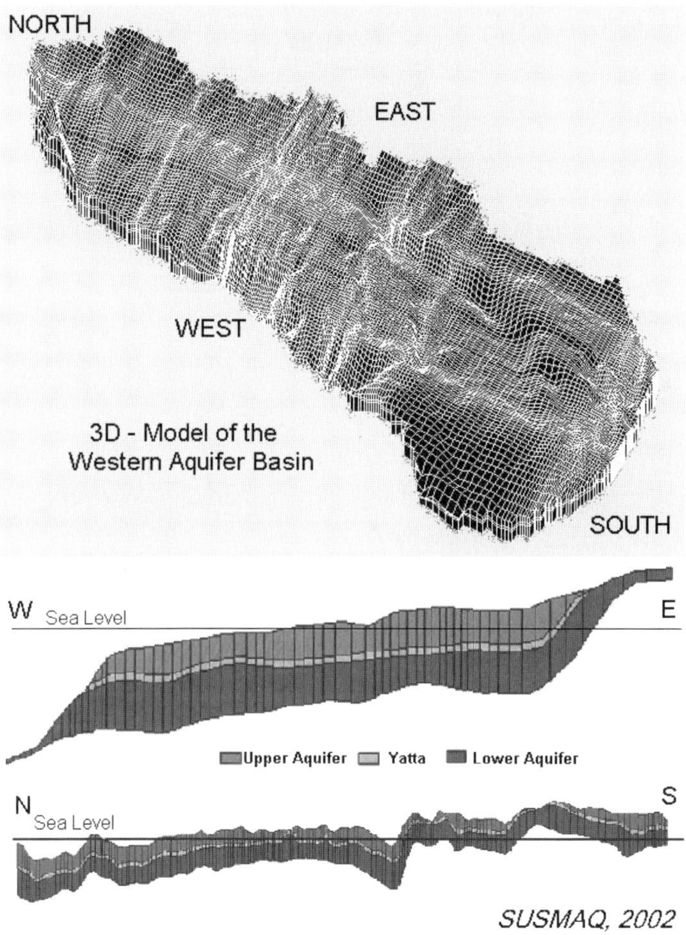

Fig. 3.4.2. Cells and Sections of the Western Basin (Geo Modelling Software)

The middle aquitard, so far was inserted at known locations (well logs, etc.) and interpolated in between. Major faults yet have to be included in the model set-up.The average aquifer thickness is around 800m, including some 100m of the middle aquitard. In the south, the thickness is reduced to only 500m. Under the coastal plain and due to the truncation by the Saqiye group, the thickness is also reduced, sometimes strongly.

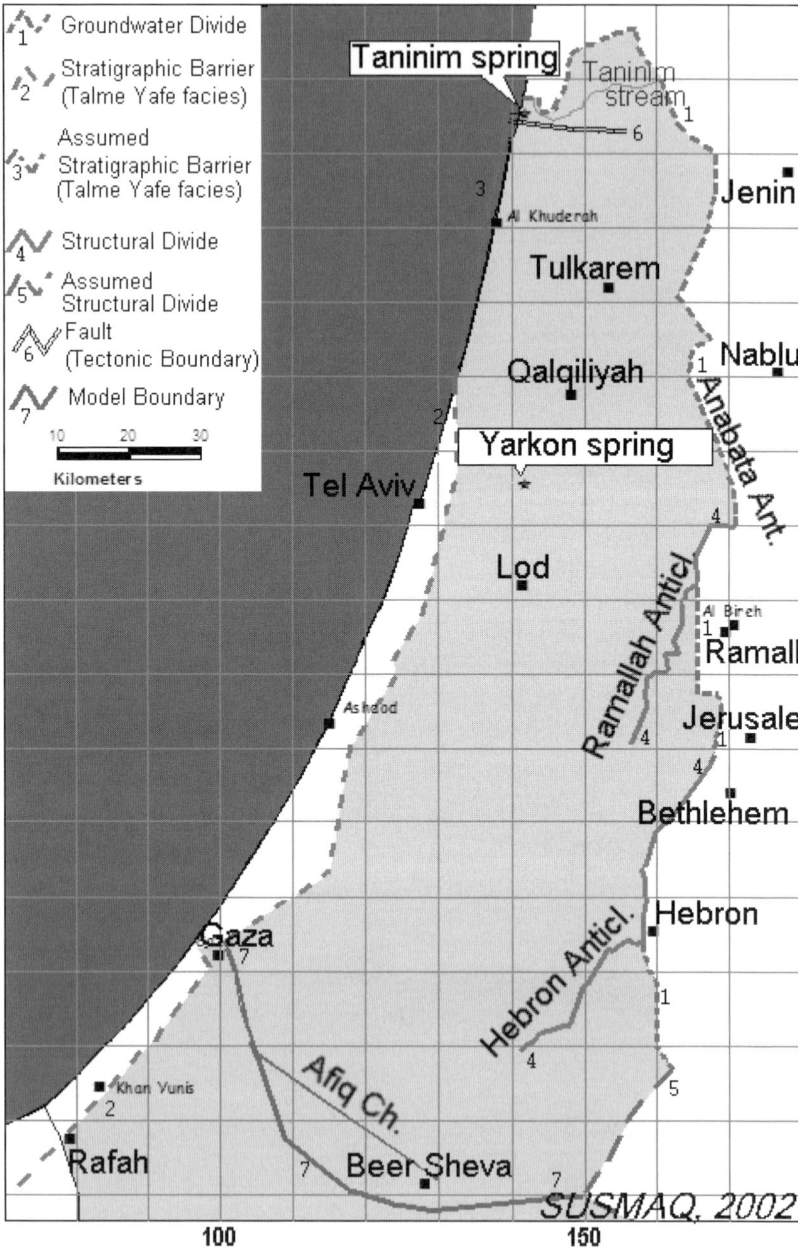

Fig. 3.4.3. Boundary Types of the Western Basin

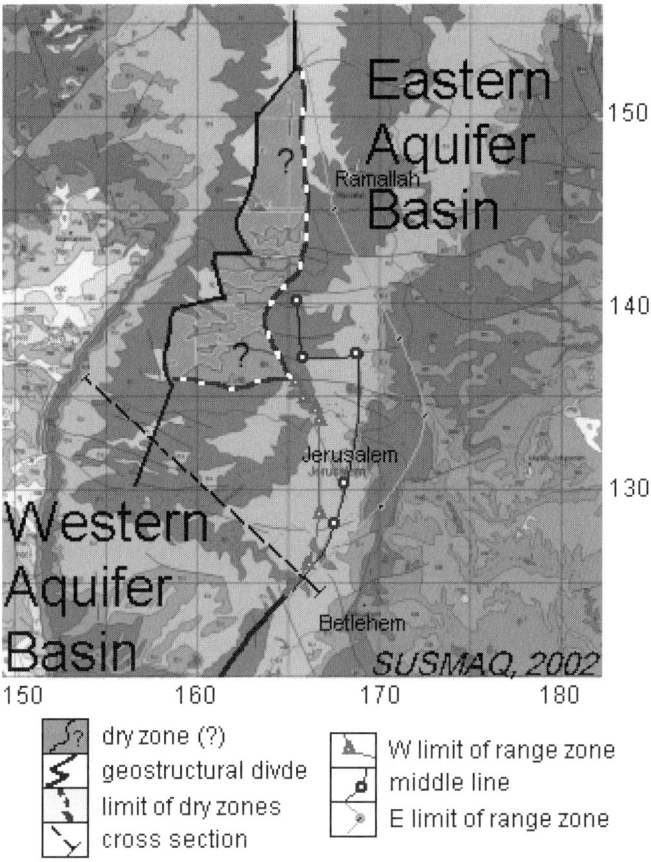

Fig. 3.4.4. Boundaries of Ein Karem Sub-basin (W-Jerusalem)

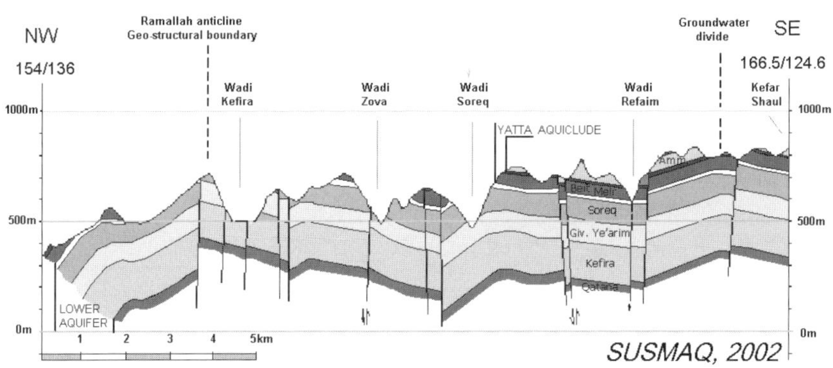

Fig. 3.4.5. Cross Section through Ein Karem Area (W-Jerusalem)

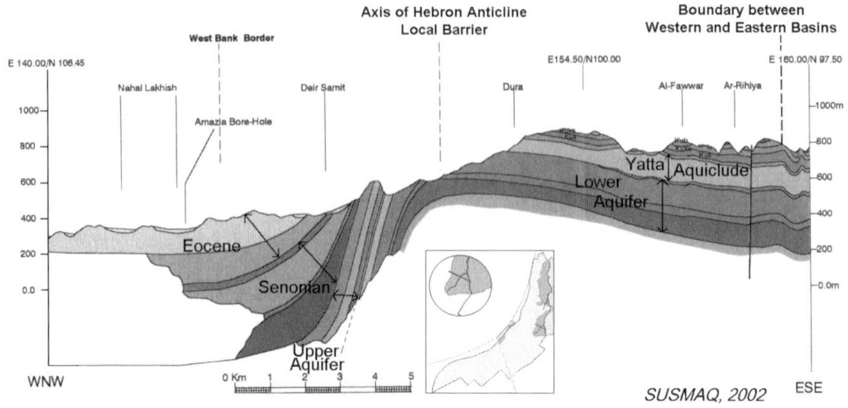

Fig. 3.4.6. Hebron Anticline (West); Boundary between Western and Eastern Basins (East)

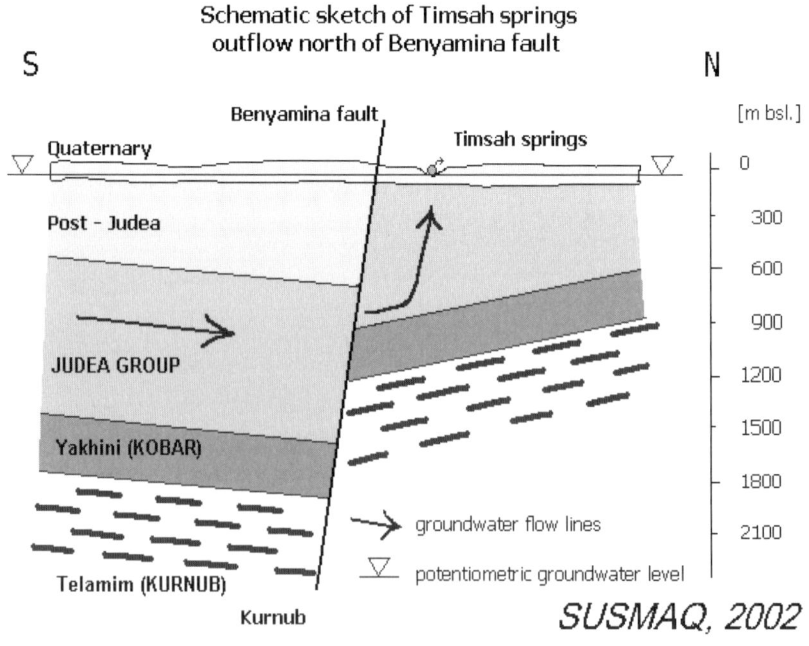

Fig. 3.4.7. Taninim (Timsah) Spring Mechanism north of Benyamina Fault

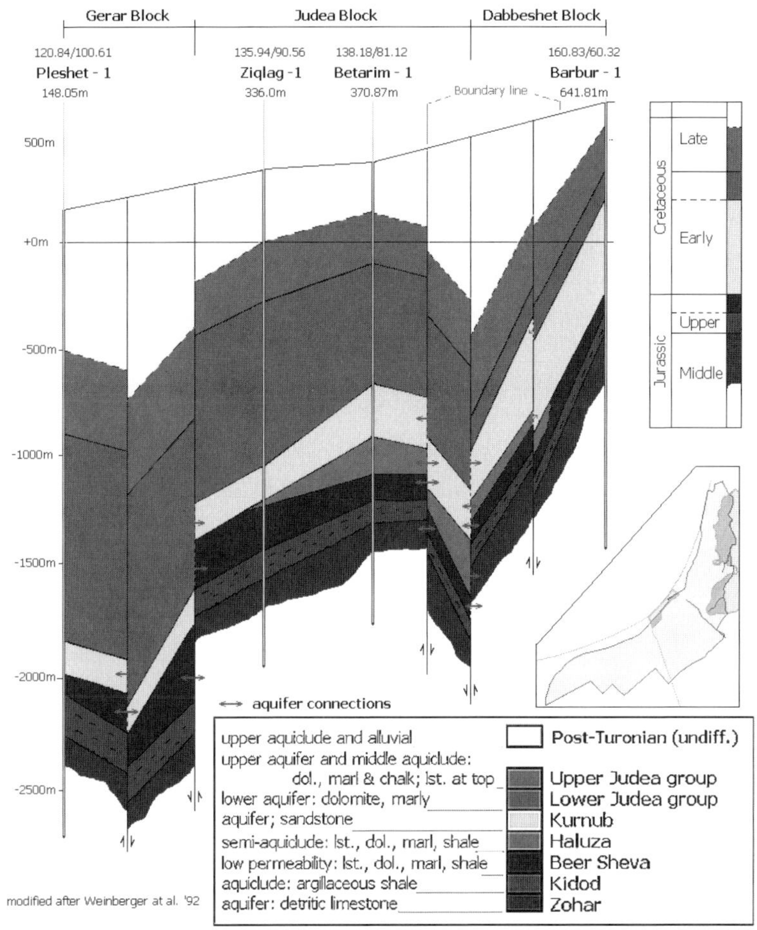

Fig. 3.4.8. Aquifer Connections in the Northern Negev

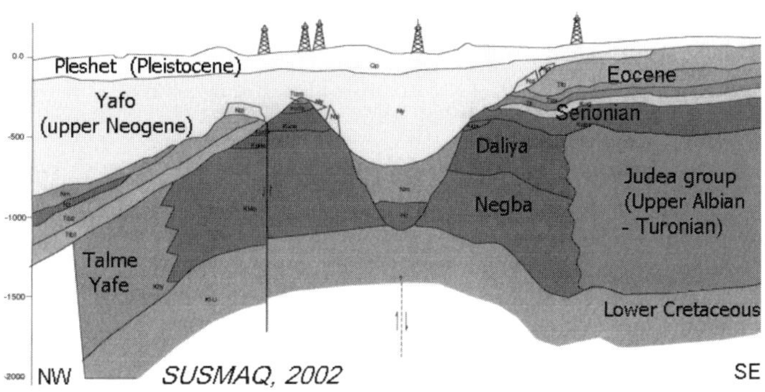

Fig. 3.4.9. Afiq Channel near Gaza dissecting the Aquifer

3.4.4 Boundaries

Detailed studies were executed to assess position and type of the aquifer boundaries. In addition to geological maps and field experience, geological cross-sections were constructed at different areas. The variety of boundary types reflects the structural and lithological diversity of the area (Fig. 3.4.3).

GW divides are found in the North, NE and in parts of the Jerusalem and Hebron anticlines. Most of the 2 anticlines (E) and presumably the Negev area (SE) are structural divides. The West is bounded by Talme Yafe stratigraphic divide. The Southern end of the Model area is built by Afiq channel. The strongly throwing Benyamina fault (700m) constitutes an internal tectonic boundary near the Northern natural outlet – Taninim (Timsah) springs.

Near Jerusalem, a sub-basin is opened by the southern continuation of the anticline (Fig. 3.4.4). The cross section (Fig. 3.4.5) shows, that SW flow is enabled, assuming water levels not below 400m asl., as found in Ein Karem wells.

Also in the Hebron anticline, a SW-trending thorn opens a sub-basin (Fig. 3.4.6). Near Carmel, the Benyamina fault connects the aquifer to the Pleistocene, with subsequent outflow in low areas near the shore (Fig. 3.4.7). Deep faults in the Negev connect the aquifer to Lower Cretaceous Sandstone and partly Jurassic aquifers with saltwater intrusion due to over-pumping (Fig. 3.4.8).

Afiq channel over 10.5km completely and for another 20km partly dissects the aquifer (Fig. 3.4.9). It is modelled therefore as a specified flow boundary.

3.4.5 Aquifer characteristics

Aquifer characteristics have not been determined in a detailed maner. They were described by the assumption of general parameters. The steady state model calibration will target on transmissivity and k-values. For the transient conditions modelling, specific yield and storativity have to be calibrated as well. Horizontal and vertical conductivity was defined according to existing studies and own estimates. Horizontal conductivity is locally increased around the main springs (Yarkon or Al-'Auja and Taninim or Timsah). It is regionally decreased in the West Bank Mountains recharge area and along the coastal Talme Yafe barrier.

3.4.6 Well files

Well abstractions increased drastically in the 1950ies and sixties, with a peak in 1999 with 572MCM (Fig. 3.4.10). This exceeds the agreed Oslo shares by far. Conversely, natural spring flow dropped to a small fraction of its former discharge; i.e. from over 300MCM/yr before extensive pumpage to now less than 50MCM/yr. Israeli artificial recharge through injection wells fluctuates with rain. Over 90% of the pumpage comes from Israeli wells. Palestinians only have an increment average share of 21MCM/yr. The abstraction rates and their spatial distribution (Fig. 3.4.11) are crucial for running a transient flow model.

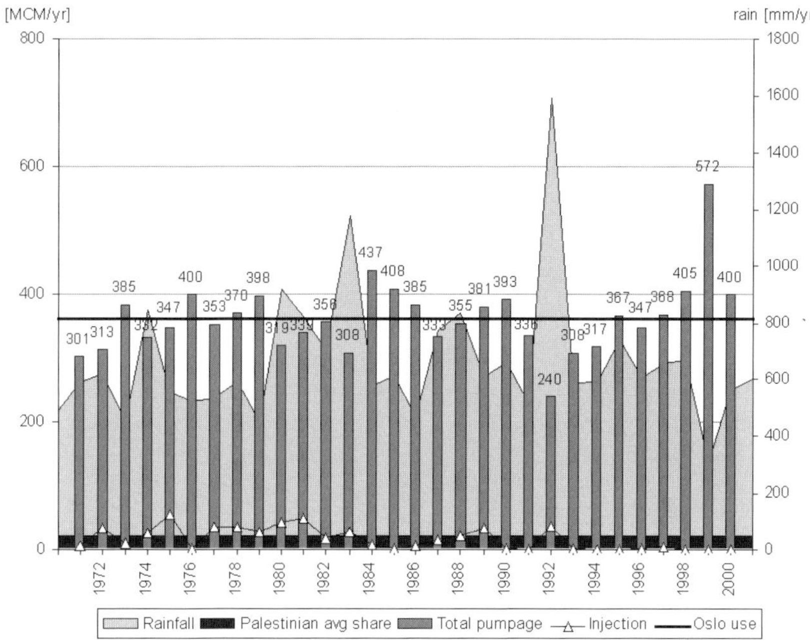

Fig. 3.4.10. Pumpage Schemes, Rain and Injections in the Western Basin

3.4.7 Rain, Recharge and Water Levels

The rainfall has its maximum (>600mm) in the Northern West Bank and sinks towards S and SW (<200mm). Recharge is correlated to the amount, intensity and duration of rainfall, as well as to the outcrop lithology, soil, vegetation and land use. Slopes rule the run-off. Less important are temperature and other climate factors, because rain almost exclusively occurs in the cold season. Both, Upper and Lower aquifer crop out predominantly in the West Bank, where >80% of recharge takes place as direct rainfall recharge (Fig. 3.4.12).

Time series on water levels (Fig. 3.4.14) will be studied for the transient run numerical model. The water levels will have to be differentiated into the Upper and Lower aquifers, respectively. For this, a better understanding of the aquifer connections is required. Therefore, the target of the steady state calibration will include the vertical aquifer connections, besides the above mentioned aquifer parameters. For this purpose, a list of target points (Fig. 3.4.13) has been chosen, where relatively good and reliable data exist.

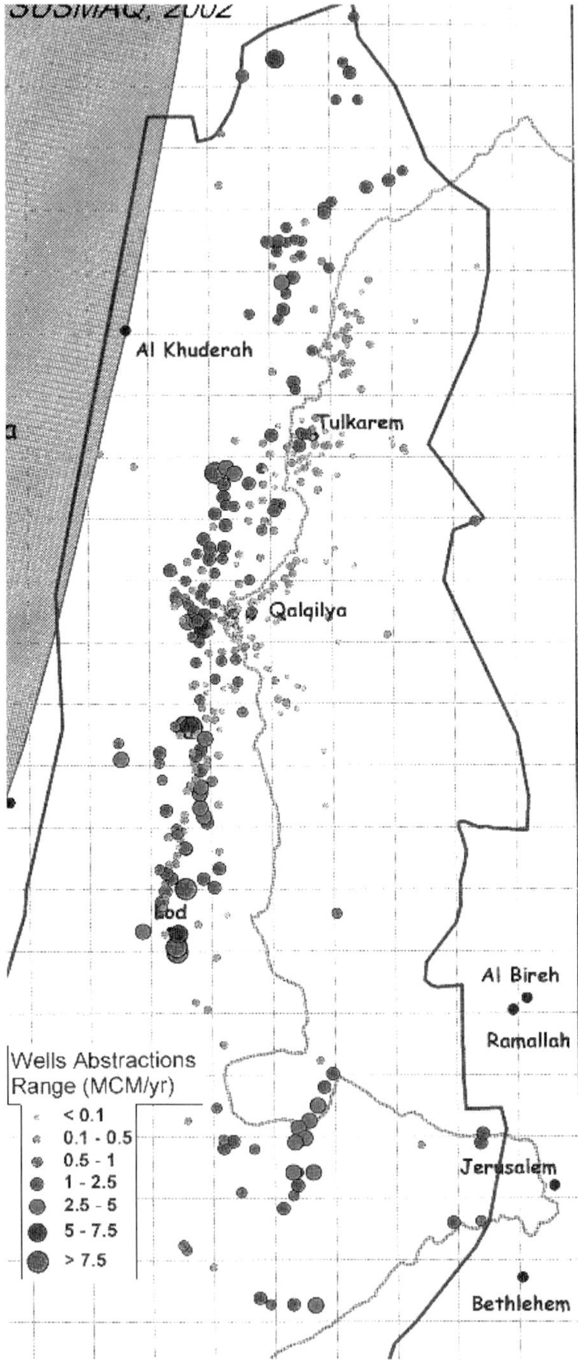

Fig. 3.4.11. Well Abstractions 1996

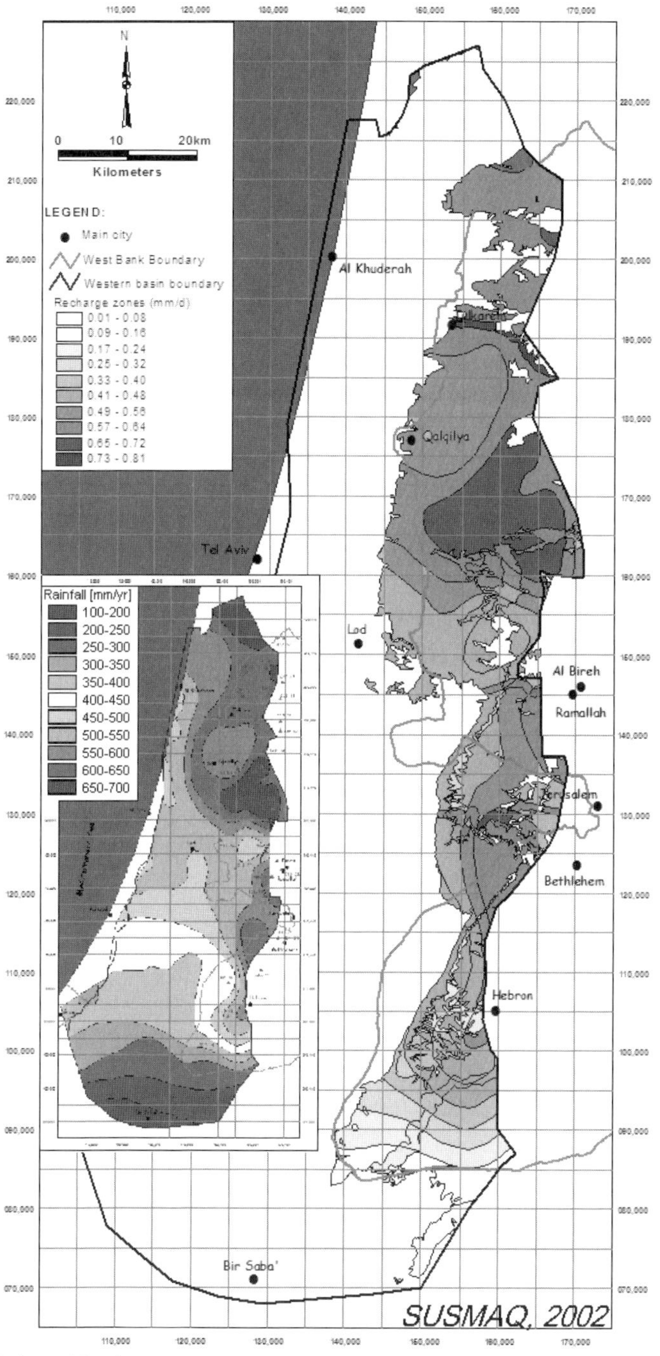

Fig. 3.4.12. Rain and Recharge

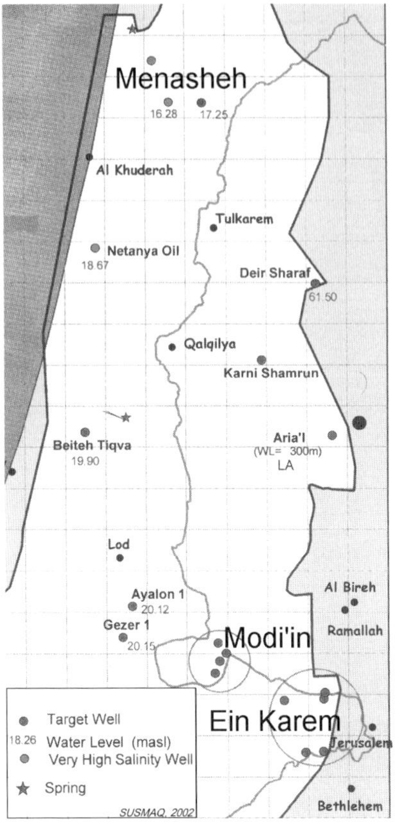

Fig. 3.4.13. Calibration Target Points

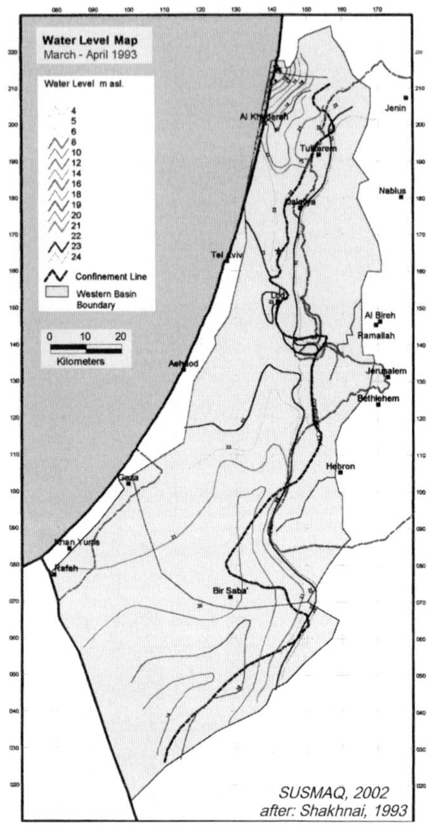

Fig. 3.4.14. Water Levels 1993

References

Antea, PHG and PWA (1998) Well Development Study of the Eastern Aquifer Basin, Vol.1; ANTEA No. A11930; Ramallah.

Braun et al. (1990) Guide Book; International Geological Correlation Program; Jerusalem.

Gelbermann, Fleischer, Wolff (1992) Structural Map on Top Judea Group; The Institute for Petroleum Research and Geophysics; Jerusalem.

Guttman, Zukerman (1995) Yarkon-Taninim-Beer Sheva Basin: Setup and calibration of the flow and salinity models; TAHAL 01/95/72; Tel Aviv.

Weinberger et al. (1994) The Yarkon-Taninim Groundwater Basin, Israel, Hydrogeology, Case Study and Critical Review; Journal of Hydrology; *161*; 227-255.

3.5 Ecosan – Introduction of Closed-Loop Approaches in Wastewater Management and Sanitation – A Supra-Regional GTZ-Project

Christine Werner, Heinz-Peter Mang, Jana Schlick, Papa Abdoulaye Fall

Deutsche Gesellschaft für Technische Zusammenarbeit (GTZ) Eschborn, Germany

Abstract

"ecosan – ecologically and economically sustainable wastewater management and sanitation systems" means approaches to bring forward a new philosophy of dealing with what is presently regarded as waste and wastewater. They are based on the systematic implementation of reuse and recycling of nutrients and water as a hygienically safe, closed-loop and holistic alternative to conventional solutions. Ecological sanitation systems enable the recovery of nutrients from human faeces and urine to the benefit of agriculture, thus helping to preserve soil fertility, to assure food security for future generations, to minimize water pollution and to recover bioenergy. They ensure that water is used economically and is recycled in a safe way to the greatest possible extent for purposes as irrigation or groundwater recharge. The main objectives of GTZ research and development project "ecosan" include broad knowledge management via a global ecosan network as well as the initiation, implementation and monitoring of appropriate pilot projects in cooperation with local and international partners.

Keywords

Closed loops; ecological sanitation; nutrient recycling; wastewater reuse; wastewater management.

3.5.1 Introduction

Water supply and treatment often receive more priority than wastewater collection and treatment, yet it should be recognized that sanitation deserves greater emphasis because of the effect that poor sanitation has on everyday lives, especially on those of the poor. It is the poor who suffer most from the growing scarcity of water and their degrading quality and by the burden of water related diseases and the degraded and dangerous environment.

Untreated excreta and wastewaters contain organic matters, plant nutrients, trace elements and micronutrients as well as pathogenic bacteria, viruses, helminths, endocrine substances and medical residues. If they are badly managed they are a major source for the spread of deseases and environmental harm; yet if well-managed they can make a positive contribution to local resources.

Currently more than 90 % of wastewater and excreta worldwide is either only poorly treated or not treated at all at discharge. In addition to the problem of pollution of water sources, such as rivers and groundwater aquifers, poor wastewater management also often lead to pools of stagnant water which may become breeding sites for insects, children playing on wet ground or near such pools may be exposed to dangers of infection, and the pools may become evil-smelling and unsightly. Badly designed or operated on-site sanitation is also contributing to groundwater pollution and contamination of the local environment. Sludge emptying is often ignored or the sludge is disposed of in the surrounding environment without precautions for hygienic safety.

Large investments have been made in water supply and sanitation over the last two decades, but the resulting health benefits have been limited by an inadequate focus on hygiene and sanitation and have often even been contraproductive as the improvement in the water supply has resulted in larger wastewater streams.

3.5.2 The problem with conventional wastewater management

Conventional forms of central wastewater management, i.e. a combined system with post-connected multistage wastewater treatment facilities, are still standard in developed industrialized nations today. Increasing criticism has, however, been levelled at these methods for ecological and economic reasons.

Increasing investment costs, high operating and maintenance costs and high water consumption as a result of misusing valuable drinking water just for transport give grounds to question such methods in rich nations, let alone their widespread application in developing countries.

Particularly for arid and semiarid zones, the search for appropriate solutions has become a pressing problem. With increasing population density and the resultant groundwater pollution, conventional decentralized disposal systems such as latrines and seepage pits are not a viable alternative either. Despite great efforts by many international and local organizations to improve water management in developing countries, the mortality rate due to water sanitation hygiene-associated diarrhoeas and some other water/sanitation related deseases (schistosomiasis, intestinal helminth infections etc.) was estimated about 2.2 million in 2000. Worldwide, over 2 billion people were infected with schistosomes and helminths, of whom 300 million sufferred serious illness, most of them children under the age of 5 [UNESCO-WWAP, 2003].

In addition to this, conventional wastewater disposal systems directly impair soil fertility as the valuable nutrients and trace elements contained in human excrement are not usually rechanneled into agriculture. Even where sewage sludge is put to agricultural use only a small fraction of the nutrients are reintroduced into the living soil layer. Most are either destroyed (e.g. nitrogen elimination) or enter the water balance, where they pollute the environment.

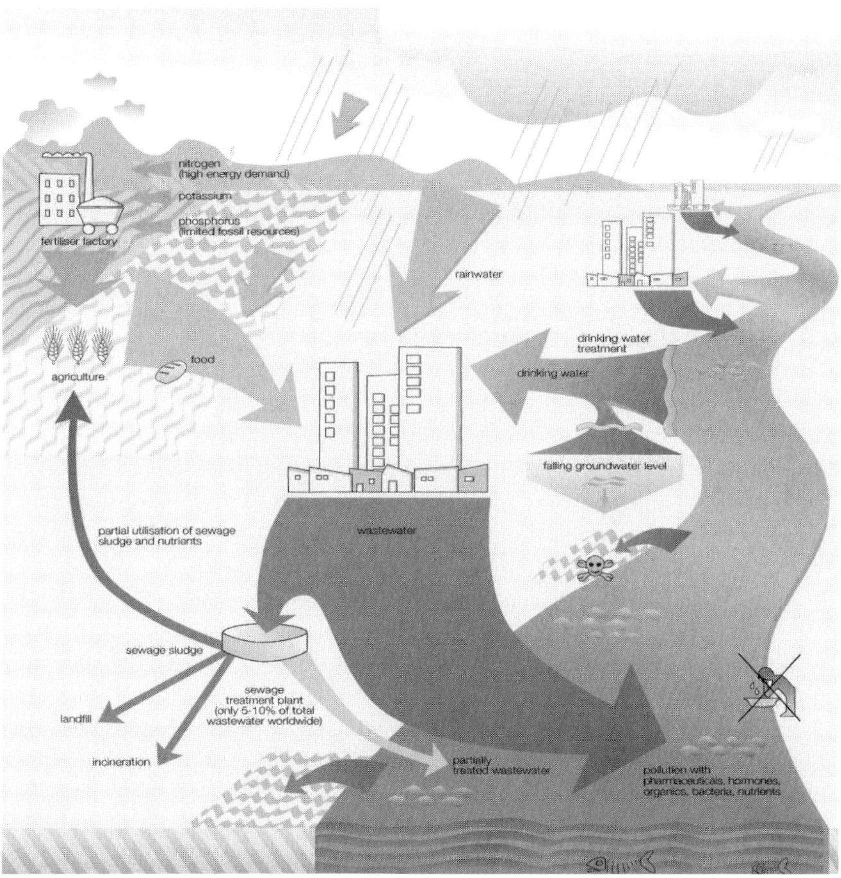

Fig. 3.5.1. Drawbacks of conventional wastewater management systems

- unsatisfactory purification or uncontrolled discharge of more than 90 % of wastewater worldwide
- pollution of waters by organics, nutrients, hazardous substances, pathogens, pharmaceutical residues, hormones etc,
- unbearable health risks and spread of disease
- severe environmental damage and eutrophication of the water cycle
- consumption of precious water for transport of waste (water carriage waste disposal systems)
- high investment, energy, operating and maintenance costs

- frequent subsidization of prosperous areas, neglect of poor settlements
- loss of valuable nutrients and trace elements contained in excrement due to discharge into waters
- impoverishment of agricultural soils, increased dependence on fertilizers
- combined central systems are predominant in organized wastewater disposal, resulting in problems with contaminated sewage sludge
- linear end-of-pipe technology

Frequently, the use of sewage sludge from central wastewater systems is also restricted as it contains too high a concentration of heavy metals and other hazardous substances, often as a result of intermixing household with commercial/industrial wastewater and with rainwater from contaminated streets. In fact, our conventional wastewater systems are largely linear end-of-pipe systems where drinking water is misused to transport waste into the water cycle, causing environmental damage and hygienic hazards [Werner et al., 2002].

3.5.3 Advantages of ecological sanitation

An alternative approach to avoid the disadvantages of conventional wastewater systems is ecological sanitation, 'ecosan' for short. This is based on an overall view of material flows as part of ecologically and economically sustainable wastewater management systems tailored to local needs. It does not favour a specific technology, but constitutes a new philosophy in handling substances that have so far been seen merely as wastewater and water-carried waste for disposal.

Systems based on this approach are used for the systematic closure of local material flow cycles and thus ultimately enable recycling systems as are already in common use for solid waste [Esrey, 2000]. Ideally, ecosan systems enable almost complete recovery of all nutrients and trace elements in household wastewater and their reuse in agriculture - after appropriate treatment. This way, they help preserve soil fertility and safeguard long-term food security.

As an integral alternative, a hallmark of ecosan is its interdisciplinary approach that goes beyond the narrow domestic water supply and technological aspects to subsume agricultural use, sociology, hygiene, health, town planning, economy/small-enterprise promotion, administration, etc. in system development.

In practice, the ecosan strategies of the separation and separate treatment of faeces, urine and greywater for example minimizes the consumption of valuable drinking water and treats the separate wastewaters at low cost for subsequent use for soil amelioration, as fertilizer or as service or irrigation water [Werner et al., 2002].

Diverse technologies can be used, from simple low-tech to sophisticated high-tech systems. These currently range from compost toilets or urine-separating dry toilets to water-saving vacuum sewage systems, possibly with separate collection and subsequent treatment of urine, faeces and greywater through to membrane technology for material separation and hygienization. Generally, precedence is given to appropriate modular and decentralized facilities, but in very densely populated areas centralized systems may still be needed [Lange and Otterpohl, 1997].

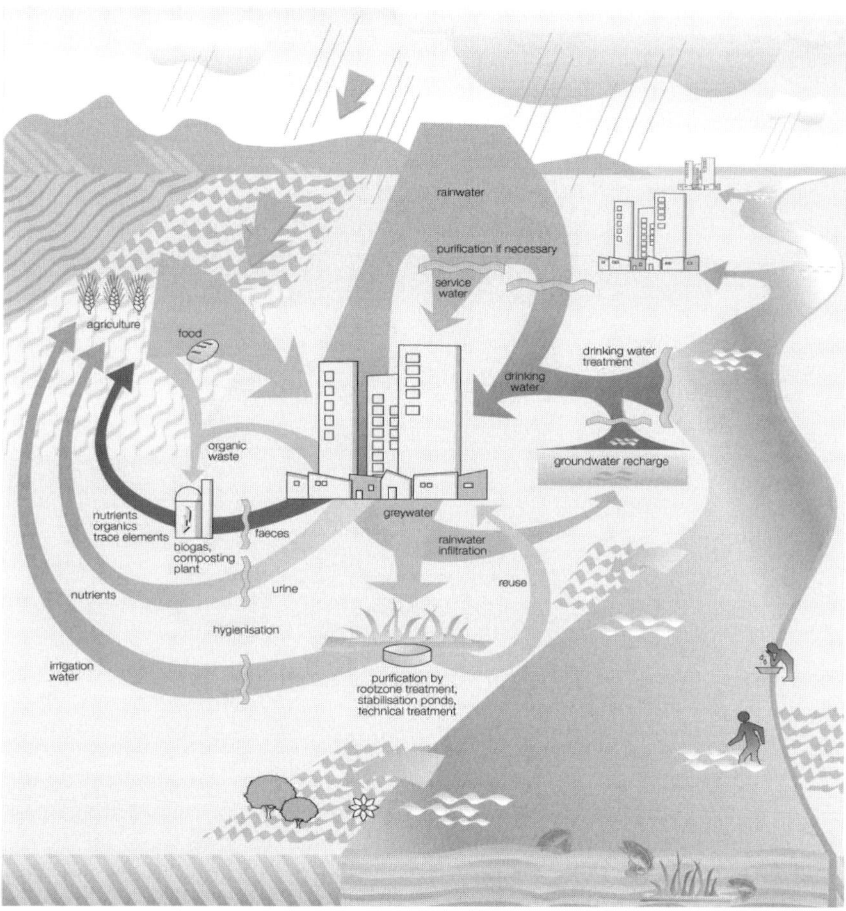

Fig. 3.5.2. Advantages of ecological sanitation

- improvement of health by minimizing the introduction of pathogens from human excrement into the water cycle
- promotion of recycling by safe, hygienic recovery and use of nutrients, trace elements, water and energy
- conservation of resources through lower water consumption, substitution of chemical fertilizers, minimization of water pollution
- preference for modular, decentralized partial-flow systems for more appropriate, cost-efficient solutions

- possibility to integrate on-plot sanitation into households, increasing user comfort and security for women and girls
- preservation of soil fertility
- improvement of agricultural productivity and hence contribution to food security
- promotion of a holistic, interdisciplinary approach (hygiene, water supply and sanitation, resource conservation, environmental protection, town planning, agriculture, irrigation, food security, small-business promotion etc.)
- Material-flow cycle instead of disposal

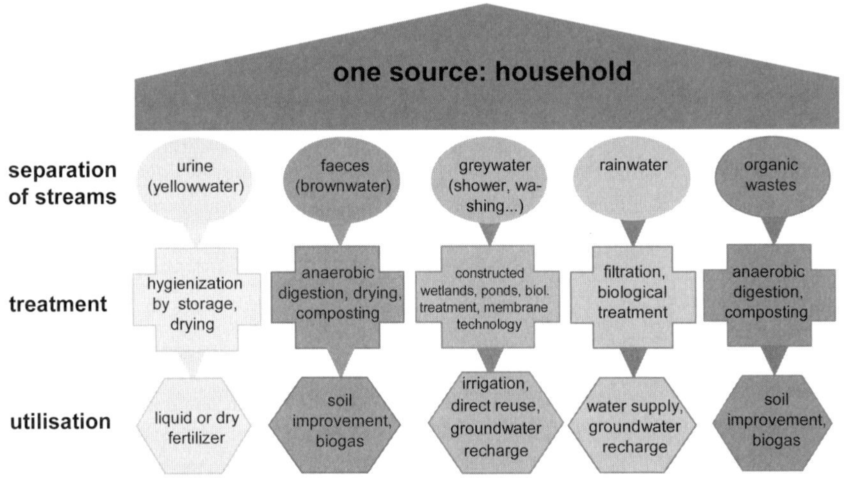

Fig. 3.5.3. Examples of possible ecosan elements

Of key importance particularly here are also innovative logistics to return nutrients to farmland, marketing strategies for the recovered nutrients and directions for their safe application in agriculture. New ecosan schemes may also entail setting up service enterprises and hence implementing income-generating measures for the construction and easy and safe operation of the installations as well as the collection, treatment and marketing of recyclates.

Closing local nutrient cycles by retrieving and using nitrogen, phosphorus, potassium, trace elements and organic components contained in excrement is even more important considering some of the disadvantages of fertilizers. For one thing, they are too expensive in many parts of the world or are unavailable to local farmers, and their effects on soil and food quality are in dispute. For another, large amounts of energy and finite fossil resources are used to produce them.

3.5.4 ecosan concepts in operation

In May 2001, GTZ started a supra-regional research and development project, ecosan, financed by the Federal Ministry for Economic Cooperation and Development (BMZ). It aims at promoting the development and application of integral ecologically, economically and socially sustainable recycling-based wastewater and sanitation concepts in developing countries by broad knowledge management, global networking and the development of pilot projects for testing and demonstrating ecosan-solutions as working examples. Ultimately it is intended to contribute to the global dissemination and application of ecosan approaches and establish these internationally as state-of-the-art techniques – also in industrialized countries [Werner et al., 2002].

3.5.5 Examples

On-plot ecosan systems for the treatment of faeces, urine and greywater in Mali

Koulikoro, Mali, has a central potable water supply system dating from the 1970s, but as yet no sewage system. In an arid sub-Saharan country like Mali, where financial and water resources are scarce, a water-carrier sewage system resembling those used in Europe would be inappropriate and too expensive. Mali is also faced with the steadily worsening problem of soil degradation, up to and including desertification, chiefly as a result of agricultural overuse and insufficient return of nutrients.

An affordable means of proper wastewater disposal is needed. GTZ is therefore developing an on-plot household ecosan system in which faeces, urine and greywater are separately collected and treated. This offers major advantages over conventional latrinebased systems, as it enables the hygienic recovery of soilamending substances from faeces and of nutrients from urine and purified greywater. The ecosan system is also in harmony with local traditions.

In 2002, the National Sewage and Solid Waste Department at the Malian Ministry of the Environment incorporated the greywater gardens and separating toilets developed by the ecosan initiative into its program. Together with GTZ, the department is now examining their suitability for widespread introduction. Ultimately, however, the success of greywater gardens depends solely on the degree to which they are accepted by women for growing vegetables, bananas and papayas.

Municipal ecosan concepts in a Chinese suburb

Located in one of Beijing's three river basins, Yang Song covers a little more than three square kilometers and is home to some 21,000 people. With its intensive livestock farming and grain and vegetable production, the region is a major source of food for the city of Beijing. The community currently produces roughly 15 tonnes of solid waste each day. Less than 10 % of the town's wastewater is treated prior to being discharged into the rivers or groundwater. Within the scope of a local ecosan project, the community is to be provided with a modern, material-separating disposal and recycling concept for wastewater and organic wastes that is in line with the principles of closed-loop wastewater management and sanitation. GTZ, Chinese and German scientists and companies are working together to analyze and compare different sanitation, wastewater treatment and recycling options in various harmonized systems. The cost-effective recovery of useful materials and energy is the main objective.

There are also plans to use water-saving vacuum technology and urine separation systems. Organic waste from kitchens and markets will be collected, shredded and, finally, fermented in a bioreactor system. The resultant fertilizer and hygienized urine will be suitable for use in growing flowers and vegetables. Greywater will be used for watering public parks and gardens.

Soilization of sewage sludge in Egypt

In many countries, the use of sewage sludge in agriculture is thwarted either by the complexity of the processing technology or by the poor quality of the sludge, which awakens very little interest among farmers for its use as a soil conditioner. In Egypt, the GTZ has therefore supported a large scale field test, carried out by IPP Consult, of a process of sewage and faecal sludge upgrading, or soilization, by means of sewing sludge polders with grass or common reed. The results are promising and the process will be introduced in other ongoing ecosan pilot projects of GTZ in Kafr el Sheikh in Egypt and in Mali:

- process technology is easy to manage and economical
- structural, aesthetic and hygienic attributes of the soiled sludge are superior to those of dried sludge
- soilized products find more acceptance and market potential is improved.

ecosan research into noncentralized applications in Cuba

Throughout Cuba, and particularly in urban areas, the wastewater management and sanitation systems lack capacity and are in urgent need of rehabilitation. Most notably in periurban areas with considerable agricultural activity, the soil, groundwater and watercourses are heavily polluted. As a result, health conditions and odor-nuisance levels are critical in many places. Moreover, many households do not have access to electricity. This forces many people to use ecologically questionable forms of fuel for their everyday needs. To address the situation, a GTZ-supported ecosan research project is conducting field tests on various household sanitation systems and looking for appropriate-technology solutions which generate cooking-energy yields. For example, on several city farms in two different project regions, the disposal or utilization of household sewage and organic waste is being integrated into the in-house production of fertilizer and cooking energy. In a third region, prefabricated components are being designed and developed for diverse decentralized disposal systems, and in a fourth region, different ecosan systems are being implemented in urban centers. The four regions in question are located in different parts of the island to ensure the study is representative of the island's diverse climatic, structural and social conditions.

Sanitary systems in Ouagadougou, Burkina Faso

The climatic and social conditions in Ouagadougou, the capital of Burkina Faso, are typical for a country in the sub-Saharan savannah belt: surface water, which represents the city's major source of drinking water, is stored in open reservoirs and therefore extremely vulnerable to pollution. The limited groundwater resources are very important, both as a relatively uncontaminated source of water and as a reserve supply in particularly dry years. The simple pit latrines used by 93% of the population infiltrate into the groundwater. Once discarded, they contaminate the receiving water body and contribute to the high levels of water-related diseases. Thus, conventional sanitary systems have created a dilemma between the need for sanitary disposal of human excreta, the protection of water resources, and the increased agricultural activity needed to meet the rising demand for food. That dilemma can be addressed if human excreta are recognized as a potentially valuable resource. The extent to which ecosan sanitation systems can be introduced in Ouagadougou as part of a closed-loop, or material-floworiented recycling process is currently being investigated by GTZ.

ecosan as an element of sustainable regional resource management in Botswana

In many countries of Africa, including Botswana, conventional forms of wastewater disposal have drawbacks for the general population. Most households located outside of the major urban centers are not connected to any existing waste management and sanitation system. Droughts and inadequate water resources make an already unsatisfactory situation even worse.

Over the next five years, a project devoted to sustainable regional resource management will be cooperating with local authorities, the International Union for the Conservation of Nature (IUCN) and the German Development Service (DED) in developing, testing and demonstrating sustainable noncentralized wastewater management and sanitation systems and methods. Initially, private households in the districts of Ghanzi, Gaborone and Serowe are to be tied into the research activities. Later, the approach will be extended to the municipal level. One of the aims of this GTZ-project is to recover nutrients and trace elements from domestic wastewater, faeces and urine for use in agriculture. This not only contributes toward long-term food security, but also provides the people with an opportunity to earn extra money.

3.5.6 Further projects in preparation

Other projects are presently being prepared with the support of GTZ-ecosan in several other countries, where advocacy workshops, baseline and feasibility-studies are running or bociseing organised at present:

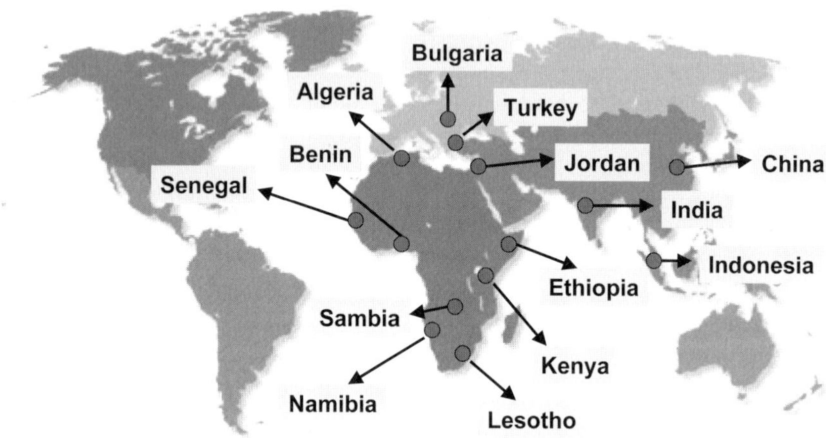

Fig. 3.5.4. Further projects in preparation

3.5.7 Conclusion

Human excreta and domestic used water have not to be considered as wastes but rather as an important natural resource to be safely reuse for agriculture. Efforts should be concentrated on developing and implementing new approaches on ecological wastewater treatment and sanitation for a variety of suitable closed-loop systems in urban areas including the efficient agricultural reuse of organics, nutrients and water.

References

Esrey SA (2000) Towards a recycling society, ecological sanitation. In: Werner C, Schlick J, Witte G, Hildebrandt A (eds) Proceedings of the International Symposium "ecosan-closing the loop in wastewater management and sanitation". Universum Verlagsanstalt, Wiesbaden, pp 34-44.

Lange J, Otterpohl R (1997) Abwasser Handbuch zu einer zukunftsfähigen Wasserwirtschaft, Jörg Lange und Ralf Otterpohl, ISBN 3-9803502-1-5.

UNESCO-WWAP (2003) Water for People, Water for Life, UN World Water Development Report (WWDR), ISBN 92-3-103881-8.

Werner C, Schlick J, Witte G, Hildebrandt A (eds) (2000) ecosan-closing the loop in wastewater management and sanitation, proceedings of the international symposium, pp 30-231, October 2000, Universum Verlagsanstalt, Wiesbaden.

3.6 I M P E T U S West Africa [1] An Integrated Approach to the Efficient Management of Scarce Water Resources in West Africa - Case Studies for Selected River Catchments in Different Climatic Zones -

P. Speth and M. Christoph

Institut für Geophysik & Meteorologie, Köln

Abstract

In the IMPETUS project thorough investigations of all aspects of the hydrological cycle are carried out within two river catchments in North West and West Africa: the wadi Drâa in the south east of Morocco and the river Ouémé in Benin. This choice is motivated by the possibility that the climates of Africa and Europe interact through atmospheric teleconnections, and the evidence that since the 1970s the droughts north and south of the Sahara have probably been related.

Since the late 1970s Morocco has experienced a number of extremely dry winter seasons, the causes of which are not fully understood. They are assumed to be related to changes of the large scale circulation on interannual and interdecadal time scales as manifested in the North Atlantic Oscillation or the El Niño-Southern Oscillation. Against this background, the development of sustainable water resource management is a strong necessity. In this context the water-balance of the Drâa catchment area and socio-economic implications are investigated. In order to address a number of imminent problems limiting the availability and allocation of water along the wadi Drâa 12 measurement sites were installed along a gradient of elevation and aridity. Monitoring of the thickness and the extent of the snow cover in the High Atlas mountains is essential to enable the competing water users (power generation, irrigation, domestic consumption) to have adequate supplies. In addition to seeking a better understanding and prediction of the geospheric, atmospheric and biospheric components of the hydrological cycle, the IMPETUS activities centre around the questions of the influence, risks, and resulting conflicts of human activities in the context of the specific social and economical structures encountered in the area. A hierarchy of nested meteorological and hydrological models have been developed to assess the effects of environmental and anthropogenic change on the hydrological cycle and to analyse likely 'future scenarios'.

1. IMPETUS is the acronym for „<u>I</u>ntegratives <u>M</u>anagement-<u>P</u>rojekt für einen <u>E</u>ffizienten und <u>T</u>ragfähigen <u>U</u>mgang mit <u>S</u>üßwasser"

3.6.1 Introduction

Fresh water is an essential component of life on Earth. It plays a vital role in the maintenance of the natural environment of the world and its continuous availability is indispensable for virtually every human activity. Shortage of fresh water is expected to be the dominant water problem of the forthcoming century and one that, along with water quality, may well jeopardise all other efforts to secure sustainable development, and even in some cases lead to social and political instability. Fresh water has already become critically scarce in many regions. The global mean water withdrawal per capita has shown a significant decrease during recent decades; there are now 22 countries that have renewable fresh water resources under 1000 m^3 per capita per year, a value commonly accepted as a benchmark for fresh water scarcity. It is forecast that, for the first quarter of the 21st century, about one-quarter of the world population will suffer from severe water scarcity. Some estimates suggest that already the amount of fresh water available for each person in Africa is only about a quarter of that in 1950 (Obasi, 1999), and that fresh water supply could become problematic especially in West Africa, where about 30 years of drought have been observed. Although the climates of West Africa are still relatively poorly known and understood, it is recognised that North West and tropical West Africa have experienced the most pronounced inter-decadal variability of climate in the world during the 20th Century (Ward et al., 1999). The possibility of human-induced climate change adds additional serious aspects to the challenging water-related problems already encountered in many parts of the world.

3.6.2 Motivation

The available fresh water is controlled by the hydrological cycle. Climate, in particular the spatial and temporal distribution of precipitation and evaporation, plays a significant role in the hydrological cycle, and climate data are therefore of the utmost importance in the analysis of ground and surface water supply for domestic and industrial users, irrigation, hydropower generation and ecosystems. Dealing effectively with the hydrological cycle and its impacts demands not only a strong co-operation between different disciplines within the natural sciences (e.g. hydrology, meteorology, botany, agriculture, geology, remote sensing), but also consideration of socio-economic and medical issues; all disciplines involved have to interact in a complex and co-ordinated manner. Hence, in order to solve possible future problems with regard to fresh water supply, a clearly interdisciplinary approach is necessary. This is done in the present initiative for West Africa and it is the purpose of this project to offer concrete ways of translating into action scientific results through scientifically-based strategies. This approach will provide a reliable basis for political measures and international agreements. In the first three-year phase the focus is set on the identification and analysis of influencing factors regarding different aspects of the water budget. Based on this, in the second three-year phase methods will be developed to predict changes during the coming decades. In the final two years the collected insights of all the disciplines will be coupled in order to assess management options and to install operational tools for the decision-making process (so called "Decision Support Systems").

3.6.3 Choice of Catchments

West Africa was chosen because (i) it has experienced the most pronounced inter-decadal variability of climate in the world during the 20th century, (ii) relations to the climates of Europe might exist via complex atmosphere-ocean interactions, and (iii) the regions north and south of the Sahara might be linked via atmospheric tele-connection processes with regard to precipitation anomalies; first results presented below give evidence for the existence of such a link by atmospheric moisture trans-ports out of the West Sahel zone across the Sahara towards the Atlas mountains.

Since the 1970s both subtropical Northwest Africa and tropical West Africa have experienced a general rainfall decline which have probably been related (cf. figs. 3.6.1a and 3.6.1b). For this reason it is of advantage to consider both areas north and south of the Sahara desert in a combined approach, realised by means of a transect between the Atlas mountains and the Gulf of Guinea (fig. 3.6.2). This transect contains two reasonably sized river catchments (< 100.000 km^2) which are representative in the following sense: the Drâa catchment in the south east of Morocco is typical of a gradient from humid/sub-humid subtropical mountains to their arid foothills; the Ouémé basin in Benin is typical of an alternating sub-humid climate ("Guineo-Soudanien") of the outer tropics embedded within a transect from the Sahelian to the Guinean Coast climate.

The feasibility of the presented initiative has been guaranteed by the good avail-ability of data of both the natural and the human sciences and by politically stable conditions in the respective countries.

3.6.4 Past and Present Situation

Moroccan precipitation is strongly related to the large-scale atmospheric circula-tion over the subtropical and extratropical North Atlantic and the Mediterranean Sea, with the bulk of precipitation occurring in winter (November - March). Since the late 1970s, Morocco has experienced a number of extremely dry winter seasons (cf. fig. 3.6.1a), the causes of which are not fully understood. Against this back-ground, the development of sustainable water resource management is even more a necessity. The considered wadi Drâa possesses two main tributaries, the wadi Dades and the wadi Ouarzazate which drain the south-eastern and the south-west-ern parts of the Atlas and confluence near the city of Ouarzazate thereby forming the wadi Drâa.

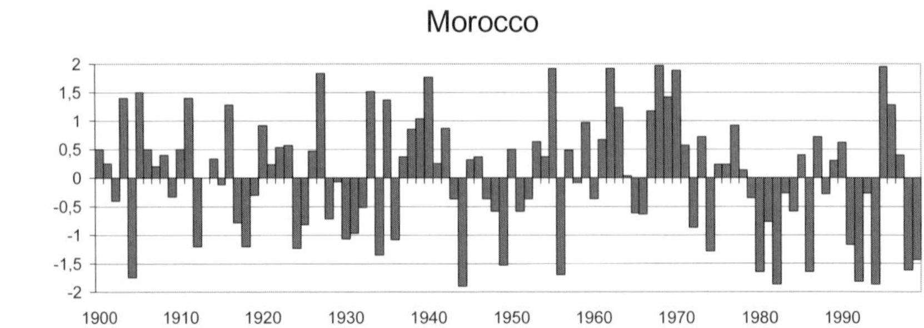

Fig. 3.6.1a. Annual precipitation variability in Morocco throughout the 20th century.

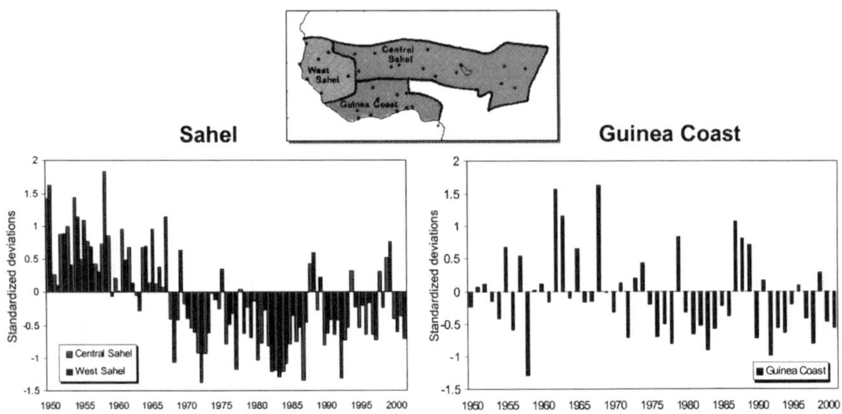

Fig. 3.6.1b. Precipitation variability in West Africa for the period June - September 1950-2002.

At the site of confluence construction of the Mansour Ed Dahbi dam (60 m height) was completed in 1972 with an original storage capacity of 560 million m^3. Due to strong sedimentation a capacity of only 440 million m^3 remains today. Approximately 250 million m^3 of stored water is released in normal years for irrigation purposes. Release is done periodically in several waves called 'lâchers' which consist of an average discharge of 25 m^3/sec during approx. 3 weeks. The first lâcher is usually planned for August and an ideal number of 7 such releases from the dam take place per hydrological year. The actual number of lâchers carried out, however, strongly depends on the temporal evolution of the filling situation. Once a critical margin of 50 million m^3 is reached irrigation is no longer possible. The main irrigation structures downstream of the Mansour Ed Dahbi dam consist of five smaller dams and a complex network of traditional and modern canals. The irrigated perimeter covers a total area of approximately 26.500 ha.

Since the snow melt in spring contributes significantly to the annual discharge of the main storage lake tributaries, diagnosing the spatial distribution of accumulated snow water equivalent in the elevated areas of the catchment is particularly desirable. An effective and sustainable management of water in the Drâa valley is essential to enable the competing users (water power generation, irrigation, domestic consumption) to have adequate supplies, and to prevent social tensions related to water resources. Figure 3.6.3a shows the dramatic low levels of the reservoir at the end of the rainy seasons of the mid 1970s, mid 1980s, and of the very recent years.

Since the early 1970s tropical West Africa has suffered from a prolonged drought that reached its first climax in the first half of the eighties (cf. fig. 3.6.1b). The average rainfall deficit over 1971-1990 was of the order of 180 mm/year compared with the interval 1951-1970. All climatic zones, from the semi-arid Sahel and the subhumid Sudanese zone down to the humid Gulf of Guinea, have been affected. The prolonged West African drought has already brought about a profound deterioration in the economic and social development of the West African countries. As a consequence river discharges in West Africa have decreased by about 40-60% in recent decades, causing shortages in river water available for domestic and agricultural purposes. For instance figure 3.6.3a shows the decrease in run-off of the Ouémé at Bétérou which reflects the integral for the southern part of the upper Ouémé catchment. This has led to extensive migrations in the past. During the rain-rich fifties, water power stations were built in the Guinea coast zone to supply a substantial amount of energy to Ivory coast, Ghana, Togo, Benin and Nigeria.

Apart from the decreasing availability of fresh water per capita both in Morocco and in Benin the current situation north and south of the Sahara is also characterized by increasing population (population growth rate more than 3% per year), increasing degradation of the natural vegetation due to overgrazing (Morocco), demands in fire wood, and shifting cultivation (Benin). As a consequence soils quickly erode in Morocco (to a lesser degree also in Benin) and salt contents rise due to intensive irrigation practices. In combination the aforementioned factors are likely to accelerate the degradation and desertification processes for the coming decades.

3.6.5 Methodology

Due to the importance of the hydrological cycle regarding the availability of fresh water, its different components and their interactions (cf. fig. 3.6.4) are identified in its complexity and quantified in the course of this project: the atmospheric variability, the continental hydrosphere, and the land surface processes from the natural science perspective. However, focal point of the investigations are human activities related to fresh water, e.g. economic behaviour, migration, construction of wells, water related rights and conflicts, etc.

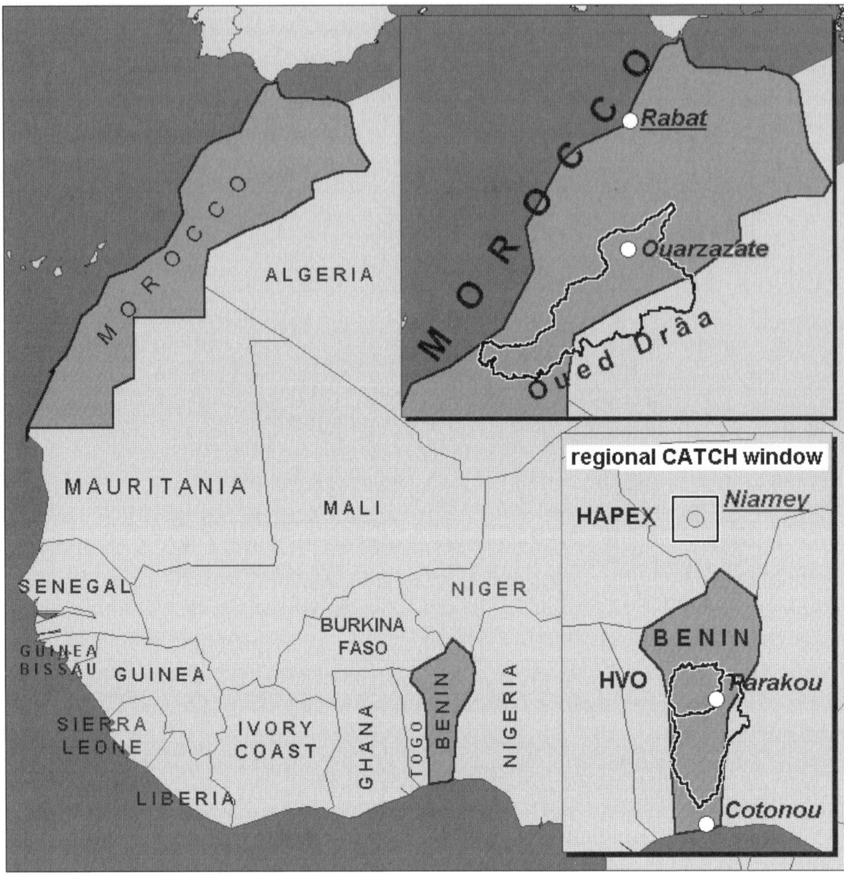

Fig. 3.6.2. The two catchments of consideration. The DRÂA catchment in Morocco and the OUÉMÉ catchment in Benin are boldly bordered. A sub-catchment of approx. 100x100 km west of Parakou (Haute Vallée de l'Ouémé: HVO) has been chosen as an area of focused investigations.

In an integrated approach a sequence of existing models for the individual components have been adapted in the first project phase in order to describe the relationships and dependencies within the hydrological cycle in its present state. Basic research was only carried out if existing competence proved to be insufficient. These adapted models will then be coupled in the subsequent phase with the aim of developing and computing likely scenarios of change for the coming decades. This will also serve as a basis for assessment of management options and the installation of operational tools for decision makers during the third phase of the project.

A measurement network of essential parameters has been set up in data sparse areas in order to fill gaps in the existing national data bases and networks. In the Drâa catchment 4 water-level gauges and 11 climate stations were installed along a representative north-south height gradient from the High Atlas to the pre-Saharan desert (see fig. 3.6.5a and 3.6.5b) thereby representing also the transition from the Mediterranean forest species to the semi-desert species. The selection of test sites also covers the most important grand units of the natural landscape (e.g High Atlas mountains, basins of the Atlas foothills, pre-cambrian mountains of the Jbel Sarrho, paleozoic ridges south of the Anti Atlas, basins of the pre-Sahara, etc.).

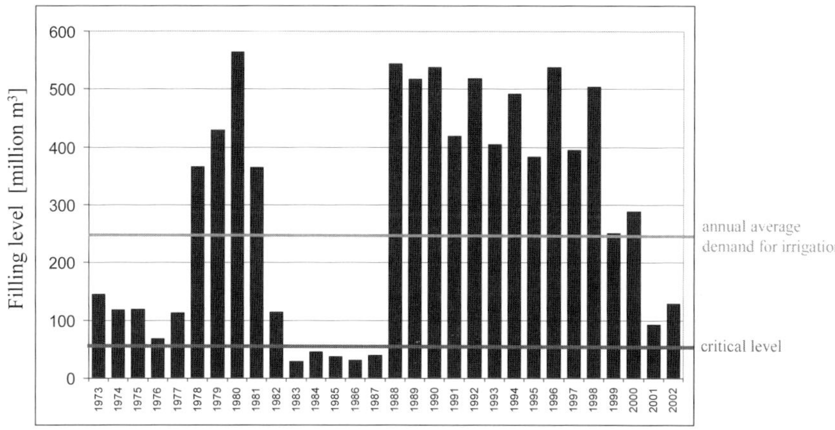

Fig. 3.6.3a. April filling levels of the "Mansour Ed Dahbi" reservoir near Ouarzazate, Morocco (1973-2002).

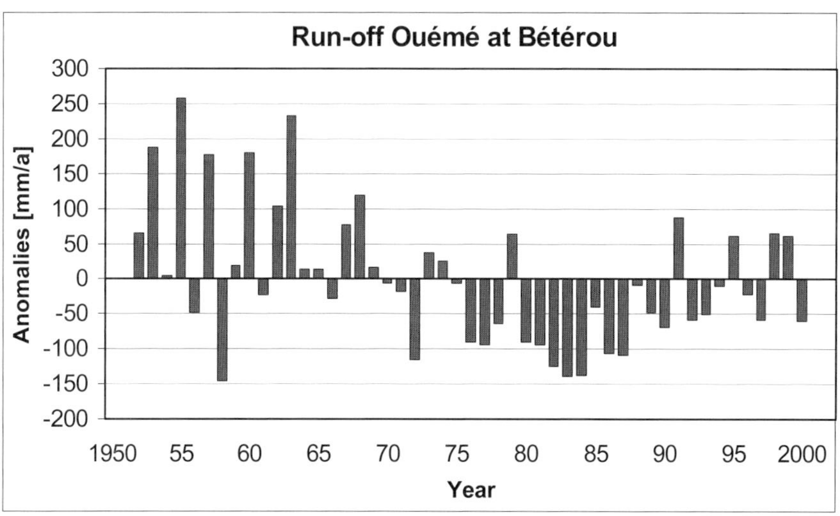

Fig. 3b. Annual run-off anomalies of the Ouémé river at Bétérou for the period 1952 through 2000. Units are in mm/year, thereby taking into account the size of the upstream catchment area.

One additional mobile climate station is located within the oasis near Zagora in the southeast of the catchment. These stations measure in 10 minute intervals air temperature, air humidity, dew point, wind speed, wind direction, precipitation, air pressure, short wave global radiation, reflected radiation, net radiation, soil temperature, soil moisture, soil heat flux, and snow height (above 2000 m). In addition a fenced vegetation plot was installed in the vicinity of each climate station (10 all together, no plot close to the M'Goun summit and within the oasis) in order to prevent live stock from grazing and thus determine the undisturbed development of vegetation. All test sites serve to characterise the hydrological processes and to develop and verify the hydrological models. For the Ouémé catchment we concentrated on the upper Ouémé valley where the existing national and IRD hydrometeorological networks have been enforced with focus on a super test site 'Aguima' near Doguè in central Benin (no figure).

3.6.6 First Results

As an example of first results achieved we chose to present the improved understanding of the relations between precipitation events in different parts of Morocco, and patterns of the large-scale atmospheric circulation and synoptic activity on the basis of observational data. The associated physical processes were examined on different time-scales. The relations found were used to explain different aspects of the spatial, seasonal, interannual and decadal precipitation variations and will serve as a basis for the evaluation of the output from climate change experiments with a global general circulation model in the second phase of IMPETUS.

In the analysis, we distinguished between tropically and extratropically induced rainfalls. Three regions with different precipitation regimes were defined, whose deviations are mainly caused by the orography. The precipitation in the northern and western parts of Morocco reveals a clear dependence on the strength and position of the North Atlantic storm track (represented by baroclinicity, storm track intensity, frequency of surface cyclones). Rainy synoptic situations are often accompanied by an upper-level trough and/or a shallow cyclone west of the Iberian Peninsula as well as predominantly westerly weather types and water vapour advection from the Atlantic Ocean (Knippertz et al., 2003 a). In this area, a strong negative correlation to the North Atlantic Oscillation (NAO) was observed, particularly, when the subtropical NAO-centre is at a relatively eastern position (see fig. 3.6.6, left). The region close to the Mediterranean coast in northeast Morocco/ northwest Algeria reveals a distinct relation to the frequency of upper-level troughs and the storm track and cyclone activity over the western Mediterranean, that is accompanied by enhanced moisture transports from northwesterly or northerly directions. The region south of the Atlas Mountains in Morocco and Algeria is affected by extratropically induced precipitation only in cases of far southward stretching upper-level troughs and/or surface cyclones, that lead to a humidity advection from the Atlantic Ocean along the southern flank of the Atlas. In these cases precipitation often appears to be connected to orographic lifting.

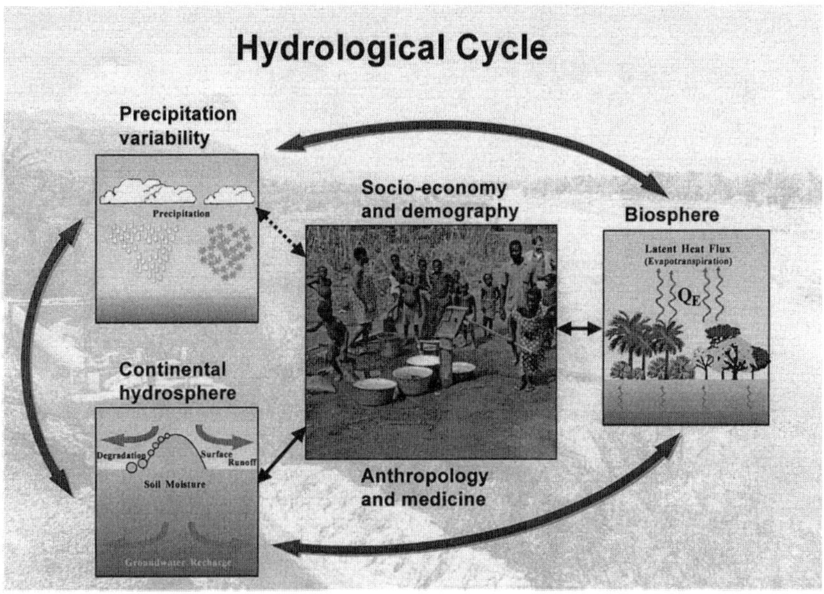

Fig. 3.6.4. Main components of the hydrological cycle and interactions considered in IMPETUS.

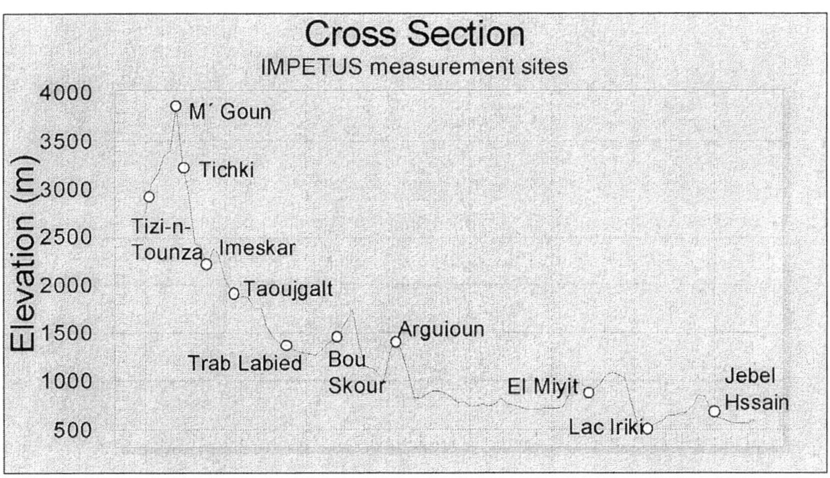

Fig. 3.6.5a. Height transect of the IMPETUS measurement sites in the Drâa catchment, Morocco.

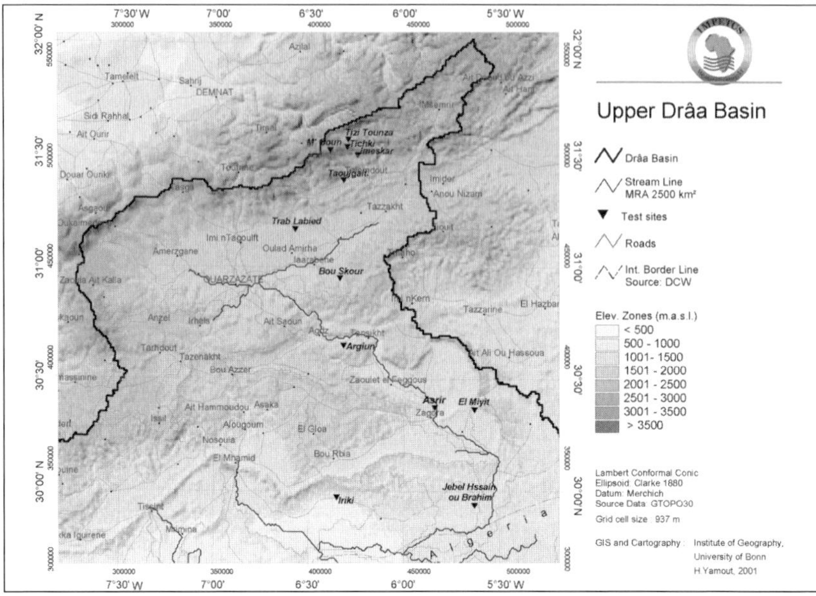

Fig. 3.6.5b. Geographical location of the IMPETUS measurement sites in the Drâa catchment in Morocco. Catchment boundaries marked by black solid line.

The spring precipitation in all three regions shows a weakly negative, temporally variable correlation to the El Niño/Southern Oscillation. Preliminary results suggest that the predicted northward shift of the entire storm track system over the North Atlantic due to anthropogenic greenhouse gas emissions implies a reduction of winter precipitation in the northern and western parts of Morocco. In particular for the region south of the Atlas, a regionalization of the climate change impacts as proposed for the second phase is necessary to determine future precipitation conditions.

In contrast to former studies a considerable influence of precipitation in connection with tropical-extratropical interactions could be demonstrated, particularly in the region south of the Atlas (Knippertz et al., 2003 b). This phenomenon was observed throughout the year, but is most pronounced in the transition seasons. The typical synoptic evolution and involved physical mechanisms of such tropically induced rainfalls were analysed by investigating in total 12 rainy episodes in late summer/early autumn. Results demonstrate that a moisture input from convective clusters or squall lines over tropical Africa and the adjacent Atlantic Ocean, which is transported to northwest Africa on the eastern side of a subtropical upper-level trough to the west of northwestern Africa, is a decisive factor. The involved tropical convection is often triggered by African Easterly Waves. The precipitation over northwest Africa is predominantly caused by upper-level divergence ahead of the trough and/or by surface heating of elevated terrain in the Atlas Mountains (see fig. 3.6.6, right).

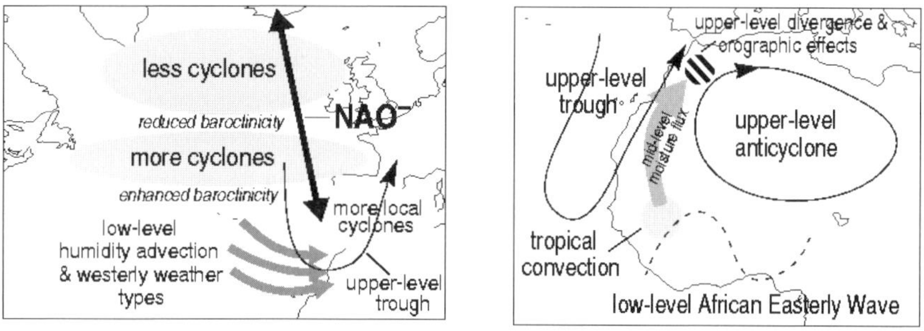

Fig. 3.6.6. Schematic overview over the most important mechanisms of precipitation generation in northwest Africa. Precipitation caused by extratropical synoptic disturbances (predominantly in winter; left) and in connection with tropical-extratropical interactions (predominantly in the transition seasons, right) [taken from Knippertz, 2003].

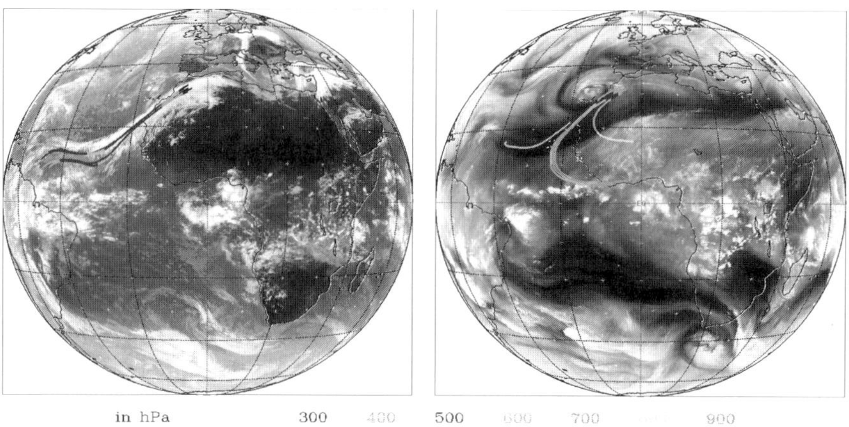

Fig. 3.6.7. Four-day backward trajectories starting from five different points over the southern margin of the High Atlas at 12 UTC, 31.3.2002. Different gray shadings indicate the height of the trajectory (see figure legend). In the left panel the infrared image of 12 UTC, 31.3. (corresponding to the beginning of the backward trajectories), and in the right panel the water vapour image of 12 UTC, 27.3. (corresponding to the end of the backward trajectories) is underlaid. The starting level is 300 hPa (left) and 400 hPa (right) [taken from Fink and Knippertz, 2003).

In addition, an extreme rain event in the region south of the High Atlas on 31st March and 1st April 2002 was investigated, among others based on half-hourly measurements from the 12 IMPETUS climate stations (Fink and Knippertz, 2003). Precipitation totals (of up to 77 mm in 23 hours) range in the order of magnitude of more than half of an average annual sum and constitute the heaviest storm of the last 25 years in this region. The immediate run-off caused flooding and damage to buildings. Besides, a substantial filling of water reservoirs (+23.6% of the total capacity of the great storage lake Mansour Eddahbi) and a storage of water in the High Atlas snow cover (up to 1 m) and in the soil was observed, which positively impacted on the region's water supply until the summer. The precipitation event reveals similarities to the late summer/early autumn cases, but additionally shows the formation of a 'tropical plume' (TP) on the eastern side of the upper-level trough. A trajectory analysis revealed that on the equatorward side of the TP, mid-level moisture transports from tropical West Africa occurred, while the high clouds connected to the actual TP originated close to tropical South America (fig. 3.6.7). In contrast to the cases of tropical-extratropical interactions in late summer/early autumn, large-scale dynamical and frontogenetic effects seem to dominate over local factors as the triggering of convection in the moist tropical air through the daytime heating of elevated terrain.

A climatological attribution of rainfall to tropical or extratropical sources on the basis of the origin of mid-level trajectories that reach the Atlas region revealed that rainfalls in connection with tropical-extratropical interactions account for up to 40% of the annual precipitation.

Literature

Fink AH, Knippertz P 2003 An extreme precipitation event in southern Morocco in spring 2002 and its hydrological implications. *Weather*, 58 (10), pp 377-386

Knippertz P 2003 Niederschlagsvariabilität in Nordwestafrika und der Zusammenhang mit der großskaligen atmosphärischen Zirkulation und der synoptischen Aktivität. *Mitteilungen aus dem Institut für Geophysik und Meteorologie der Universität zu Köln*, Vol. 152, 136 p

Knippertz P, Christoph M, Speth P 2003a Long-term precipitation variability in Morocco and the link to the large-scale circulation in recent and future climates. *Meteorol. Atmos. Phys.*, 83, pp 67-88

Knippertz P, Fink AH, Reiner A, Speth P 2003b Three late summer/early autumn cases of tropical-extratropical interactions causing precipitation in Northwest Africa. *Mon. Wea. Rev.*, 131/1, pp 116-135

Obasi GOP 1999 Hydrology and water resources: a global challenge for WMO. Lecture at the 14th Conference on Hydrology – 79th Annual Meeting of the American Meteorological Society. Dallas, Texas, USA, 16 p

Ward MN, P-J Lamb, DH Portis, M El Hamly Rachid Sebbari 1998 Climate variability in Northern Africa: Understanding droughts in the Sahel and the Magreb. To appear as chapter 6 in: Beyond El Niño – Decadal variability in the climate system, ed. A Navarra, Springer-Verlag

3.7 Capacity Building in Water Management in Palestine - Experience of InWEnt Palestinian Water Sector Training Programme (1995 – 2000)

Ismail Al Baz
InWEnt, Capacity Building International Germany

3.7.1 Introduction

The West Bank/Gaza has a total of about 3.2 million inhabitants: 48% of the population lives in cities, 35% in rural areas and 17% in refugee camps. The West Bank covers an area of 5,682 square kilometres; Gaza covers an area of 365 square kilometres.

The average annual rainfall is about 600 to 800 mm on high land while in Gaza and Jericho it is only between 200 and 400 mm.

In Palestine the water situation is in a very bad shape due to military occupation and military restrictions, absence of Palestinian central body for water resources management and lack of institutional co-ordination and financial resources.

Qualified human resources who are able to deal with the water problems and to improve the situation were not sufficiently available in Palestine when this project started in 1995. Training institutions lacked practical training facilities, professional trainers, practical experience and networking. Therefore, the training needs for technical and managerial personnel in different fields of the water management sector in Palestine were very high. With the beginning of the Oslo negotiations, international donor organisations started offering technical and financial aid for the Palestinian community and capacity building was one of the core issues. of international development co-operation with Palestine.

In this paper, I am going to introduce the results of a training project of InWEt (a German Foundation for International Capacity Building) which was conducted in Palestine from 1995 to 2000.

3.7.2 Background information

The water situation in Palestine

In the Oslo II agreement, article 40 of Annex 3 "Israel recognises the Palestinian water rights in the West Bank. The Israelis accept to transfer authority of water management to the Palestinians which include responsibilities of water and sewage infrastructure activities in the West Bank and Gaza". The future water demand of the Palestinians according to Oslo II are estimated to be between 70 - 80 MCM/year.

Israel agreed to supply the Palestinian population in the West Bank with additional 28.6 MCM/year of freshwater for domestic use during the interim period . The remainder of the estimated quantity of the Palestinian future needs (41.4 - 51.4 MCM/year) shall be developed by the Palestinians from the Eastern aquifer and other sources in the West Bank as agreed upon. According to Jad Isaac, the Palestinian Water Authority received only 12 MCM of this agreed amount.

Institutional framework

In 1996 the Palestinian Water Authority (P.W.A.) was founded and it started its responsibility as the regulatory body for water resources management for Palestine. P.W.A. is responsible for the legislation, monitoring and human resources development. Water distribution and wastewater treatment still lie under the responsibility of municipalities. In the same year, the P.W.A. started also a reformation in the water sector in Palestine and the first water contract was signed for the Gaza strip where 26 municipalities are now served through private companies. This private system will be extended in other parts of Palestine.

Water resources

The Western Aquifer System, the largest, has a safe yield of 362 MCM/year (of which 40 MCM/year are brackish). This basin is mostly located within the West Bank boundaries, whereas 80% of the storage area is located within Israeli borders. Groundwater flows to the coastal plain in the West, making this a shared basin between Israelis and Palestinians. The water is mainly used for municipal supply because of its good quality. Israelis exploit the water through 300 deep groundwater wells to the west of the Green Line. The Palestinians consume only about 7.5% of its safe yield through 138 wells (120 for irrigation and 18 for domestic use) The cities Qalqilya, Tulkarm and West Nablus. exploit the water through 34 springs with an annual discharge of about 2 MCM/year (Jaad Isaac).

The Northeastern Aquifer System has an annual safe yield of 145 MCM (of which 70 MCM are brackish). 100% of the water is recharged by rainwater falling within the West Bank area. Palestinians consume about 18% of the safe yield of the aquifer through 86 wells in the Jenin district and East Nablus for both irrigation and domestic purposes. There are 24 springs with an annual discharge of about 3.3 MCM (Jad Isaac)

The Eastern Aquifer System has an annual safe yield of 172 MCM (of which 70-80 MCM are brackish). It lies entirely within the West Bank territory. Before 1967 the eastern aquifer was under the control of Palestinian farmers. Thereafter, Israel expanded its control over this aquifer and began to extract the water to supply Israeli settlements who settled in the area. after 1967. This aquifer is mainly drained by a group of springs. There are 56 springs with an annual discharge of about 55.5 MCM (Jad Isaac)

The Palestinians extract their water through 122 wells (109 for irrigation and 13 for domestic use). Most of these wells have been over-pumped which has led to a drop down in the water table and deterioration of its water quality.(Jad Isaac)

The coastal basin of Gaza is a shared aquifer with Israel. More than 2200 wells tap this aquifer with depths mostly ranging between 25 and 30 meters. The annual safe yield is 55 MCM while water extraction is 110 MCM. Due to this extensive use, the groundwater table is dropping below sea level with the result of sea water intrusion into groundwater. Additionally, there is a return flow from very intensive irrigation activities contributing to further contamination of the groundwater. According to the Palestinian Environmental Authority, nitrate concentrations in some sites in the Gaza Strip reach an amount of 250 – 1000 mg/L.

Table 3.7.1. Natural resource management indicators in Palestine (after Jad Isaac in Co-operation on Transboundary Rivers, Al Baz et.al., 2002)

Indicator	Palestine
Population (million)	3.2
Total water consumption (MCM)	286
Agriculture water consumption (MCM)	174
Agriculture water consumption per capita and year (CM)	58
Domestic water consumption per capita and year (CM)	30
Daily domestic water consumption per capita (in litres)	63 – 104

3.7.3 InWent Capacity Building Programme in Palestine (1995 ñ 2000)

InWEnt, Capacity Building International Germany, organises and implements on behalf of the German Ministry for Economic Co-operation and Development international training and dialogue programmes all over the world. InWEnt is a merger between the former Carl Duisberg Gesellschaft (CDG) and the former German Foundation for International Development(DSE). Both organisations have 50 years of experience in capacity building and more than 35.000 professionals take part in InWEnt programmes every year.

InWEnt's Department for Environment, Natural Resources and Food, Division for Environmental Policy and Environmental Management, plans and conducts training programmes in the fields of water management, environmental protection, natural resources, renewable energy, rural development and food in order to promote environmental protection and resource conservation. The programmes are designed for junior executive personnel and decision-makers from developing countries who deal professionally with water and environmental issues.

Main Objectives of our training programmes aim at, developing human resources, enhancing intercultural dialogue, transfer of practical experience in integrated and sustainable water resources management, strengthen potentials for co-operation on international water resources and improving efficiency of water and wastewater utilities

To achieve this, different tools are applied like, Practical Training on the Job in *Germany and abroad,* networking, workshops, international conferences and *web-based training*

From 1995 – 2000 we conducted a training programme in Palestine with the main objective to upgrade Palestinian professionals to be able to plan and implement a better and more efficient management of their very limited water resources. The training started with a planning workshop which was held in March 1995 in the city of Bethlehem in the West Bank with all stakeholders from the West Bank and Gaza. During the planning workshop the targets, target groups, training modules, contents and the steering committee were selected and set up.

In the planning workshop two different target groups have been identified and allocated in different training programmes:

Target groups

- Technicians who are responsible for operation and maintenance of water and wastewater treatment from municipalities, universities and NGOs (Middle Management)
- Engineers who are the heads of water and wastewater departments at the different municipalities(Upper Management)

Training modules

Different training programmes have been suggested in the planning workshop and modified by the steering committee :

- Management training with focus on financing management, construction management, personnel and project planning management
- Operation and maintenance of water treatment plants, water distribution, water networks, water losses detection, water metering and water billing
- Groundwater management: groundwater modelling, quality control, GIS, well design and construction
- Environmental impact assessment
- Sludge treatment
- Solid waste management
- Water quality management

Organisation and implementation of the training modules

The training courses were organised in very close co-operation with the steering committee in Palestine which was representing the main Partner of InWEnt : Jerusalem Water Undertaking (JWU, private water company supplying more than 200.000 inhabitants in Ramallah and Jerusalem area), both Palestinian Association of Engineers in the West Bank and Gaza, the Palestinian Water Authority (PWA), representatives of universities (Birzeit University, Al Najah University/ Nablus and Islamic University Gaza), representatives of NGOs (PHG- Palestinian Hydrology Group), representatives of Municipalities(Gaza Municipality). The steering committee met 4- 5 times a year and decided on the contents of the training courses and selected the participants. The courses had been previously announced in the local newspapers and the application procedures were managed through the offices of the Association of Engineers in the whole of Palestine by distribution of application forms and organisation of interviews. The training courses have been offered in different 3 - 7 day modules to adapt several target groups and different needs and conducted in several parts of Palestine(Al Biereh Municipality, Birzeit University, Al Najah University/Nablus, Hebron city/Association of Engineers, Islamic University /Gaza).

Most of the training modules have been organised locally in Palestine and with Palestinian experts. During the period of project implementation, some training courses were conducted abroad:

- Two 6 months training courses on water and wastewater management in 1996 and 1997 were conducted in Germany
- One 6 weeks training course in water supply management was conducted in 1998 in Rabat/ Morocco in co-operation with Office National de l Eau Potable (ONEP)
- One 4 weeks course was conducted in 1999 in Tunisia on wastewater management in co-operation with Tunisian Wastewater Authority (ONAS)

3.7.4 Results

- Around 600 Palestinian engineers, more than 25 % of all engineers working in the water sector in Palestine, were trained
- A relatively high number of women (more than 12 %) participated in the training courses
- NGOs, academic and governmental representatives were involved in the design and implementation of the training courses and in the selection of participants
- Most training courses were conducted in Palestine with local experts
- Experience was shared with neighbouring Arab -countries (Tunisia, Morocco)
- Most training courses were practice-oriented and lecturers came from water and wastewater companies.

3.7.5 Selected InWEnt Capacity Building Programmes in Water Management that are being offered for the MENA Region

Sustainable Water and Wastewater Management (SWAMP Project) (2002-2005)

- integrated management of water supply and wastewater disposal
- (long-term training courses, 9 months in Germany, Oct. 2002 - July 2003, Oct. 2004 – July 2005)
- water and wastewater management Jordan , short-term, 1-26.06. 2003, wastewater management Tunisia, 8-26 Sept. 2003
- water management in the Middle East, conference, 16-24.03., Kyoto, Japan,Third World Water Forum
- experts assignment, Yemen 2002, Jordan and Palestine 2003, 2004
- water supply management, 4 weeks course, 2005 in Morocco
- water management in the MENA Region, conference, Morocco 2005
- web-based training course 2003, 2005
- follow-up workshop 2005

Euro-Mediterranean Regional Programme for Local Water Management (MEDA Project):

EU-funded project (2003 – 2006):
Efficient Management of waste water treatment and reuse:
Countries: Palestine, Lebanon, Turkey and Jordan. 12 Different local training courses 8 weeks each, 3 regional training courses 4 weeks each, 3 web-based training courses, 4 country studies, 4 pilot plants, 1 handbook on innovative solutions, 1 regional conference, several regional and evaluation workshops.

Policy Guidelines for Wastewater Management in the Gaza Strip, Palestine:

EU-funded project, Life Third Countries (2000 – 2002)
Draft proposal for policy guidelines for wastewater treatment and reuse, published in December 2002. It included several workshops, training courses on sampling techniques in Gaza 2000, and Morocco 2001, a country study on water situation in Gaza and a regional study on wastewater treatment and reuse standards in the region.

3.7.6 Conclusions

Capacity building is very important in order to upgrade the water situation in Palestine by supporting the Palestinian engineers in solving their water and wastewater problems. During the 5 years of this training project, many Palestinian professionals have become accustomed with different aspects of the integrated water resources management, wastewater treatment, water pricing and solid waste management. Many of these engineers have established contacts with water experts from abroad, Germany, Tunisia and Morocco and were able to transfer some of this knowledge to Palestine. In spite of the political situation in Palestine which has been deteriorating during the last years, many developments and improvements have been achieved in the water sector:

- the establishment of the Palestinian Water Authority in 1996 as a central body for water resources development. Since that time the PWA is doing an excellent job in Palestine
- privatisation of part of the water management sector in the Gaza Strip. Management contract for 26 communities in Gaza
- establishing different water training centres at the Palestinian Universities (Islamic University Gaza, Birzeit University West Bank)
- construction of a wastewater treatment plant in Al Biereh city to serve 20.000 inhabitants and for wastewater reuse
- similar treatment plant projects are planned for Hebron, Nablus and Sulfit
- rehabilitation of the wastewater treatment plants in Gaza City
- solid waste management project in Gaza (construction of dumping place in Der Al Balah and User Association)
- rehabilitation of water network in Ramallah and Bethlehem.

These changes indicate that within a short time a huge development took place in Palestine in the field of water and environment. We can be sure that in the near future when the political situation will become more stable and the peace track will be opened again, the process of development and improvement in the water sector will continue with success.

In the light of the information I have presented in this paper, we can safely say in summary that the water reserves of Palestine - which for natural reasons are in a very short supply - are being used up excessively and becoming exhausted. The course the peace negotiations is taking is having a severe impact on this development

Living together in peace means being willing to share the limited water resources available. This willingness includes relinquishing the role of the "super power" in the region and starting to take practical steps towards joint development, which basically means sharing the limited water resources which are vital for everyone.

References

Ismail Al Baz, Volkmar Hartje, Waltina Scheumann (2002) Co-operation on Transboundary Rivers, Nomos Verlag, Baden Baden.

Yasser El Nahhal (1995) Water situation and problems-A case study from Gaza Strip.Environmental Protection and Research Institute.Gaza/Palestine.

Bernd Zuppke (1994) Wasserprobleme im Nahen Osten, Wasser und Boden *8*. pp 63-66.

Jamal M. Safi (1993) The State of Environment in the Gaza Strip Environmental Protection and Research Institute, Gaza, Palestine.

Hisham Zarour and Jad Isaac (1993) Nature's Apportionment and the Open Market: A Promising Solution to the Arab Israeli Water Conflict.Water International 18: pp 40-53.

P. Wolf (1992) Durstiges Israel.Eine kritische Betrachtung zur Wasserwirtschaftlichen Situation Israels. Berichte Nr.28 der Gesamthochschule Kassel, Fachbereich 21, Internationale Agrarwirtschaft, Fachgebiet Kulturtechnik und Wasserwirtschaft.

R. Abdulhadi, Abdellkarim As'sad, Karin Assaf, Marwam Hadad, Reinout Koning, Taher Naser Eddin and Philip Rorak (1994) Water Conservation in Palestine, Centre for Engineering and Planning, Ramallah, Palestine.

3.8 Efficient Groundwater-Management for Organizations with a Small Financial Budget – the Continuous Improvement Method

Kurt von Storch
Jenaer Straße 16, D-65205 Wiesbaden

Abstract

All over the world, the majority of groundwater resources are managed by small and medium-sized companies. Despite having limited personnel and financial resources, these companies have to ensure an adequate groundwater management. This is an objective that any company can achieve with the means at its disposal by using the continuous improvement method.

In contrast to the technical approaches mainly used, which focus on the aquifer and the groundwater, the method presented here is specifically geared towards the individual conditions of the company extracting the water. A method of sustainable groundwater management is steadily built up in this manner by starting with the resources a company has available. The advantage of this approach is that any company, irrespective of its financial and personal resources, can, before long, achieve suitable groundwater by doing as much as it can within the bounds of its possibilities. Instead of calling in expensive external experts at the start, the continuous improvement method initially calls for the participation of the company's employees and other individuals and organizations who are involved. A systematic continuous improvement means that the desired perfection of the groundwater situation is followed in a stepwise manner. This methodology covers three scopes: groundwater abstraction, groundwater protection and an appropriate business management regime.

Key words

Water resources management, groundwater protection, risk management, quality management, natural tracers, groundwater overexploitation

3.8.1 Introduction

Most countries in the Middle East suffer from groundwater-scarcity and an increasing deterioration of the general water quality. Sustainable management of the resource water is – as elsewhere in the world – of essential relevance both economically and socially. The objective is to reach a form of water abstraction which could, in the long term, provide all participants in a region with adequate quantities of water of an appropriate quality. The basic requirement, therefore, is the realization of the principle of sustainability as it is described on the World Commission of Environment and Development and is being applied to groundwater use. Nevertheless, the realization of a sustainable groundwater use is a process, which often needs years and which will be hindered, for example, by the following problems:

Problem: Frequently insufficient knowledge about the real groundwater situation

Adequate knowledge of the real groundwater situation is necessary for optimal exploitation of the available groundwater. However, in most cases there is insufficient knowledge concerning recharging rates and their fluctuation as well as of the structure and the vulnerability of the aquifer and the predictable development of the groundwater under the existing abstraction conditions. In general, it is not possible – despite progress in modelling techniques – to predict the water development over years in a manner which yields sustainable development. Custodio (2002) pointed out the requirement for progressive refinement of the knowledge about the groundwater and the aquifer on account of the many hydrogeological uncertainties.

Problem: Generally scarce financial resources of the organizations concerned

A large proportion of the water resources of the Middle East is managed by small and medium-sized companies. Technically (and therefore also financially) intensive solutions are not usually appropriate for these organizations. Hence, the major proportion of the groundwater resources cannot be managed adequately. Sustainability starts with the finances: this means first of all, that an organization has to be able to manage the right form of groundwater abstraction over the long term within its own financial resources. External financial funding by third parties can only be used as a catalyst. An efficient groundwater management has to be developed from the point of view of the organization concerned and its capabilities and not – as often occurs – from the scientific-technical perspective alone.

Problem: Groundwater management with single projects

An important problem is rooted in 'classical', often relatively cost-intensive information procurement. Attempts are made to improve the supply of water by conducting individual projects (surveys, mapping, computer-based modelling, abstracting new water resources, etc.). However, this discontinuous, project-like approach entails the company in having to mobilize (often considerable) additional personnel and financial resources for a certain period of time. This is something the many small water supply companies all over the world in particular often only carry out once the quality problem has become relatively serious.

Further Problems:

Inadequate legislation, national and/or external funding policies and limited national infrastructure are hindering adequate groundwater management (Lloyd 1994). Also federally funded engineering and federally administered technological regulation are ill-designed to address complex, site-specific water resource problems (Lant 1999).

3.8.2 General demands made on groundwater management systems for organizations with a small financial budget

A groundwater management, that focussed on the above described problems, cannot be based on technical solutions – anyway these are often "island solutions". Examples are the techniques of groundwater modelling, which are, of course often an important tool in groundwater management. Computer based groundwater models cannot displace a groundwater management system, because they do not recognize the "human factor". This means the human management aspects influence groundwater development. Examples are:

- The willingness to adapt the abstraction rates durably on the natural conditions – i.e. in the case of a change in management or economical constraints;
- The general political conditions, which could change;
- The willingness and capability to recognize future industrial developments in the peripherals;
- The technical knowledge of the personnel and their motivation;
- The behavior of neighboring companies abstracting groundwater or potentially contaminating it.

An effective groundwater management has to be orientated on this "human factor". The quality management system according to ISO 9001 as an established management system applies, for example, to this point of view. Technical solutions could only be an instrument within a management system. Their usability depends on the resources an organization can apply to them.

A groundwater-management system, that will operate efficiently under the constraints of a low financial budget, has to be designed in a way that it is suitable, especially for organizations, exhibiting the following features:

Small water utilities: They provide small regions or towns. They are independent organizations or they belong to larger nationwide or international organizations.

Agricultural organizations with the demand on additional irrigation: Companies, which are dependent on water from their own wells.

Food producers: Companies using groundwater for food production, as, for example, the beverage industry or bottlers of packaged drinking water (e.g. "water cooler"). Often they are dependent on a few wells.

Baths and spas, if the supply is from their own wells.

Local water authorities (regional): As a superordinate institution inside a town or district, they are responsible for the water quality and availability in a region. They have to control a group of groundwater users often with relatively small personnel resources. They play a central role in the realization of a sustainable water supply. Standardized management systems are a helpful tool for supervisory authorities.

A groundwater management for these organizations needs to be designed according to the following principles:

Comparability and simplicity

It should accord to a unique standard, so that it is comparable. Additionally it should be easy enough to apply to each organization. The comparability is an important tool for assessing a system in relation to others. In case of an objective assessment, the willingness for an improvement of the groundwater situation could be made visual. Further advances are:

- Funding institutions (as banks, third-party-funds) have better possibilities to check if the organizations concerned really fulfil the organizational requirements for a sustainable groundwater abstraction. Also they can check the long-time effect of the success of their benefits. Better conditions for verification could increase the readiness of banks to give financial support for investments in water supplies.
- Water authorities or water associations could transmit to the groundwater users such a standardized management system in form of a burden. Water authorities can oblige companies to use a standardized management system, in the case of receiving an official admission for groundwater abstraction.
- Organizational defaults of individual companies could be verified more easily in the case of a widespread groundwater deterioration.
- Start-up funding with governmental funds could be provided with a smaller administration effort.
- It is easier to promise "water awards" or to accord "groundwater quality seals" which provide an additional motivation for groundwater users.
- It is possible to rank the groundwater management systems of different organizations. This could provide additional motivation for local water authorities, in particular.

Low initial threshold and frictionless functionality

The less the effort required to start implementing such a standardized system the more organizations will be encouraged to take this route. Hence, a low initial threshold is essential for a successful groundwater management system. Ideally every company ought to be able to start with a groundwater management system independent of its financial and personnel resources. It has to be frictionless in day-to-day operation and the additional inputs required must be acceptable. Inexpensive planning and monitoring instruments which still function reliably, despite the limitations on company resources, are important tools for meeting these heavy demands.

Integrated Management

The groundwater management system must be integrated into the internal-organizational processes. Groundwater for production has to be regarded like other production-components and, as such, is subject to materials administration, of risk management and of quality assurance. If a management system is appended and not integrated there is the risk that it will not be updated consequently. If this is the case then an important condition of continuity for the improvement of the groundwater situation is missing.

Networks and long-distance support

Small organizations are particularly dependent on external support. Involvement in networks helps especially in the transfer of know-how and technology.

Continuous improvement

The system has to enforce the continuous, economically viable improvement of groundwater management.

Inclusion of the participants

Groundwater problems should be solved only by experts – that is the general opinion which causes disintegration of sustainability in daily business operations. The predominant reliance on the know how of experts is a barrier to the implementation of an integrated management system. The involvement of external experts is necessary in special situations, but an integration into the organizations can only occur if the staff has an adequate basic knowledge about groundwater aspects. Sophocleous (2000) has pointed out the importance of an extensive understanding of the nature, complexity, and diversity of ground water resources as an influence on the attitudes of the various stakeholders.

3.8.3 The continuous improvement method

Instead of the discontinuous process in single projects as described above, the principle of continuous improvement is applied in small steps. This "Kaizen" philosophy formulated in Japan is a key element of the generally recognized management systems for product quality (e.g. ISO 9001:2000) and for environmental protection (e.g. EMAS, ISO 14001). The principle of the continuous improvement can be used at any time by any water supply company, even under difficult financial or hydrogeological conditions. Continuous improvement can be started from any level of quality. Over time, the many small steps accumulate to create a significant improvement in water quality. In addition to the technically possible standards to solve problems in the water supply, above all, the utility's resources (personnel, infrastructure, finance, etc.) are taken into account by the method.

However, achieving high water quality through continuity is only possible if a suitable organizational management system is in place. Hence, the continuous improvement method is a combination of techniques from the spheres of geology and business management. A systematic continuous improvement means that the desired perfection of the situation with regard to water quality and availability is followed in a stepwise manner.

Therefore, the following requirements are obligatory:

A quality-policy with regard to groundwater is necessary. There the long-term objective, the achievable perfect situation in groundwater aspects is documented. This quality policy must be known inside the organization as well as by all the participants involved. It is the red thread by which the measurements to groundwater assurance are orientated.

The long-term objective will be realized step by step as a result of reaching short-term objectives. The short-term objectives could come from very different areas: e.g. the improvement of the quality of the collection of data, of the maps or of the monitoring system. Short-term objectives should be measurable and reachable within limited timespans (up to max. 3 years) and – that is very important – they have to be set out in writing and personal responsibilities must be allotted.

Financial resources must be provided for short-term objectives and budgeting must be transparent. The short-term objectives must be communicated and the advantages of reaching them must be outlined to the staff and participants affected within the region. Externally (e.g. funding institutions) the achievement of short-term objectives can be combined with financial incentives. It is necessary to have a manual in which the inner-organizational structure and the processes for the realization of the objectives are documented. Ideally the realization of short-term objectives will be verified by neutral auditors.

3.8.4 Examples of technical measures

The continuous improvement of groundwater quality covers three fields:

- Groundwater extraction: Extracting groundwater in a way that the quality and availability is constant (entailing appropriate monitoring);
- Groundwater protection: Continuously reducing any internal and external risks, which could impair the groundwater quality and availability;
- The management system: Having a business management regime, which systematically sustains the continuous improvement process.

Continuous improvement of groundwater extraction

When applied to groundwater extraction, the continuous improvement method has two objectives:

- Optimizing knowledge about groundwater;
- Refining the monitoring programme.

The interaction between the two improvement cycles "improving groundwater knowledge" and "refining the monitoring programme" results in an upward quality spiral (Fig. 3.8.2).

First constant improvement cycle: "improving groundwater knowledge"

Collection of basic data. The first step is to recollect systematically the organization-owned documents concerning the aspects listed below. The particular knowledge must be continuously refined in the future.

Table 3.8.1. Examples of indicator parameters

Natural protection of the groundwater (influence of young water)	e.g. tritium, CFCs, oxygen-18 (time series)
Anthropogenic influences	e.g. nitrate, borate, pesticides, CHCs, phenols, PACs etc.
Information about the groundwater recharge area	e.g. deuterium/oxygen-18
Extraction tied to certain geological units	Mineralization typical of aquifers, characteristic trace substances
Mining of a groundwater resource	e.g. via the alteration of the mean retention time with the help of carbon-14
Influence of highly mineralized deep groundwater	e.g. chloride, boron, Br/Cl

- Information on the geology, hydrogeology, topography, meteorology, etc.;
- Information on water quality: physics, chemistry, isotopes, microbiology;
- Previous quality problems (at the organization's own water abstraction or a neighboring one);
- Written interpretation of groundwater genesis, taking into account all the available data including neighboring wells (if possible). The main role is played by interpretation of the chemical parameters and the environmental isotopes, since they provide the clearest reflection of groundwater genesis;
- Outlining a draft conceptual hydrogeologic model using maps and profiles;
- Deciding action priorities: concrete aims to clarify outstanding questions in order to refine the conceptional model.

Second constant improvement cycle: "monitoring programme"

The monitoring concept is drawn up on the basis of the hydrogeological conceptual model:

- Online well parameters;
- Parameters indicating anthropogenic influences;
- Natural tracers representative of the groundwater genetic type;
- Extensive chemical analysis as a reference ("baseline");
- Establishing the reasons for any deviation in subsequent water analyses from reference analysis (if possible, with the involvement of external experts) and additional investigations;
- Comparison with the conceptual model and adaptation of monitoring.

Continuous improvement in groundwater protection

Alongside groundwater extraction, the second pillar of water management is groundwater protection. Each water operation is affected by a local individual risk situation, which may, for example, be influenced by:

- Industry and agriculture (e.g. fertilizer, soil-salinzation as a consequence of intensive irrigation, pesticides);
- Usage competitors, such as waterworks or other plants;
- Own production wells, monitoring wells, external wells;
- Road and railway traffic;
- Future construction projects and land use;
- Natural problems, such as flooding (e.g. germs, chloride) or geogenic contamination (e.g. fluoride, arsenic, chloride);
- Influence from lakes, seawater intrusion etc.;
- Influence from settlements and populations (e.g. sewage, hydrocarbon spills);
- Earthquakes or other shocks.

Groundwater, especially in the case of the public water supply is usually protected by protection zones in which certain activities which could put the resource at risk are banned (in Germany: DVGW 1995). In other words, these regulations constitute protection via prevention. If, however, groundwater contamination within a protection zone is caused by an accident, risk management measures are necessary. The snag is that risk management affording genuine protection is often only implemented to a limited extent at the companies concerned. Continuous improvement comprises warding off hazards in order to genuinely protect water resources. This includes a whole range of measures which need to be integrated into the company's organization:

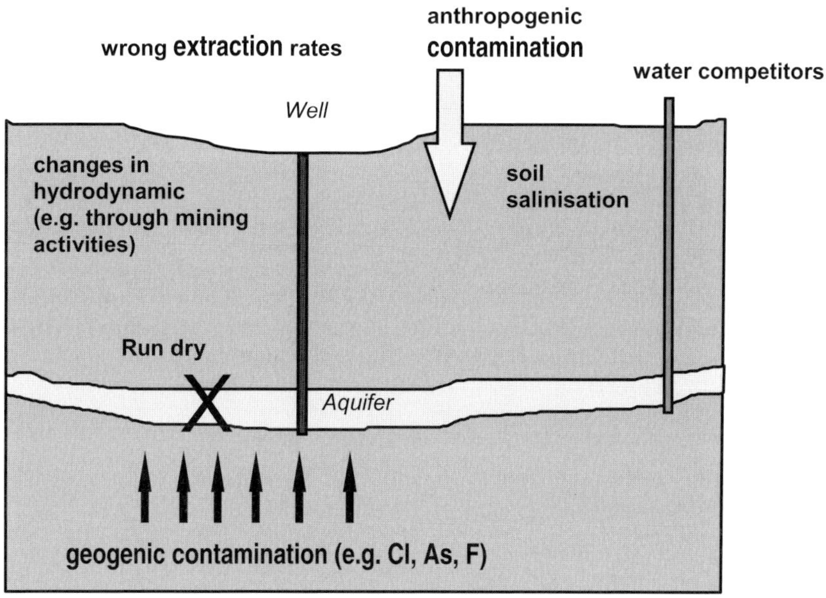

Fig. 3.8.1. Risks for groundwater deterioration, examples

- Continuously increasing knowledge about the natural protection function of the soil and the geological conditions;
- Continuously refining the gauging of internal risks and local external risks, along with their assessment;
- Continuously adapting and improving systems of measures to prevent acute dangers;
- Continuous training for staff; providing information to staff and possibly external organizations, such as the fire brigade.

In order to determine a suitable protection and monitoring concept, the company has to differentiate between two types of risk:

Gradual contamination: if contamination is caused by anthropogenic contaminants due to a long time input such as nitrate from agriculture or – geogenic – for example due to an upwelling of chloride-rich deep waters, larger areas could often be affected. In this case, long-term measures need to be conceived to improve groundwater quality over larger areas.

Sudden contamination: caused by an accident or natural hazard (flood, earthquake etc.). The actual harm caused to a company by an accident involving substances suddenly contaminating the water depends not only on how badly affected the water is theory but also the time required for the internal defense mechanisms to take effect.

Therefore, the continuous improvement process covers:

- The knowledge of the risk situation in the catchment area;
- The effectiveness of prevention / mitigation and the speed of defense measures.

First constant improvement cycle: knowledge of the risk situation "knowledge risk"

Risk identification:

- Evaluating previous groundwater quality problems as a result of internal and external influences and their causes – comparison with the hydrogeological conceptual model;
- Recording and written documentation of (initially the most serious) potential risks and their mapping (e.g. using a GIS).

Knowledge of the individual risk situation must be continually refined as time goes on. This is a task, which is required to be carried out by all company employees, neighboring companies, and the supervisory water authorities.

Risk assessment:

- Assessment of the consequences of the damage concerned;
- Inclusion of information from the natural environment (soil, geology, hydrogeology, groundwater genesis, etc.);
- Risk classification ("ABC"-rating) taking into account the natural protection properties of the overlying strata.

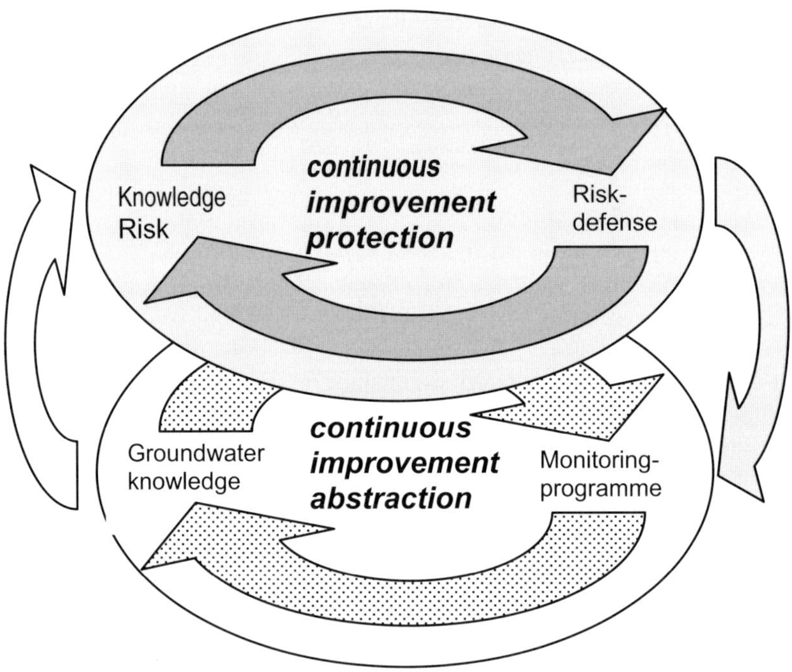

Fig. 3.8.2. The interaction of continuous improvement: Groundwater abstraction and protection

Second constant improvement cycle: improving prevention and defense measures "risk defense"

- Constantly adapting the monitoring program to the current state of knowledge about the risk situation;
- Prioritizing preventive measures;
- Compiling and updating emergency plans with the involvement of public authorities and the fire brigade; improving process and working instructions; adapting communication management;
- Staff training to recognize and prevent danger.

This method requires co-operation among all those affecting groundwater in any way (companies using groundwater, potential contaminators and supervisory authorities). The water authorities can play a central role since they can include a number of different groundwater users.

3.8.5 The organizational measures – the management system

The obvious choice is to base the standard orientated on the international quality norm ISO 9001:2000 (International Standardization Organization 2000). This standard enables global comparability. Four areas in ISO 9001:2000 need to be emphasized:

- Responsibility of the management;
- Management of resources;
- Product implementation;
- Measurement, analysis and improvement.

Documentation takes the form of a management manual, which forms the heart of a quality management system. However, the area of risk aspect (danger of groundwater contamination) must be described in more detail than is stipulated in ISO 9001:2000. Finally, the management system needs to be reviewed by a neutral body in order to make sure it is completely objective.

References

Custodio E (2002) Aquifer overexploitation: what does it mean?.- Hydrogeology Journal *10*: pp 54-277

DVGW (1995) Protective Areas, Part 1: Protective Areas for Groundwater - Code of Practice W *101*: 23 p ISSN 0176-3490

International Standardization Organization (1996) Environmental management systems – Specification with guidance for use (ISO 14001:1996).- Beuth Verlag Berlin, 36 p

International Standardization Organization (2000) Quality management systems – requirements (ISO 9001:2000).- Beuth Verlag Berlin, 63 p

Lant (1999) Human dimensions of watershed management. Journal of American Water Resources Association *35* (3): pp 483-486

Lloyd JW (1994) Groundwater-management problems in the developing world.- Applied Hydrogeology *4*: pp 35-48

Sophoclus M (2000) From safe yield to sustainable development of water resources – the Kansas experience.- Journal of Hydrology *235*: pp 27-43

WCED (1987) Our common future. World Commission of Environment and Development (Bruntland Commission). Oxford Univ. Press, No. *4*

4 Water as Source of Conflict

4.1 Water Conflict and Water Management in the Middle East

Jamill Sabbagh

RegiComun-Institute for Stratetic Regional Management, Department of Geography, University of Mainz

"Water is Life": this is often forgotten in countries with abundant water ressources. As a ressource it only gains importance once it is in short supply or once its quality declines. In the case of shortage the battle for water becomes inevitable. At the moment we have up to 200 conflicts of border-crossing water systems in the world. The Middle East is considered to be a typical example of this politically significant problem

4.1.1 Introduction

For the Near East problems and conflicts, resulting from water abundance, will be one of the greatest challenges in the near future. Most of the countries in the respective area have already to undertake immense efforts to ensure their water supply. This problem is just right to worsten the cross-national relations in this region anyway full of ethnic, religious and political strains. Indeed, the problem of water abundance could motivate the individual countries to work together in case of recognising their only chance in it. Probably, a first vague step towards a regional cooperation has already been made. The plans for the Near-East peace process envision bilateral dialogues of negotiations to be added onto the multilateral level. One of the working groups, established on the first multilateral conference in Moscow in January, will be engaged in the issue of water in Vienna from the month of May on (Sabbagh 1999).

The main conflict region is build up by the abutting states of the Jordan River network (Lebanon, Syria, Israel, the Palestinian districts and Jordan).

The River Jordan is a complex river system, whose four main inflows (the Dan, the Hasbani, the Banias and the Yarmuk) spring of different countries. The river bed contains of very little water to supply the five abutting states. Except of the Lebanon all of these countries suffer of too little water.

At a first glance

In Israel green avenues, many lawn sprinklers on private and public ground, private and public swimming pools etc. convey - due to most modern watering technologies and, if compared with other regions, high water ressources - the impression that water shortage does not exist.

In the Palestinian areas the contrast is most evident next to the Israeli settlements, which give the impression of "water abundance", the Palestinian villages seem to suffer from "water deprivation." Barren and dry open spaces, trees craving for water, and defective water pipes determine the first impression of the Palestinian villages.

In Jordan, too, one encounters empty hotel swimming pools, and one's eyes are directed towards unaccountable water containers on the roofs as well as towards water merchants crossing the towns in their tank trucks. ***All this evidence testifies to a situation of "water stress."***

Historical Outline of the Water Conflict

The history of earlier societies shows the importance of water in this arid region. The Nabateans in the 2nd century B.C. as well as Romans and Byzantines until the 6th century, but also the Arabs in the Middle Ages developed forms of settlement that were adapted to the water shortage. Their forms of settlement were designed to avoid evaporation when transporting or storing water and thereby to secure their survival during dry periods.

Only at the beginning of the 20th century and particularly with the political reordering of the Middle East after the First and the Second World War, "water" gained its potential for political, economic, and ecological conflict. The initial causes for the politicization of this conflict can without doubt be ascribed to the emerging Zionist state on the one hand side and, on the other hand side, to the outbreak of Arab nationalism, and also to the altered forms of economy. The Zionist movement strove for a state territory adapted to the water ressources. That way the claim for the water-abundant north up to the river Litani was - pointing out the high immigration rates - for the first time pronounced and economically justified (cp. Vallianatos Grapengeter 1996). Moreover, the rivers Jordan and Jarmuk, which contain a lot of water, were for the first time since 1917 declared an international border which separated different mandated territories from each other.

After the foundation of the state of Israel in 1948, "water" became the central issue in the military conflicts between Israel and its neighboring countries. In the 50ies there was a "Cold War for water" (McDonald-Plan or Johnston-Plan 55). Israel started draining the Hula-valley and constructed afterwards the "National Water Carrier" from the Sea of Galilee up to the Nordnegev. The withdrawn amount of water (420 million m^3/a during the 80ies) was regulated through the construction of floodgates where the Jordan exits the Sea of Galilee. This step had serious consequences for the Jordanian, but also for the Palestinian side.

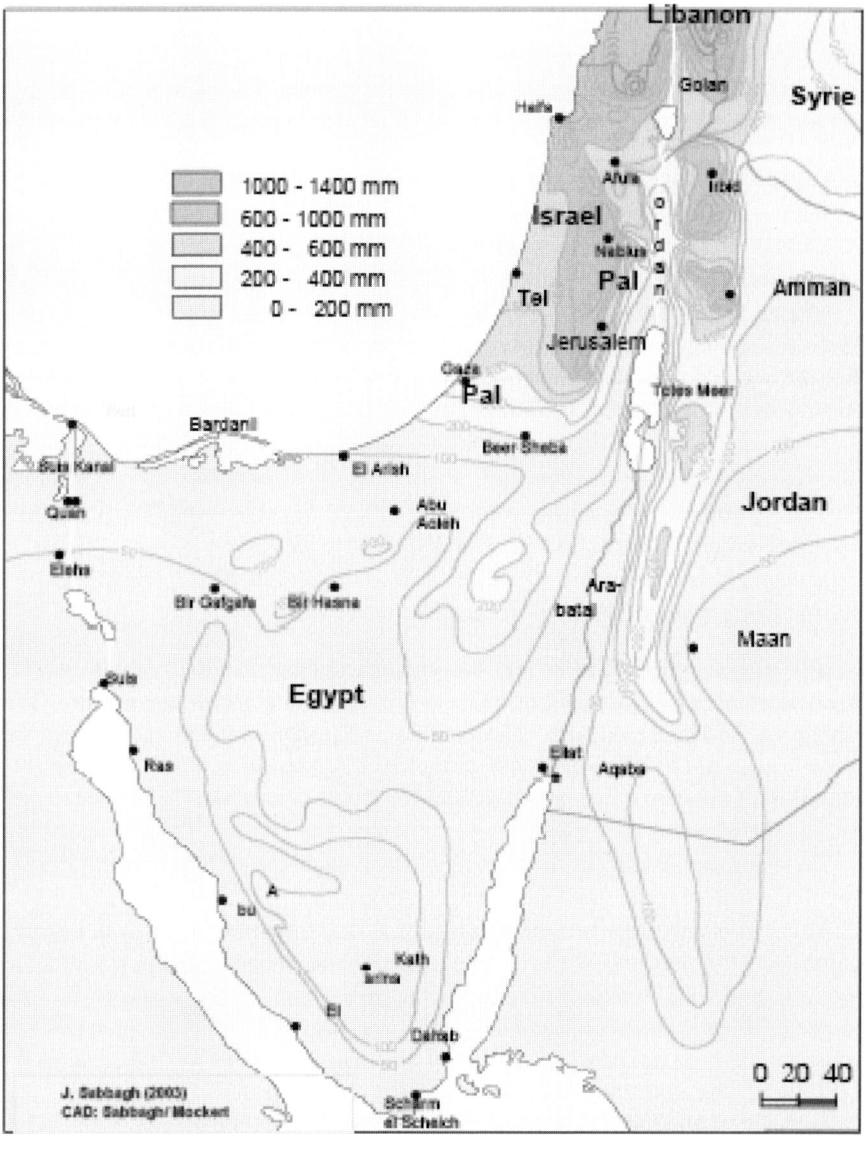

Fig. 4.1.1. Annual rainfall in Israel and Neighbour Countries

From the Israeli point of view the Six-Days-War in 1967 did not only mean territorial expansion, but also the conquest of precious water ressources, so that 60% of today's water in Israel come from those areas that had been occupied in 1967. This fact emphasizes the great potential of water to become a matter of conflict.

The Jordan lost almost its entire water potential south of the Sea of Galilee. Its annually carried amount of water decreased from 650 million m^3/a to 40 million m^3/a. After that, an Arab plan was designed to tap the rivers in Lebanon and Syria that were tributaries to the Jordan (Hasbani and Banias). This plan was not carried out due to a military threat from Israel. These events launched the construction (1958) of the "East Ghor Canal" in Jordan, that allowed to divert 140 m^3/a of Jarmuk-water southwards. In Syria, more than 19 dams at the Jarmuk and its tributaries were constructed until the 80ies; their storage capacity was estimated at almost 90 million m^3 until 1987. Future plans intend to enlarge the storage capacity up to 215 million m^3.

After the 1967th War, the Israeli government declared the water in the occupied areas as a strategic resource and kept it under military control. Since then, Palstinians are not able to expand their water supply while Israeli began to dig deep fountains to support the jewish settlements in the occupied districts. Some of them are located such nearby natural springs that the fountains of Palestinian farmers dried up. In this context it is important to annotate that, according to Israeli occupation law, Palestinian land that hasn't been worked on for two years can be dispossessed and furthermore declared as Israeli state possession (Sabbagh 1999).

Water Supply and Demand in Israel

Israel's geographical location and the geological characteristics of the various regions are primarily held responsible for the water shortage in this country. The largest part of Israel is located in areas of an arid to semi-arid climate. 80% of the water supply are to be found north of the river Jarkon in an area that represents only 30% of the state's territory. Israel's water supply is estimated to be about 2.100 million m^3/a *(cp. Figure 4.1.1)*.

The mountain aquifers of the Palestinian West Bank cover more than 22% of the Israeli demand of water, the Golan Heights about 7%.

At the moment Israel is working on several plans to increase the water supply and reduce its consumption. On the one hand side, they hope to find new sources in foreign countries on the basis of regional initiatives, on the other hand side, they try to decrease agricultural consumption. Until the year 2000 the agricultural water consumption is to be reduced from 1.000 million m^3/a to 740 million m^3/a (SOFFER 1993). The construction of two separated water pipelines for households and for agriculture and industry is also being discussed. Particularly remarkable is the fact that the aquifers are supplied with the surplus water from the rainy winter months, which in the past years happened more and more often.

Water Supply and Demand in the Palestinian Areas

The Palestinian areas West Bank and the Gaza Strip are separated from each other by the North-Negev (Israeli area). The Gaza Strip is located at the south-eastern shore of the Mediterranean and comprises an area of 363 km^2. The difference in water ressources of those two areas is due to their different geographical location and their orographic characteristics. The Gaza Strip is marked by its arid to semi-arid climate; the annual precipitation in the south hardly amounts to 200 mm and in the northern part to about 400 mm. It is considered a very arid region, whose entire water ressources are gained from a sandy, 10 to 15 meter deep aquifer with an estimated regenerable storage capacity of about 60 million m³ *(cp. Table 4.1.2)*. Today's demand of water in the Gaza Strip amounts to 110 million m³. Such a large withdrawal leads to an annual fall of the ground-water level (up to 20 cm). This inevitable over-use has as its consequence in sea water penetrating into the lower aquifers and oversalts and pollutes the unstable stratum of ground-water (chlorite 400 mg/l).

Table 4.1.1. Water potential in Israel (Mio. m³)

Origin	Ground water	Surface water	Saltwater
Lake Genezareth		490	
Flood water		160	
Treated waste water		241	
Banias/Hasbani (Hulavalley)		122	
regenerative groundwater			
- coastal aquifer	283		
- Karmel und - Galilee Aquifere	237		
WESTBANK-AQUIFER	330		
groundwater with high salt concentrations			232
Salty water at Lake Genezareth			20
non-regenerative ground water (Fossil water)	KA		
Total	850	1013	252
Total Waterpotential			2.115 Mio m³

For the year 2010 the demand for water is estimated at 180 million m³ (GTZ), of which only about 19% are intended to be used in households. The water quality in the Gaza Strip gets worse every year. The Israeli authorities did not make any efforts during the period of occupation to find a lasting solution, and for the Palestinian authorities it is simply too much at the moment. Therefore the ground water is still gained in uncontrollable amounts from deep down by the Palestinians as well as by 3.700 Israeli settlers. The annual consumption of the Israeli settlers (892 m³/a) is eight times higher than the one of the Palestinians (122 m³/a). Also the untreated effluents have been polluting the ground water for years. At the moment in almost all big cities, supported by German aid, modern systems for effluents and their purification are built to reprocess the water.

On the West Bank, however, the situation is completely different. The water ressources are estimated at 600 - 650 m³ *(cp. Figure 4.1.2)* and supplied by three aquifers. The annual precipitation shifts - with the exception of the Judean desert and the Jordan valley - between 600 - 900 mm, spreading out mainly over the West exponation and the main mountain crests. Beside, 1.3 million Palestinian inhabitants and about 140.000 Israelian (with out east Jerusalem) settlers live here.

Table 4.1.2. Water Potential and distribution in the palestinean areas (Mio. m^3)

	Potential	Palestinean consumption	Israelean consumption
Lake Genezareth	-	-	-
Flood water	176	25	-
treated wast water	-	-	-
Jordan River	190	-	190
regenerative ground water			
Western Aquifers			
Eastern Aquifers	335	30	310
NE Aquifers	125	25	110
Gaza-Aquifers	140	60	65
	65	110	-
non-regenerative ground water (Fossil water)	ND	-	-
Total	1.031 Mio m^3	250 Mio m^3	675 Mio m^3

Data-base: Schuval 1983, Assaf 1993, Sofer 1998

Thus the water problem on the West Bank seems, on the surface, not to be caused by the scarcity of water, but by its distribution.

Still more problems are created by an obsolete network of water pipes in the Palestinian areas. During the entire period of Israeli occupation the maintenance of the water pipes had been utterly neglected. As a consequence, about 50% of the supplied water in 1994 were estimated to be lost.

The watered plain of the Palestinians in the Westbank sank from 27% before the Six-Days-War in 1967 down to 3.7% in the year of 1991. In the meantime the proportion of the watered areas in the Isreali settled part of the Westbank increased up to 70%.

Water Supply and Demand in Jordan

More than 90% of the kingdom's area have desert-like climatic conditions, and the precipitation amounts to less than 200 mm/a. Only in the mountains of Ajlun, Moab, and Adom it is as high as 400 - 650 mm/a. For this reason, Jordan depends on the running waters in the northern and western parts of the country.

Estimations of Jordanian water ressources differ quite a bit: while SOFFER (1993) estimates them to be 900-1.100 million m^3, ALLAN (1991) estimates them at 84 million m^3 *(cp. material 4c)*. Jordan's water balance has been categorized as negative for more than 10 years; the annual deficit is at 60 million m^3 according to

ABU TALEB (1992). At the moment a great increase of the population (3,8%) pushes the demand for water to a critical limit - despite a very low consumption per person. Thus, Jordan's situation regarding its supply and distribution of water is next to the Palestinian considered the most alarming of that region. In many regions in Jordan, water for household purposes must be supplied in rotation (only once or twice per week).

The political conflicts with Israel and also Syria about water - that originate in the McDonald and the Johnston Plan from 1949 and 1955 which legally entitled Jordan to use the rivers Jarmuk, Jordan, and the Sea of Galilee - led to a severe shortage of Jordan's water supply in the course of geo-political developments.

This was the reason to build many dams on the west exponating sides of the mountains, that today have a storage capacity of 110 million m³. In addition, Jordan has been exploiting the valuable, not regenerable fossil water since the late 80ies. The consumption of this water is estimated at about 210 million m³ today (Table 4.1.3).

Table 4.1.3. Waterpotential in Jordan (Mio. m³)

	Ground water	Surface water
Jarmuk rever		120
Eastern triputary to Jordan Jordan (all dammes)		110
Southern Regine und souther-eastern destert		70
treated waste water		52
regenerative groundwater	275	
non-regenerative groundwater (Fossil water)	210	
Total	485	352
Total Water potential		837 Mio. m³

Therefore Jordan won't be able to find medium- or long-term solutions to the water problem by its own efforts. That means that apart from national possibilites only supra-national solutions can be considered for a just distribution of the water and for tapping new water ressources.

4.1.2 Over-Use of Water Ressources

Due to its excellent technologies, Israel has often been thought of as the international leader in water management. However, its hegemonic position in water consumption is also undeniable. Israel's agriculture uses annually about 1.400 million m³ (about 75% of the supply; the remaining 530 million m³ are intended for the demand of households and industry) *(cp. Table 4.1.4).*

Fig. 4.1.2. Agro-business Farming in the Desert. (Wadi el Diseh south Jordan)

The Palestinians use 79% of the water for agricultural purposes, the Jordanians 77%. It becomes clear that the agricultural sector represents almost the only cause for the over-use of water. The cultivation of irrigation-intensive plantations and of the deserts in Israel and Jordan made the demand for water increase since the 1970's. For example, in El Diese in the Wadi Rum (southern Jordan), agro-business is done on a large scale with valuable fossil water (cultivation of grain, potatoes, raisins, or apples). Gigantic, computer-directed irrigation systems sprinkle the fields at 45 degrees Celsius according to demand *(cp. photography).*In Israel's deserts, too, for example, in the Wadi Araba, grain, cotton, and various tropical fruit that require a lot of water are cultivated. In the Gaza Strip the over-use of water has reached the critical limits a long time ago: while the citrus fruit plantations and the vegetable fields eat up 190 million m^3, only 45 million m^3 are left for household purposes.

Water Management and Water Diplomacy

Water management refers to supply, demand, and the economical processing. It comprises all those factors that permit an optimal economic use. In this region it includes, for example, the general shortage of water, the border-crossing water resources, the disadvantage suffered by the Palestinians regarding the water distribution, etc.

Water diplomacy means the necessity of regulations by contract, aiming at a just distribution of the water, a mediation of potential conflicts among the neighboring countries, or the implementation of existing treaties.

While dealing with this subject it has become obvious that solutions require the inclusion of both elements.

To prevent future shortages Jordan and Syria planned together the construction of the Maquarin barrage on the Yarmuk. In 1967 however, Israel occupied the Golan Heights and with it the planned location for the project. A new location was then found and the Syrians and Jordanians agreed in 1986 on a new barrage project on the Yarmuk. However, influential investors, e.g. the US Agency for International Development, insist on the agreement of all affected countries to the plans before the construction will go on. Israel didn't do that so far.

Table 4.1.4. Water ressources, Population Growth and Water Consumption in Countries of the Middle East

Country	Land area	Inhabitants		Availible FWR			AA /iAA	Water cons.			
	1000 km²	Mio		Mio m³/a	m³/Inh./a		1.000 km²	%			
		1997	2025	1995	1995	2025	1994	%	A	H	I
Egypt	995,4	64,5	95,8	60.000	1.000	500	32,56	90	86	6	8
Palestine	6	2,5	5	200	100	40	0,25	48	76	23	1
Iraq	437,4	21,2	41,6	42.800	2019	1028	97,8	36	92	3	5
Israel	20,6	5,8	8	1.500	300	150	5,81	34	79	16	5
Jordan	88,9	4,5	9,3	880	250	90	11,96	6	75	22	3
Lebanon	10,3	3,1	4,4	9.000	3.000	2.100	3,22	27	68	28	4
Syria	183,8	14,9	26,3	15.000	1.250	580	137,86	8	94	4	2
Turkey	401,5	63,3	85,8	250.000	4.500	3.000	401,49	10	72	16	11

Data-base: own Composing of: Allan, 1996; Wolf, 1995; FAO, 1998; CIA, 1998; Government of Israel, 1996

AA Agriculture Area
iAA irrigated Agriculture Area
A, H, I Agriculture, Households, Industry
FWR Fresh Water Resource
Cons. Consumption

4.1.3 The Water Issue in the Peace Process

In the bilateral and multilateral negotiations of the Middle-East peace process the distribution and use of water play a central role. In the course of the negotiations it has, however, become clear again and again how difficult it is to reconcile the different positions regarding the water issue (Palestine/Jordan in the summer of 98). The agreement to extend autonomy into the West Bank (Oslo-II, Sept. 28, 1995) would have almost failed because of the water issue. The clarification of the water rights, i.e. the question, whether the Palestinians possess sovereignty over the water ressources of the West Bank, had to be defered to later negotiations about the final status of the occupied areas (cp. Sabbagh 1996).

4.1.4 Conclusion

Actual water politics are, on the one hand side, very nationalistic, on the other hand, extremely hostile to peace. Israel has made the water ressources of the occupied Golan Heights and of the Palestinian areas part of its water budget a long time ago. While the Palestinians complain about shortage of water, hundred millions cubic meters of water are annually transported from the Palestinian mountain aquifers to Israel. Jordan's capital has to content itself with water twice a week. In Lebanon, on the contrary, there is water in abundance. Syria has permanent conflicts with Turkey, which built its dams up to three kilometers to the Syrian border.

The only way to avoid conflict in this region is to find supra-regional regulations for questions of water management and water politics. Concerning the water management, for example, one would need to bring in line the supply and demand of water at the lowest economical and ecological cost. Another future task would be to decentralize and control the demand in some areas, which can only be obtained by introducing special tarifs or incentives. Regarding water politics or diplomacy, the makers would have to abandon their nationalist attitudes and approach a just sharing of the border-crossing water ressources. Moreover, house cisterns, very much appreciated in earlier times and meant to catch up the rain water and thereby cover the need of water, should be reintroduced by law.

References

Abu Taleb, MF et al 1992 Water Ressources Planning and Development in Jordan: Problem Future Scenarios and Recommendations. World Bank, Washington, D.C.

Allan, JA 1992 Substitutes for Wate rare Being Found in the Middle East and North Africa. In: Geo Journal, 28 (3), S. 375- 385

Dombrowsky, I 1995 Wasserprobleme im Jordanbecken. Perspektiven einer gerechten und nachhaltigen Nutzung internationale Ressourcen. Peter Lang Verlag. Frankfurt am/Main.

Isaac, J Hillel, S 1994 (eds) Water and Peace in the Middle East. Amsterdam, Elsevier.

Sabbagh, J 1999 Der Kampf um das weiße Gold – Wasserkonflikt zwischen Israel und seinen Nachbarn. In: Geographie Heute, Jg. *20*, H. *169*, S. 28 -32. Seelze.

Sabbagh, J 1996 Israel und die PLO auf dem Weg zum Frieden. In: Geographische Rundschau, Jg. *48*, H. *3*, S. 183-189

Soffer, A 1993 River of Fire – the conflict of water in the Middle East. Tel Aviv

Vallianatos-Grapengeter, IM 1996 Der Nahostkonflikt im Prisma der Wasserproblematik: Wasserpolitik im Jordantal 1882-1976. Münster.

4.2 Syria and Turkey in Water Diplomacy (1962-2003)

Marwa Daoudy
Development Studies Department, School of Oriental and African Studies, London

Summary

The complexity of water as a multi-dimensional strategic resource has been extensively highlighted in recent years. Its vital importance is exacerbated in arid and semi-arid regions such as the Middle East. In the case of Syria and Turkey, and at the core of the political and strategic interaction, lie the Euphrates and Tigris waters– the famous Al Furat and Dijleh (Arabic) or Firat and Dicle (Turkish) Rivers. Both rivers have seen the rise of ancient civilizations and the early development of irrigation practices dating back to the Sumerian and Akkadian periods (4000-5000 BC). Meetings have taken place, information on both sides has been exchanged over more than forty years, and experts have established a water-related modus vivendi based on bilateral agreements. This relationship has evolved over the years depending on the regional context, the strategies that were deployed and the link between the water dispute and wider strategic issues. Our purpose will be to analyze bilateral dynamics while highlighting context-related linkages and the importance of water for each actor. We will consider the core problem (the disputed sharing of common water resources in the Euphrates Basin) through the optic of Syria and Turkey's historical and political interaction, within a range between cooperation and non-cooperation. The minutes of proceedings of the negotiation rounds held between 1962 and 1993 –which have been gathered from the Syrian part- will serve as primary sources for the analysis of face-to-face official positions and exchange of information. These documents constitute an important source of information since they were not formulated in order to sway public opinion, and since their validity has been confirmed by the signature of Turkish negotiators at the time.

4.2.1 Sharing the Commons

The Euphrates and Tigris Rivers both originate in the mountains of Eastern Turkey. They flow into Syria and Iraq and join the sea at the head of the Arabic-Persian Gulf. The Euphrates enters Syria at Karkamis which is located downstream from the Turkish town of Birecik. Joined by major tributaries, the Balikh and the Khabur, the river flows Southeast before entering Iraq. The Tigris flows through Turkey until it reaches the border city of Cizre and eventually forms the border between Turkey and Syria over 39 km before crossing into Iraq. The Tigris and Euphrates unite near Qurna in Iraq to form the Shatt-Al-Arab and empty into the Gulf. The Euphrates is the longest river (3000 km) in Western Asia, distributed between Turkey (1230 km), Syria (710 km) and Iraq (1060 km). The Euphrates Basin lies 28% in Turkey, 17% in Syria, 40% in Iraq and 15% in Saudi Arabia. According to Allan, 62% of the Euphrates watershed is fed from Turkish territory and 38% from Syria (Allan 2000). The Tigris is the second longest river in Southwest Asia (1,840 km), it is divided between Turkey (400 km), Syria (32 km) and principally Iraq (1408 km). Its basin lies between Turkey (12%), Syria (0.2%), Iraq (54%) and Iran (34%).

Flow estimates appear to vary depending on which source one is using. It is worth recalling that some experts mention the natural runoff and others observed flows, or seasonal records instead of mean average flows. The Euphrates and Tigris Rivers' natural runoff is respectively estimated between 28,4-31,8 billion cubic meters (bcm) and 44,4-52,6 bcm –inclusive of tributaries. It is estimated that Turkey contributes 88% of the annual flow of the Euphrates while Syria contributes 12% (Beaumont 1978). These figures are however disputed, some authors suggest that Turkey contributes 98,6% of the flow (Kolars and Mitchell 1991). It is also estimated that Turkey contributes 51% and Iraq 39% of the Tigris flow.

4.2.2 Setting the Stage: The Turkish Great Anatolia Project, Objectives and Consequences

The GAP Project: 1980-2003

An upstream-downstream relation is characteristic of the interaction between Turkey and Syria on the Euphrates and Tigris Rivers. The geographic asymmetry is emphasized by economic and military advantages that favor the upstream riparian, Turkey. A water conflict exists when a riparian feels constrained in the realization of its national goals and objectives by the upstream unilateral exploitation of the water resources. Turkey has been building since 1980 a huge development project called the GAP (Great Anatolian Project or Guneydogu Anadolu Projesi), consisting of 22 dams and 19 HEPP (hydro-electric power plants). The ultimate purpose is to irrigate about 1,7 million hectares in the Harran Valley (9,7% of Turkey's total surface) and produce 27 billion kilo watt hours annually, thus the equivalent of 19% of the irrigable area throughout the country and 22% of its total hydro-electric potential. The ultimate objective is the expansion of irrigated agriculture but also socio-economic development through transports, enhancement of non-agricultural work opportunities, education and sanitary services. Turkey aims at compensating its lack of oil resources while developing one of its most under-developed regions, Southeast Anatolia, which represents 9,5% of the country (Republic of Turkey 2003; Unver 1994). The Turkish government has framed the project in terms of "sustainable development" of various fields such as irrigation, agricultural production, environment and society, with a long-term objective towards economic growth, social stability and the enhancement of exporting capacities (Republic of Turkey 2003).

Implementing the GAP: Socio-Economic, Environmental and Downstream Consequences

If the socio-economic objective objectives of the GAP are clear, implementing the project has been far less obvious So far, only 201,080, out of the 1,6 million, hectares are being irrigated in the GAP region (Republic of Turkey 2003). And to date, merely 6, out of 22, dams (the Atatürk, Karakaya, Krakilzi, Karkamis and Birecik) have been completed (Republic of Turkey 2003). The Keban Dam was constructed in 1975, before the launch of the GAP. Original projections suggested the GAP would be finished by 2013 (Kolars and Mitchell 1991). The accumulation of financial problems, combined with geographic hurdles and the lack of qualified workers delayed the project (Kibaroglu 1998) until 2050 (Al Mustaqbal 1999). Today, 44% of the GAP has been achieved, an estimate which corresponds to 12,8% of the agricultural plans, 75,4% of the energy-oriented schemes and 58% of the social projects (Republic of Turkey 2003). Financial difficulties have resulted from the enormous pressure that has been put on Turkey's national budget, including reduced international credit, where the latter is mainly due to the campaigns led by non-governmental coalitions in the field of the environment and human rights, notably the Ilisu Dam Campaign or the Export Credit Campaign. As a result, several European countries (United Kingdom) and companies (Swiss UBS, British Balfour Beatty and Italian Impregilio) withdrew their support for the project between 2001 and 2002 (Ahmad 2001; International Rivers Network 2001, 2002).

The socio-economic, environmental and archeological impact of this mega-project has been criticized worldwide. As a result of the flooding of numerous villages within the Gaziantep, Urfa, Adiyaman, Malatya, Elazig, Tunceli, Diyarbakir, Mardin, Siirt, Batman and Sirnak provinces of the GAP region (which are also called the "Kurdish" areas of Turkey), about 100'000 villagers have been displaced, a majority of who are Kurds, (Biegala 2001). The destruction of cultural sites has either actually happened like in the case of the city of Zeugma (and its impressive Greco-Roman mosaics) and the Kurdish city of Halfeti (dating back to 1000 BC) or has been imminent like in the case of Hasankeyf, a city on the Tigris River dating back to the 7th century BC, in this latter case the Turkish Government eventually agreed to change the location of the Ilisu Dam (Council of Europe 2001; International Friends of Kurdistan 2000). In addition, recent reports show plans to flood 52 villages and 15 towns, with a potential displacement of between 30'000 and 70'000 villagers (Carr-Brown 2001).

At the regional level, the impact on downstream countries is no less alarming. Although Turkey considers this project to be a "domestic" enterprise, inspired by the founder of the Turkish Republic, the consequences surpass national borders. According to international experts, a full implementation of the GAP will ultimately withdraw a maximum of 70% of the Euphrates natural flow (Kolars and Mitchell 1991) and about 40% of its observed flow (Kliot 1994; Picard 1993). The consequences for Syria would be highly problematic in light of her dependence on external water sources (80%) and the preeminence of the Euphrates Basin, in particular, for the country's water supply (65% of overall resources). Initially, Syria planned to irrigate 640'000 hectares in the Euphrates Basin and 210'000 hectares in the Khabur-Tigris Basin, but this scheme is contingent upon regular flows from the Euphrates, the Tigris, and their tributaries. Thus, these figures have been adjusted downwards. Revisions have also occurred as a result of the irrigation that has taken place over many millenia, and the accumulation of gypsum, which has led to the deterioration of both water and soil. The drastic reductions in water supply from the Euphrates, following the filling of the Atatürk Dam (1990) and the opening of the Sanliurfa Canals (1994-1995), have already started to affect Syria (Kolars, 2000: 253). Official sources show that there has been an overall drop in the average flow of the Euphrates at the Turkish-Syrian border, with an initial decline in 1982 (from 1063 m3/sec.), and in 1984 (698 m3/sec.), followed by a slight increase between 1987 (725 m3/s.) and 1989 (795 m3/s.), and a plunge in 1995 to 591 m3/s. (Syrian Arab Republic 1999). Later, the Syrian and Iraqi Irrigation Ministers complain about a decrease of the Euphrates flow to an average of 450 m3/s. between 2000 and 2001 (Syrian Arab Republic 2001; Tishrin 2001). A similar process has characterized the fate of tributaries such as the Balikh and the Sajour (Daoud 1999). In addition to these quantitative aspects, environmentalists highlight the risks of severely polluting the Tigris and Euphrates Rivers, principally, because the GAP authorities have not arranged for treatment of the drainage waters (Bosshard 1998). The risks of downstream flooding and water-logging are therefore very high and rising. It is thus estimated that 40% of the Euphrates waters from Turkey to Syria will also eventually be polluted, and 25% of the Tigris waters from Turkey to Iraq (Kliot 1994). Turkish experts evaluate the rate of pollution of return flows at 700 ppm – a reasonable amount for some experts only if Turkey is able to restrain the input of uncontrolled additional return flows (Kolars 2000). On one hand, the Balikh and the Khabur are principally put at risk. On the other hand, some sources show that pollution of the underground resources, lying across the common border, has already taken place (Al Hayat 1996). In turn, one envisages Syria will pollute 50% of the Euphrates waters flowing to Iraq (Kliot 1994).

4.2.3 Cooperating Over Water

A History of Water-Related Conventions

It is here worthwhile to uncover the underlying interests at play in finding, a regime to distribute the Tigris and Euphrates waters, the legacy dates back to the European Mandates in the region. In 1916, France and Great Britain held secret negotiations on the Middle East; the two powers adopted a clause regarding "a guarantee for a defined quantity of water from the Euphrates and Tigris" (Hokayem and Bittar 1981). The 1921 Convention, signed between the French Mandate in Syria and Turkey, introduces the notion of equitable distribution of the Kweik waters, a tributary of the Euphrates, mainly for the sake of the city of Aleppo (League of Nations 1926-

1927). This clause is reiterated five years later in another treaty between the French Mandate and Turkey (League of Nations 1926-1927). Another treaty signed in 1939 also had a significant impact on future hydro-strategic relations between Syria and Turkey. In exchange for Turkey's participation to the Allied war effort and the prevention of German occupation of Syria, the French Mandate officially handed over the Syrian Sandjak of Iskandaroun – later named Hatay Province by Turkey- in which lived an important Turkish minority (League of Nations 1940-1944). Obviously, Syria considered the territorial split perpetrated by France to be illegal, especially since part of the national territory and population was simply handed over without any consultation of the largely Arabic local population (Mardam Bey 1994). In effect, Turkey acquired an enhanced upstream position towards Syria on the Orontes River, which Syria does not, and will possibly never ever, recognize. Today's official maps still include Iskandaroun within national borders. Felt as a territorial "loss" for Syria, Iskandaroun and the Orontes waters are thus a considerable obstacle to improving the countries' bilateral relations (Sanjian 1956) and the forty-year long water negotiations on the Euphrates and Tigris. Ever since the two countries started their 'water talks', Turkey would put the issue of the Orontes on the negotiation table. The strategy would be perceived by Syria as a tactical maneuver (4th Bilateral Session 1971) aimed at forcing Syria to a de facto recognition of Turkey's sovereignty on disputed territory.

Negotiating on the Waters: Bilateral and Trilateral Meetings

In effect, Syria and Turkey have been meeting on a regular basis to discuss issues relating to respective water developments in the Euphrates and Tigris Basins. The process will be predominantly conflictual with periods of cooperation and peaks of crisis that will be examined in the next section. The actual rounds were launched on a bilateral basis between Syria and Turkey (1962-1971) and Syria and Iraq (1962-1974). Four phases will follow the initial rounds: tripartite (1972-1974; 1983-1992) and bilateral (Syria/Turkey in 1993; Syria/Turkey, Turkey/Iraq & Syria/Iraq in 2001). Both Syria and Turkey highlight the need, from 1962 to 1971, to exchange water, climate-related data in order to favor cooperation and "reach an equitable distribution of the Euphrates waters" (2nd Bilateral Session 1964). As of 1962, Turkey suggests to study "all the shared resources" which Syria refutes on the ground of "known political reasons such as the Orontes (Asi) and the Sandjak of Iskandaroun" (1st Bilateral Session 1962). This proposal by Turkey will be regularly reiterated over the years (3rd Bilateral Session 1969; Turkey's Verbal Note to Syria 1995; Republic of Turkey 1996) and Syria will continue to emphasize the fact that the Orontes River "is not part of the three countries' common waters" (3rd Tripartite Session 1983).

Turkey also emphasizes her will to pursue negotiations with both Syria and Iraq on the question of common projects to be built on the Turkish side of the Euphrates. A technical Protocol signed in 1969 states the will "to search for a distribution of water rights" (4th Bilateral Session 1971). Later bilateral sessions coordinate the filling of the Keban (30 km3) and Tabqa (11,6 km3) reservoirs to be built respectively in Turkey and Syrian (4th Bilateral Session 1971). Syria emphasizes her intention to irrigate 640,000 hectares with the Euphrates waters. The five tripartite sessions held around the filling of the upstream dams reveal an escalation of tension. Syria considers that downstream countries were put on a "*fait accompli*" by Turkey and looks into "unifying positions with Iraq" (First Tripartite Session 1972). Turkey's reply is to envisage a distribution of the waters only after analyzing all projects, each country's needs and possible water transfers from other rivers (2nd Tripartite Session 1972). The discussions are quickly deadlocked as Turkey and Syria start filling their reservoirs at the same time, thus impacting severely on Iraq's water volume. The ensuing crisis will be tackled in the next part.

Cooperation similarly characterizes the start of the third phase of the process (1982-1992), following downstream countries' concerns about the launch of the GAP project. Starting on a bilateral Turkish/Iraqi basis (1982), the sixteen bilateral and trilateral sessions later include Syria and are held alternatively in each capital twice a year. Parties convene on carrying field missions, exchanging data and establishing common measures of the rivers. Positions however quickly clash over water evaluations (10th Tripartite Session 1988) as well as legal interpretations of the international status of the rivers. As shown in the next part, periods of tensions are followed by cooperative moves that are punctually concretized by the signing of three bilateral agreements. These agreements should however be put in a regional context and in a power perspective.

4.2.4 Conflicting and Cooperating over Water: Crisis, Agreements, Positions and Power Relations

Crisis and Bilateral Agreements

The first hydro-political crisis erupted in 1974 between Syria and Iraq. A combination of planned upstream extraction in Syria and Turkey, severe drought and political tension brings the two countries to the verge of war. Iraq accuses Syria of withdrawing a third of the river's flow while Syria argues that the level of extraction corresponds to her annual consumption (12,8 km3). In addition to the exceptional precipitation levels of March 1974, a Saudi and Soviet mediation in the region, prevent the conflict from escalating further (Bari 1977; Kienle 1990). Syria and Iraq agree orally on a minimum flow of 450 m3/second reaching the downstream riparian, while filling the Tabqa Dam (Farah 1995). Water here appears to be the catalyst for a crisis, which has its origins in political and ideological rivalries, separating these two regional powers. This later favored a political rapprochement between Iraq and Turkey over trade in oil (Yumurtalik pipeline), the Kurdish file (pursuit agreement) and an interruption of water negotiations with Syria until 1983. In April 1974, a fifth trilateral session gathers all actors to the dispute (5th Tripartite session 1974). During the session, Turkey announces a plan to fill the Keban reservoir from early July 1974 until 1975. The upstream riparian commits to let an average flow of 100-170 m3/second pass through the common border to Syria. In order to satisfy her needs, Syria, in turn, claims an average flow of 370 m3/second between the months of April-July and 260 m3/second during the summer... The discussions do not give any result and the average flow varies between 100-200 m3/second. During the 10th bilateral session an agreement is reached between Syria and Iraq. Syria commits "to let Iraq have a minimum of 90 m3/second during the month of June and 110 m3/second during the first days of July" (19 May-5 June 1974). No trilateral agreement is reached and in the first days of July 1974, the two upstream reservoirs start to fill.. The following bilateral meetings between Syria and Iraq reveal a strong dissent about the quantity of water that is released to Iraq. Syria suggests a minimum amount of 250 m3/second, with a possibility to raise it to 450 m3/sec. during the month of January. No compromise is reached and the two riparians revert to non-official provisional agreements to regulate the consequent filling of the Tabqa Dam. Even though, contacts are very rare between 1974-1980, cooperation then becomes possible in the wake of the GAP project. Tensions start rising again between Syria and Turkey when the latter launches the Karakaya Dam (1984). In 1986, Turkey's commitment to let 500 m3/second pass through the common border is first rejected by Syria who stresses "the need to specify volumes in the framework of a tripartite agreement" (7th Tripartite Session 1986). In 1987, Turkey commits in writing to let this minimum volume pass through to Syria in the framework of an economic cooperation Protocol in various sectors, i.e., oil, gas, trade, electricity, telecommunications. The water clause specifies that "during the filling up period of the Atatürk Dam reservoir and until the final allocation of the waters of the Euphrates among the three riparian countries, the Turkish side undertakes to release a yearly average of more than 500 m3/sec. at the Turkish-Syrian borders, and in cases where the monthly flow falls below the level of 500 m3/sec., the Turkish side agrees to make up the difference during the following month" (Syrian Arab Republic 1987).

During the 10th Tripartite Session, Turkey announces the completion of the Karakaya Dam and gives an evaluation of expected volumes for the Atatürk Dam. Syria and Iraq proceed likewise with regards to Al-Baath and Al-Qadisiya Dams in their respective territories (10th Tripartite Session 1988). During the same session, downstream countries express a concern about the possible impact of the Atatürk Dam, a concern to which Turkey responds that "no damages will be suffered by downstream countries along the same guarantees offered during the filling of the Karakaya Dam" (10th Tipartite Session 1988). This position is reversed as of 1989. During the 14th Session, Turkey announces that she intends to fill the Atatürk Dam between 13 January and 12 February 1990, and completely cut off the flow of the Euphrates river during this period (14th Tripartite Session 1989). Syria and Iraq strongly reject this solution, suggesting some technical alternatives to "alleviate the damages encountered by the three countries" (Ibid). Syria demands that Turkey respects the 1987 Protocol. Turkey proceeds to drastic cuts of the Euphrates flow in early 1990 and the two countries are on the brink of war. The crisis has reached its zenith. Then, during the following trilateral meeting, Syria reiterates the importance of the Euphrates for its projects and also puts forth claims to the Tigris waters (15th Tripartite Session 1990). Turkey commits to "try as much as possible to respect the commitment to let 500 m3/sec. pass through the common border with Syria" (Ibid). While Iraq claims a minimum water flow of 750 m3/sec. to satisfy all downstream needs, Syria insists on the provisional dimension of the 1987 agreement before reaching a complete allocation of the waters (15th Tripartite Session 1990). Turkey refutes both propositions. Hence, prior to the 11th Tripartite Session held in April 1989, the two downstream countries convene to distribute the waters of the Euphrates amongst themselves, "Iraq's share across the Syrian-Iraqi border (...) will amount to 58% of the water volume crossing the Turkish-Syrian border" and "the rest, 42%, will go to Syria" (Syrian Arab Republic 1989).

A 16th Tripartite Session is then held in September-October 1992, followed by a brief bilateral Syria/Turkey Session in May 1993, in which the two riparians commit to reaching a final tripartite agreement on the distribution of shared waters before year's end (Syrian Arab Republic 1993). Thus was the trilateral session stillborn. Hereafter, instead of pursuing negotiations, the two riparians will exchange official notes and mutual complaints through their embassies and their official position papers. Influencing variables can be found in analyzing the nature of power relations between Syria and Turkey—an analysis provided in the next section.

Conflicting Positions on the Euphrates and Tigris

As per the above analysis, positions clash within the face-to-face negotiations and outside their strict framework. A micro-analysis of the negotiating documents and official papers brings us to summarize and examine dominant positions on both sides. A larger undertaking—beyond the space allowed by this study—would imply a confrontation of arguments and counter-arguments with the codified principles of international law for the protection and management of shared water resources and each country's interpretation of these principles. In effect, there has been a consolidation of Turkey's upstream position, which has evolved in line with the progress she has made in her national water infrastructure programs, notably the GAP.

During the first encounters with Syria, Turkey did recognize the necessity to "reach an equitable distribution of the Euphrates waters" (2nd Bilateral Session 1964). However, she then introduced a conditional understanding of water distribution, which was completely tied up with a basin-wide assessment of her long-term projects, needs, and water transfers (Ibid), as well as an "allocation of the uses of water" and not "the distribution of water between the countries concerned" (Turkey's Verbal Note to Syria, 1995). As a consequence, Turkey came to argue for an "interpretation of the two Euphrates and Tigris basins as a unique basin" and "a centralized planning of all dam projects in the two basins across the three countries" (1st Bilateral Session 1980; 10th Tripartite Session 1988; 13th Tripartite Session 1989) for the sake of "rational and optimal" utilization of the waters (12th Tripartite Session 1989) and a standardization of the data in the face of "inequitable and non-economic" irrigation of infertile lands in Syria (Republic of Turkey 1992, 1996). This approach was then reiterated in later official water-related publications (Republic of Turkey 2003) as well as in declarations about the "transboundary" nature of the two rivers (Bagis 1994). In 1990, Turkish President Özal stated that "Turkey refuses to have to share the Euphrates waters because the Euphrates is a Turkish river" (Chalaby & Majzoub 1995). Obviously, Turkey is enforcing an absolute sovereignty approach based on the fact that both rivers originate in her territory, where most of the fertile lands and ambitious projects are found. Moreover, Turkey has long defended the value of regulation and stocking provided by upstream reservoirs, supposedly to the benefit of all riparians in periods of severe drought such as in 1958-1962 and 1970-1975 (Bilen 1994). Backed by the codified principles of international law on the protection and management of international watercourses (United Nations 1997), Syria has consistently contested such arguments. Syria has also reminded Turkey of the historic Conventions signed between the two parties in 1921 and 1926, in an attempt to underline the historic basis for upstream responsibilities, and the need to respect "good neighborly" relations (Rifai 1998). The 1987 agreement is also quoted as proof of the "full recognition by Turkey of the international nature of the watercourse" (Syria's Verbal Note to Turkey 1995). In doing so, Syria highlights an inherent paradox within the Turkish position: "We may ask which water that countries can utilize jointly if it was not common water? How is it possible to distribute water utilization if it does not mean to get part of that water (…)?" (Kasm 1996). It also backs Iraq's approach based on "acquired" or "historic rights" by stating that their "utilization of the Euphrates waters for irrigation and economic needs dates back to fifty years" (Farah 1995; Kasm 1996). Syria considers Turkey's claims as a way of postponing the issue of water distribution until the completion of the GAP (Kasm 1996).

In concluding, and considering these diverging approaches to the common waters, it would be fair to wonder why Turkey who benefits from an upstream position combined with stronger military and economic endowments would agree to guarantee a minimal flow of the Euphrates at a rate of 500 m3/second. The context in which the negotiation process evolves, and the nature of the power relations between the upstream and downstream riparians, are here key variables for a better understanding of inter-national water diplomacy.

Brief Analysis of Power Relations: The Kurdish Card

Water is not negotiated in absolute terms but is rather linked through power strategies to other issues such as security concerns. Turkey's answer to the regional outreach of the Kurdish issue has been besides military suppression of the rebellion to try and develop the socio-economic structure at the heart of the Kurdish uprising, principally in Southeast Anatolia. According to a report by the Council of Europe, the dilemma lies in the official objective set by the GAP administration (to reduce massive rural inflows to burgeoning urban centers and stabilize the "Kurdish" provinces) while the Government's struggle against "terrorism", together with the Kurdistan Workers Party (PKK) 's violent operations on the ground, have directly contributed to such human inflows (Council of Europe, 2002: 2). According to Turkish sources, twenty years of war between the Turkish army and the PKK has displaced about 3 million villagers (Bayramoglu 2001).

Faced with the steady progress of the GAP project and the deadlock in negotiations, Syria started to employ, what can be conceptualized as, "tactical issue-linkage" (Haas 1980). As of 1984, Syria's support of the Kurdish PKK and its leader Abdullah Öcallan would represent an important bargaining chip, one which would enhance Syria's position within the negotiation process (Daoudy 1996, 1999). This tactic became somewhat of a success, and certainly broke the impasse, because in exchange for collaboration with Syria on security issues, Turkey became much more willing to compromise on some minimal allocation of common waters. The Kurdish issue, and its link with the water file, was a bone of contention and triggered several crises between the two countries, notably in 1993, 1996 and 1998, until the capture of Öcallan by Turkey in 1999. Typically, the crisis of 1990, 1993 and 1996 were activated by one of two factors either a significant reduction of water from the Turkish side, or a refusal to reconvene the negotiations and then followed by an intensification of Syria's support of the Kurds. In 1990, Turkey filled the Atatürk Dam reservoir, leaving "a trickle of water to downstream countries" (Morris 1990). However, the internal destabilization became much too burdensome, and so the Turkish Minister of Foreign Affairs agreed to again institute the 500 m3/s. minimum quota in signing the 1992 Security Protocol with Syria in which both countries also commit to fight terrorism (Syrian Arab Republic 1992). Yet, in 1993, the flows are again significantly reduced, and this trend will continue and culminate with the launch of the Birecik Dam on the Euphrates in 1994-1995, as well as the opening of the Sanliurfa canals. In 1996, Turkey openly accuses Syria of supporting the PKK, but, refuses to make any link to ongoing water negotiations (Bell 1998). In 1998, Öcallan is expulsed from Syria and later captured and jailed by the Turkish authorities. Later on, the two countries have harmonized their relations. This *rapprochement* was recently concretized with the signing of a second bilateral water agreement in June 2002, in which Turkey commits to stop the downstream flow of polluted waters to Syria; the two riparians also decide to actively share their hydraulic expertise (An-Nahar 2002; Syrian Arab Republic 2001). Since then, and certainly in the wake of the Anglo-American war against Iraq (2003), Syria and Turkey have witnessed a foreign policy alignment.

4.2.5 Concluding Remarks

By virtue of crossing national boundaries, water forces riparian states into a situation of interdependence. The mutual dependency is aggravated by the fact that, despite two bilateral agreements, the two immediate countries (Syria, Turkey) have not yet succeeded in reaching a comprehensive water-sharing agreement. Thus, we have analyzed the content of the face-to-face negotiation, as well as the evolution of each actor's position and strategy on the water issue. The analysis has shown the complex nature of the resource located in the Euphrates and Tigris basins with historic and political context-related linkages. In concluding, the process of negotiation has been mixed but predominantly conflictual.

References

Ahmad K (2001) UK Drops Turkish Dam Plan. In: The Observer 1 July

Al Hayat (2001) GreenPeace Reveals Water Pollution from Turkey to Syria (in Arabic). 15 August

Al Mustaqbal (1999) Israel in the GAP (in Arabic). 12 October, p. 15

Allan JA (2000) The Middle East Water Question, Hydropolitics and the Global Economy. I.B. Tauris & Co.Ltd, London

An-Nahar (2002) A Syrian-Turkish Project of Agreement around Hydraulic Cooperation and Exchange of Expertise (in Arabic). 21 June

Bagis AI (1994) Water in the Region: Potential and Prospects – An Overview. In: Bagis AI (ed) Water as an element of Cooperation and Development in the Middle East. Hacettepe University of Ankara and Friedrich Naumann Foundation, Ankara, pp 15-27

Bari Z (1977) The Syrian-Iraqi Dispute over the Euphrates Waters. In: International Studies vol 16 no 2: pp 227-244

Bayramoglu A (2001) Southeast of Turkey is Victim of Forced Rural Exodus. In: Sabah (in Turkish). Translated by Vanrie P Turkish Press Review, European Institute for Euro-Arab and Mediterranean Research and Cooperation., http:// www.medea.be/site.html?

Beaumont P (1978) The Euphrates River, an International Problem of Water Resources Development. Water Resources Development vol 5 no 1. pp 35-43

Bell A (1998) Turkey Turns Down Water Talks with Syrians. CNN Interactive, 15 October

Biegala E (2001) Mixed Blessings From GAP Project for Turkey's Southeast. In: Turkey Update, http:// www.turkeyupdate.com/features/gap.html

Bilen O (1994) Prospects for Technical Cooperation in the Euphrates-Tigris Basin. In: Biswas A (ed) International Waters in the Middle East, From Euphrates-Tigris to Nile. Oxford University Press, Delhi, pp 95-116

Bosshard P (1998) Ilisu. Bern Declaration, Switzerland, http:// www.rivernet.org/ turquie/ ilisu.htm

Carr-Brown J (2001) Blair in Drive for Turkish Dam Deal. Sunday Times 28 October

Chalaby H, Majzoub T (1995) Turkey, the Waters of the Euphrates and Public International Law. In: Allan JA, Mallat C (eds) Water in the Middle East, Legal, Political and Commercial Implications. Tauris Academic Studies, London, p. 211

Council of Europe (2002) Humanitarian Situation of the Displaced Kurdish Population of Turkey. Report by the Commission on Migrations, Refugees and Demography of the Parliamentary Assembly, http:// assembly.coe.int/Documents/WorkingDocs/Doc02/FDOC9391.htm

Council of Europe (2001) Ultural Aspects of the Project Relating to Ilisu Dam, Turkey. Report by the Commission on Culture, Science and Education of the Parliamentary Assembly, http:// assembly.coe.int/ Documents/ WorkingDocs/ Doc01/FDOC9301.htm

Daoud M (1999) Personal Communication to the author (former member of Syrian delegation to the Tripartite Sessions). Ministry of Irrigation, Damascus, 28 November

Daoudy M (1999) Water, Institutions and Development in Syria: A Downstream Perspective from the Euphrates and Tigris. World Commission on Dams, Cairo, http:// www.dams.org/submissions/sub_p_af4.htm

Daoudy M (1996) Entre le Tigre et l'Euphrate, une négociation en eaux troubles. Les Cahiers de l'Orient 44: pp 73-95

Farah Z (1995) Personal Communication to the author (former Director of Syrian Ministry of Irrigation). Damascus, 20 November

Haas EB (1980) Why Collaborate? Issue-Linkage and International Regimes. World Politics vol 32: pp 357-405

Hokayem A, Bittar MC (1981) 'Empire Ottoman, les Arabes et les Grandes Puissances, 1914-1920. Editions Universitaires du Liban, Collection : l'Histoire par les Documents, Beyrouth

International Friends of Kurdistan (2000) The Ilisu Dam and GAP, http:// international-friends-of-kurdistan.s5.com/Ilisu%20Dam.htm

International Rivers Network (2002) Swiss Bank UBS Quits Turkish Ilisu Dam Project

International Rivers Network (2001) Balfour Beatty Withdraws from Ilisu Dam Project

Kasm B (1996) The Syrian-Turkish Dispute over the Euphrates River in International Law: Recent Developments (in Arabic). Embassy of the Syrian Arab Republic, Geneva

Kibaroglu A (1998) Management and Allocation of the Waters of the Euphrates-Tigris Basin: Lessons drawn from Global Experiences. Ph.D thesis, Bilkent University

Kienle E (1990) Ba'th v. Ba'th, the Conflict between Syria and Iraq, 1968-1989. I.B. Tauris & Co Ltd, London and New York

Kliot N (1994) Water Resources and Conflicts in the Middle East. Routledge, London

Kolars JF (2000) The Spatial Attributes of Water Negotiation, the Need for a River Ethic and River Advocacy in the Middle East. In: Amery H, Wolf AT (eds) Water in the Middle East, A Geography of Peace. The University of Texas Press, Austin, pp 245-261

Kolars JF, Mitchell WA (1991) The Euphrates River and the Southeast Anatolia Development Project. Southern Illinois University Press, Carbondale

League of Nations (1940-1944) Great Britain and Northern Ireland, France and Turkey, Treaty of Mutual Assistance, Signed at Angora, October 19, 1939 and two Protocols annexed thereto of the same Date. Treaty Series no 4689 vol 200, pp 167-189

League of Nations (1926-1927) Convention of Friendship and Good Neighborly Relations, with Procès-Verbal of Signature, Signed at Angora, May 30, 1926. Treaty Series no 1285 vol *54*, pp 195-229

League of Nations (1926-1927) France and Turkey, Agreement with a View to promoting Peace with Protocol relating thereto, Protocol concerning into coming into force, and Exchange of Notes, Signed at Angora, October 20, 1921. Treaty Series no 1284 vol *54*, pp 177-193

Mardam Bey S (1994) Syria's Quest for Independence. Ithaca Press Reading, Beyrouth

Morris M (1997) Water and Conflict in the Middle East: Threats and Opportunities. Studies in Conflict and Terrorism vol *20* no 1. pp 1-13

Picard E (1993) Eau et sécurité dans le bassin de l'Euphrate. In: Géopolitique vol 43. pp 75-79

Republic of Turkey (2003) The Turkish National Policy for Utilizing the Waters of the Euphrates-Tigris Basin (The Three-Staged Plan). Ministry of Foreign Affairs, http:// www.mfa.gov.tr/grupa/ad/adg/adge/default.htm

Republic of Turkey (2003) The Criteria Which Would Satisfy Each of the Countries in Allocating Transboundary Waters. Ministry of Foreign Affairs, http://www.mfa.gov.tr/ grupa/ad/adg/adgd/Chap3.HTM

Republic of Turkey (1996) Water Issues between Turkey, Syria and Iraq. Ministry of Foreign Affairs, Department of Regional and Transboundary Waters, Ankara

Republic of Turkey (1992) Turkey and the Question of Water in the Middle East. Turkish Embassy, Department of Information, France

Republic of Turkey (1990) Press Statement on the Water Summit between Iraq, Syria and Turkey. Ministry of Foreign Affairs, Ankara, 26-27 June

Rifai M (1998) A Regime for International Watercourses and its Application to the Tigris and Euphrates Basins (in Arabic). In: Water in the Arab World. Maalumat Editions, Beyrouth

Syrian Arab Republic (2001) Al Atrash-Yilmaz Statements. Syrian Arab News Agency, Damascus, 24 August

Syrian Arab Republic (1999) Hydraulic Plans: 1995-1999. Ministry of Irrigation, Damascus

Syrian Arab Republic (1995) Hydraulic Plans: 1990-1995. Ministry of Irrigation, Damascus

Syrian Arab Republic (1993-1995) Exchange of Official Notes with Turkey.

Syrian Arab Republic (1993) Common Statement Following the Visit of Turkish Prime Minister S. Demirel. Damascus, 19-20 January

Syrian Arab Republic (1992) Common Statement Following the Visit of Turkish Minister of Foreign Affairs H. Cetin. Damascus, 2 August

Syrian Arab Republic (1989) Common Minutes of Proceedings between Syria and Iraq. Baghdad, 17 April.

Syrian Arab Republic (1987) Protocol of 1987 on Issues relating to Economic Cooperation between the Syrian Arab Republic and the Republic of Turkey. 17 July

Syrian Arab Republic (1982-1992) Internal reports by the Syrian Delegation on the Two Technical Committees between Iraq/Turkey and the Tripartite Syria/Iraq/Turkey Sessions. 1st Session of 17 to 27 May 1982; 2nd Session of 29 November to 2 January 1982; 3rd Tripartite Session of 26 to 28 September 1983; 4th Tripartite Session of 11 to 14 June 1984; 5th Tripartite Session of 5 to 8 November 1984; 6th Tripartite Session of 5 to 9 June 1985; 7th Tripartite Session of 25 to 30 January 1986; 8th Tripartite Session of 23 to 30 June 1986; 9th Tripartite Session of 10 to 16 January 1987; 10th Tripartite Session of 5 to 10 January 1988; 11th Tripartite Session of 1988; 12th Tripartite Session 13 to 20 March 1989; 13th Tripartite Session of 18 to 24 April 1989; 14th Tripartite Session of 29 November to 1 December 1989; 15th Tripartite Session of 7 to 12 March 1990; 16th Tripartite Session of 28 September to 2 October 1992

Syrian Arab Republic (1972-1974) Internal reports by the Syrian Delegation on Tripartite Syria/Iraq/Turkey Sessions 1st Trilateral Session of 9 to 10 April 1972; 2nd Trilateral Session of 26 August to 7 September 1972; 3rd Trilateral Session of 1 to 27 November 1972; 4th Trilateral Session of 9 September to 6 October 1973; 5th Trilateral Session of 1 to 17 April 1974

Syrian Arab Republic (1962-1974) Internal reports by the Syrian Delegation on Bilateral Syria/Iraq Sessions. 1st Bilateral Session of 24 September-9 October 1962; 2nd Bilateral Session of July 1963; 3rd Bilateral Session of 26 May to 2 June 1966; 4th Bilateral Session of 26 January to 9 February 1967; 5th Bilateral Session of 12 April to 7 May 1967; 6th Bilateral Session of 1-26 November 1967; 7th Bilateral Session of 6 to 22 April 1971; 8th Bilateral Session of 27 December 1971 to 3 January 1972; 9th Bilateral Session of March 1972; 10th Bilateral Session of 19 May to 5 June 1974; 11th Bilateral Session of 16 to 20 July 1974; 12th Bilateral session of 18 to 21 November 1974

Syrian Arab Republic (1962-1971) Internal reports by the Syrian Delegation on Bilateral Syria/Turkey Sessions. 1st Bilateral Session of 16 to 22 January 1962; 2nd Bilateral Session of 3 to 16 September 1963; 3rd Bilateral Session of 19 to 29 January 1969; 4th Bilateral Session of 14 to 24 June 1971

Tishrin (2001) Syria, Iraq Concerned at Turkish Use of Euphrates, Tigris. Damascus, 28 November

United Nations (1997) Convention on the Law of the Non-Navigational Uses of International Watercourses. General Assembly, International Law Commission, Document 51/229, 21 May, http://www.un.org/law/ilc/texts/nnavfra.htm

Unver O (1994) Innovations in Water Resources Development in Southeastern Anatolia Project (GAP) of Turkey. In: Bagis AI (ed) Water as an Element of Cooperation and Development in the Middle East. Hacettepe University of Ankara and Friedrich Naumann Foundation, Ankara, pp 27-37

4.3 Water Issue Among the Riparian States of Euphrates and Tigris Transboundary Rivers

Gurer, Ibrahim
Gazi University, Ankara, Turkey

Abstract

Turkey is a country located between Europe and Asia and has a surface area of 779 452 km^2 with a total population of about 68 million according to 1998 census. Annual rainfall in Turkey varies between 220 mm to 2500 mm with an average of 642.5 mm and this corresponds to an average annual rainfall of 501 km^3. Approximately 186 km^3 of this water flows in rivers as surface water. Since the Euphrates and Tigris rivers form 28.5 % of all the water resources of the country, they are the most important water resources of the eastern and southeastern part of the Turkey. After they originate in Turkey, they cross the border to Syria and Iraq and there has been always a water issue to be discussed among the neighbors. It is necessary to find an equitable and satisfactory means for allocating water of Euphrates and Tigris river system to the riparian states. At present, among the riparian states Iraq has the highest, and the Syria has the lowest amount of water consumption per capita but compared to water rich countries all three states use rather little amount of water.

Key Words

Water Resources, Transboundary rivers in Middle East, Turkey, Euphrates, Tigris

4.3.1 Introduction

Turkey is a country located between Europe and Asia. There are seven geographical regions in Turkey and she has the continental climate at the interior part and the Mediterranean climate at the Aegean and Mediterranean coastal regions. The mean annual precipitation is 642.5 mm and it corresponds to 501.0 Km^3 of total water volume. The figures for the surface water potential of Turkey can be stated 186.05 Km^3 as surface runoff with a surface runoff coefficient of 37 %, 95.0 Km^3 as consumable water volume, 25.9 Km^3 as the actual consumed volume, and 12.0 Km^3 as exploitable ground water potential with an allocated volume of 7.6 km^3, and the real consumed volume of 5.7 km^3.

Turkey has been divided into 26 river drainage basins to study the water resources as extensively as possible. The water and soil resources of each basin are given in Table 4.3.1a and Table 4.3.1b respectively (DSI, 1999). It is also possible to get information on drainage area the specific discharge expressed in l/s/km^2, volume of annual runoff total in km^3 for main drainage basins of Turkey from the table. In some of the drainage basins, the construction of dams and thus utilization of water for multi purposes cannot be possible due to topographical and geological conditions, therefore about 91 km^3 of surface water could not be utilized due to technical and economical reasons. The water, which can be utilized for various purpose will be thus 95 km^3 after the construction of all the necessary water structures to regulate the virgin rivers. In order to regulate the existing river system of turkey it is planned to build 730 dams (DSI, 1999), but it is noted that with detailed studies on the drainage basins this number may change. The Euphrates and the Tigris rivers are the most important water resources of at the eastern and southeastern Anatolia and constitute of 28.5 % of all the potential of Turkey.

4.3.2 Euphrates and Tigris rivers

The water issue among the riparian states of the river system of Euphrates and Tigris can be discussed referring to a series of paragraphs expressing the same idea by a different wording as "Satisfactory means for allocating water of Euphrates and Tigris river system to riparian states is to be devised if the scramble for water is not to become another item for further poisoning the relations among the neighboring states"(SAM, 1996).

The main river courses supplying most of the fresh water to the Middle East counties are the river Nile, the river Jordan and the Euphrates and the Tigris (MFA, 1995). Each river basin varies considerably form the others with distinct problems requiring different solutions.

It is believed that if harnessed by modern technology, the Euphrates and the Tigris rivers have the potential to make agricultural to flourish in a very short period of time. The water resources of three countries vary considerably, Iraq has the highest and the Syria has the lowest amount of water per capita (Table 4.3.2) (MFA, 1996). So Syria, Iraq and Turkey need to arrange the optimum use of the water of the river system of Euphrates and Tigris, which will be equitable and will stand the test of time.

The following points are mentioned in all meetings as the starting points for any kind of discussion, but of course they are not the preconditions to start the dialogue on the water issue.

- Water is an important theme for peace in Middle East
- Turkey is not a water rich country
- Euphrates is sometimes seen as a panacea for the water problems of the countries to the south of Turkey

Table 4.3.1a. Water resources of main drainage basins in Turkey (DSI,1999).

Basin			Water Potentials Of Basin			Impounded Water Within The Basin	
No	Name	Area (km^2)	Annual Precipitation (mm)	Runoff vol (km^3)	Specific Discharge (lt/s/km^2)	Number of Dams	Impounded Water (hm^3)
1	Meric Ergene	14560	604	1.33	2.9	21	1817
2	Marmara	24100	728.7	8.33	11	58	2894.5
3	Susurluk	22399	711.6	5.43	7.2	26	3848
4	North Aegean Basin	10003	624.2	2.09	7.4	15	797
5	Gediz	18000	603	1.95	3.6	16	3565.9
6	K. Menderes	6907	727.4	1.19	5.3	17	1697.7
7	B. Menderes	24976	664.3	3.03	3.9	22	2739.9
8	West Mediterranean	20953	875.8	8.93	12.4	25	1830
9	Antalya	19577	1000.4	11.06	24.2	14	2858
10	Burdur Lakes	6374	446.3	0.5	1.8	9	161.7
11	Akarcay	7605	451.8	0.49	1.9	3	172
12	Sakarya	58160	524.7	6.4	3.6	45	6827.9
13	Western Black Sea	29598	811	9.93	10.6	28	2784
14	Yesilrmak	36114	496.5	5.8	5.1	44	6287.9
15	Kzlrmak	78180	446.1	6.48	2.6	78	23774.3
16	Konya Closed Basin	53850	416.8	4.52	2.5	25	2800.8
17	East Mediterranean	22048	745	11.07	15.6	11	10173.5
18	Seyhan	20450	624	8.01	12.3	18	6124.5
19	Asi (Orontes)	7796	815.6	1.17	3.4	8	11086.5
20	Ceyhan	21982	731.6	7.18	10.7	27	8229.3
21	Euphrates	127304	540.1	31.61	8.3	89	112193.2
22	Eastern Black Sea	24077	1198.2	14.9	19.5	41	1491.6
23	Coruh	19872	629.4	6.3	10.1	21	7467.3
24	Aras	27548	432.4	4.63	5.3	20	4085.2
25	Van Lake Closed B.	19405	474.3	2.39	5	7	608.7
26	Tigris	57614	807.2	21.33	13.1	42	30630.5
	Total	779452	642.6	86.05	209.3	730	246853.9

Table 4.3.1b.The soil resources of main drainage basins of Turkey (DSI, 1999).

Basin			Soil Resources Of Basin	
No	Name	Area (km^2)	Land Area (ha)	Irrigable Land Area (ha)
1	Meric Ergene	14560	1095320	1077992
2	Marmara	24100	865704	729957
3	Susurluk	22399	850046	755934
4	North Aegean Basin	10003	367479	316348
5	Gediz	18000	667207	623403
6	K. Menderes	6907	223437	194799
7	B. Menderes	24976	104296	907383
8	West Mediterranean	20953	437356	406601
9	Antalya	19577	451224	448111
10	Burdur Lakes	6374	251403	249484
11	Akarcay	7605	364411	359938
12	Sakarya	58160	2814341	2681137
13	Western Black Sea	29598	855008	640557
14	Yesilrmak	36114	1617206	1401213
15	Kzlrmak	78180	4049796	3761142
16	Konya Closed Basin	53850	2182762	2134915
17	East Mediterranean	22048	438281	327790
18	Seyhan	20450	764673	714014
19	Asi (Orontes)	7796	376240	331719
20	Ceyhan	21982	779792	713670
21	Euphrates	127304	4293793	4111316
22	Eastern Black Sea	24077	712575	350717
23	Coruh	19872	326220	303362
24	Aras	27548	642017	641137
25	Van Lake Closed B.	19405	436485	433319
26	Tigris	57614	1148238	1137628
	Total	779452 (1)	28054310	25753586

Euphrates-Tigris basin is a completely separate issue from the guest for peace in the Middle East. The Euphrates and the Tigris rivers rise in the high mountains of northeastern Anatolia and flow down through Turkey, Syria and Iraq and join to form Shatt Al-Arab about 200 km before they reach the Gulf (Fig. 4.3.1) (MFA, 1996). From Turkish part of watersheds, the contributions of the long term annual mean water volumes for the Euphrates and the Tigris are 31 km^3 and 21 km^3 respectively (DSI,1999). The surface runoff contributions of the riparian states to Euphrates and Tigris river system is given in Table 4.3.3 (MFA, 1996).

Table 4.3.2. The annual amount of water consumption per capita in Middle East (MFA,1996).

Country	Water quantity (m^3/year/ capita)
Iraq	2110
Turkey	1830
Syria	1420
Israel	300
Jordan	250
Palestine	100

Table 4.3.3. The runoff contribution of riparian states to the Euphrates and the Tigris rivers

State	Euphrates	Tigris
Turkey	89 %	52 %
Syria	11 %	0 %
Iraq	0 %	48 %
Total	100 %	100 %

4.3.3 The present demands

It seems that the anticipated combined demand for water from transboundary rivers by riparian counties is actually greater than the total volume of the river system as could be seen from Table 4.3.4, and at the moment the demands of the riparian states do not look like very realistic.

Table 4.3.4. The claims by the riparian states as the percentage of annual flow (MFA, 1996).

State	On Euphrates	On Tigris
Turkey	52 %	14.1 %
Syria	32 %	5.4 %
Iraq	65 %	92.5 %
Total	149 %	112 %

It is known that, Turkey depends on the waters of this river system for both energy production and irrigation water requirement more than Syria and Iraq. Because Iraq and Syria are considered as petroleum producing countries but Turkey produces so little that, by hydropower production part of the energy need is planned to be compensated. In the eastern and southeastern part of the country, the agricultural industry and the percentage of the population involved is much higher than that of Syria and Iraq, and therefore they need more water. Turkey is said to have seven times more irrigable land in the river system basin than Syria, and out of 2.5 million ha, 1.7 million high quality land is planned to be irrigated by the water of the river system of Euphrates and Tigris within Southeastern Anatolian Project (GAP) (see Fig. 4.3.1). Syria plans to irrigate 770 000 ha of her land by the water of Euphrates but due to low lying topography and high content of salt and gypsum of soil, most of the water is believed to be wasted. Iraq has larger irrigable area, but topographical conditions would make it more difficult to build dams on the river system in Syria and Iraq. At present Turkey uses a very small portion of the water of Tigris. In Iraq part of the water of the Tigris river is diverted to Euphrates after it is used for irrigation (Fig. 4.3.2). So if the waters of Euphrates and Tigris are combined in one river system concept, it is believed that it will be possible for all three countries to apply their irrigation plans along the Euphrates river.

4.3.4 Southeastern Anatolian Project; GAP

For decades southeastern part of Anatolian peninsula was the least economically developed region of Turkey. Turkey started GAP project in 60's with the intention to harness the hydropower potential of the Euphrates and Tigris rivers. The Southeastern Anatolian Project is said to be one of the biggest irrigated water resources utilizations projects and it covers an area of 74 000 km^2, including the plains lying between the rivers of Euphrates and Tigris, and having a population of 5 275 000 which is about 9 % of the national total (See Fig. 4.3.1). GAP is a group of thirteen projects and their sub-projects which are multi-purpose; irrigation, hydropower, domestic water supply and flood control. Upon the full development of 22 dams and 19 hydropower plants, besides the irrigation of 1.7 million ha of land, with an installed capacity of 75 % of the present capacity of the country, with GAP Project Turkey will produce 27.3 TWh energy (DSI, 1999).

4.3.5 Dialogue on water

The most important institutional forum among the riparian states is the Joint Technical Committee (JTC), set in 1980 by Turkey and Iraq to discuss the water issue. Syria joined the JTC in 1983. The committee held 16 official and ministerial level meetings since that time (MFA, 1996). Turkey committed herself to provide minimum 500 m^3/s water flow to Syria with the protocol signed in 1987 and fulfilled her responsibility even in the most difficult situations like impounding water in the reservoir of Ataturk dam.

In 1990, during the filling up of Ataturk dam, Turkey notified Syria a month before according to the protocol signed and 768 m^3/sec flow was released from the Euphrates in the month prior to filling, and during the filling of the reservoir extra 60 m^3/sec flowed into Syria from downstream of the reservoir and, on average between 23 November 1989 and 13 February 1990, 509.12 m^3/sec flow was released from the river of Euphrates to Syria.

In 1995, the annual average flow of the Euphrates river was computed as 830 m^3/sec, and in the first half of 1996,Turkey released a flow varying between 1147 m^3/sec and 1684 m3/sec, which was far above the amount of water specified by the protocol. According to latest computations done by the experts of Turkish State Hydraulic Works, 500 m^3/sec flow can hardly be released in the months of August and September in normal years when the rainfall is about at the level of the long term average value, and in case of dry years June-September period becomes critical in providing this value (Gurer, I. 1996).

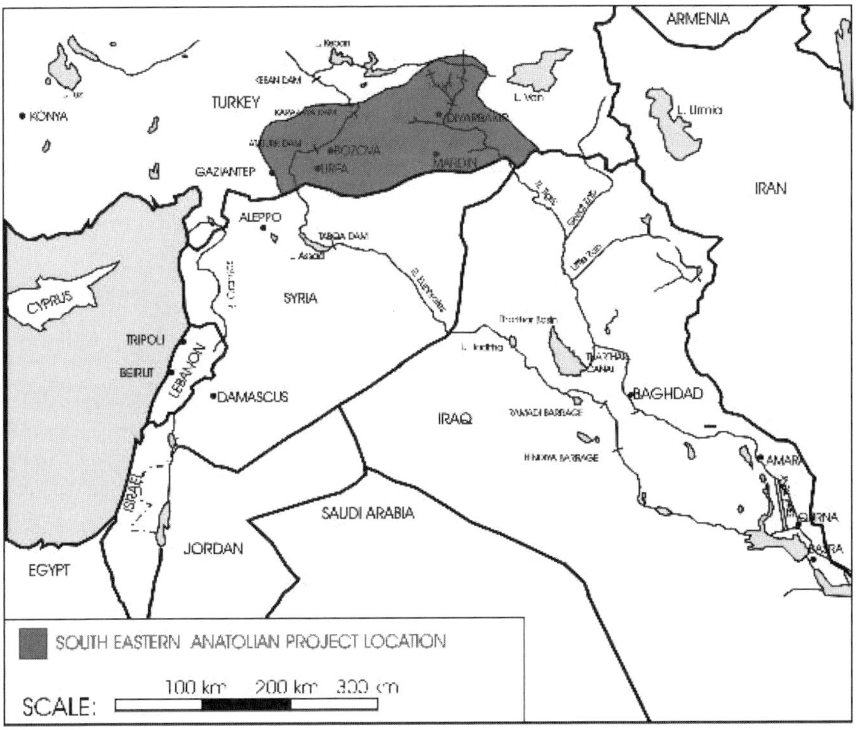

Fig. 4.3.1. The close up view of Euphrates and Tigris river system

The perspectives of Syria and Iraq

Syria and Iraq brings up the following points to the water issue related to the rivers of Euphrates and Tigris :

- Turkey failed in informing her neighbors in advance about her plans to built dams and did not follow the customary practices of international law on water courses.
- Hydropower projects built on Turkish side reduces the amount of inflowing water to Syria and therefore harm the farming on Syrian side.
- Turkey releases polluted water across the boundaries (SAM,1996).

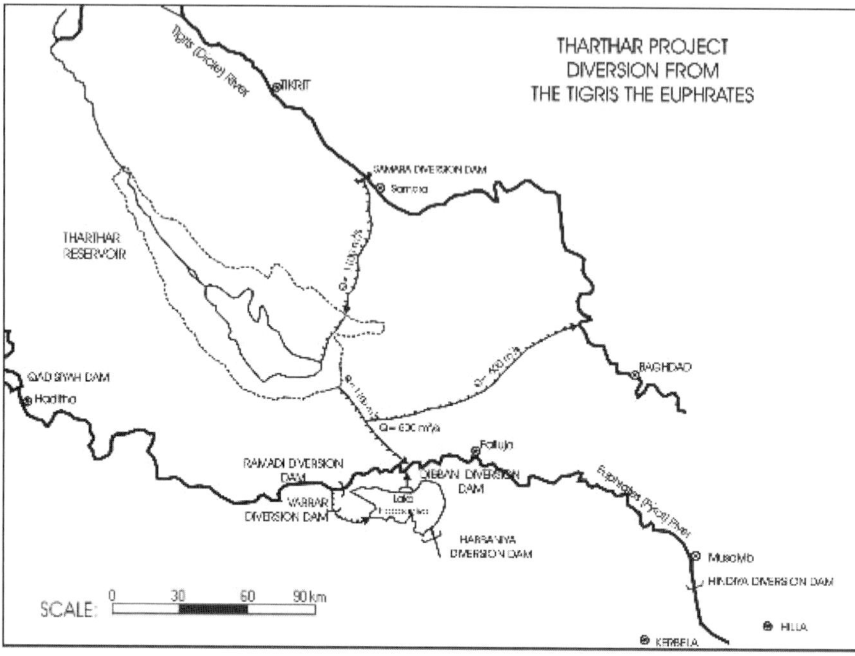

Fig. 4.3.2. The close up view of Tharthar project; diversion from Tigris to Euphrates.

The perspective of Turkey

In connection with the perspectives of Syria and Iraq listed in the above paragraph, Turkey has the following points:

- In JTC meetings Turkey always provided the complete data on Euphrates and Tigris rivers at the beginning of the meetings.
- Hydropower dams like Keban, Karakaya and Ataturk regulate the water levels and sharp fluctuations of different seasons on Euphrates river. After regulation from Euphrates Turkey provides the 500 m^3/sec flow to Syria and Iraq. The discharge figure is computed from the monthly mean values, but if this amount of water could not be released in any dry month, it has to be compensated in the consecutive month. But during 2001 Turkey could not provide the annual total volume computed from 500 m^3/sec to her neighbors due to the very dry spell. The neighboring relations improving in positive direction helped to judge the situation with understanding.
- In case of Tigris, since there is no regulation on the river almost all the water contributed from Turkey flows into neighboring countries. In this connection, it is also important to mention the use of regulating reservoirs (dams) at Birecik and Karkamis on Turkish side, Al-Baath dam just below the Tabqa dam on Syrian side and Badush dam below the Saddam Dam on the Tigris and Bagdadi dam at the downstream of Qadissiya dam on Euphrates on Iraq side. All these

reservoirs are to protect the ecology of the downstream parts of the dams during the peak flow in the reservoirs and for the regulation purpose and enable the country to release more regulated water to the downstream when needed.

- Source of pollution for Euphrates and Tigris rivers can be either from industry or irrigation return flow, but at present there are almost no industrial development which can pollute the river and also so little area is open for irrigation that the return water from irrigation is not important at all, therefore Syria's claim is unfounded because Turkey releases unused water to Syria (SAM,1996).

At a later stages of GAP, when Turkey completes all the irrigation projects planned to irrigate 1.7 million ha then rather big return flow will be put back into the river, but since Turkey is aware of this risk, so there will be dilution with 30 % fresh water. This can easily be done when one has regulation dams as it is planned.

The case of the river Orontes

The river Orontes rises in the mountains of Lebanon and flows 40 km in Lebanon and then continues in Syria for about 325 km before it arrives in Turkey for the last reach of 88 km and reaches the Mediterranean Sea (See Fig. 4.3.1) (MFA, 1995). So here Syria is upstream country and with its dams and regulatory reservoirs about 90 % of the water is used for irrigation and municipal water by Syria. The natural flow of Orontes at Turkish border is computed as 1.2 billion m^3 but before the completion of the two impoundment structures at Zeyzoun and Kastun locations only 120 million m^3 was released and furthermore when two more reservoirs on Orontes river at are completed only 25 million m^3 is expected to be released. Therefore Turkey believes that Syrian claims over the Euphrates river can also be judged against the records on the Orontes (SAM, 1996).

In June 2002, El Zeyzoun dam, located nearby Hama city of Syria, suddenly released about 70 million m^3 of water and besides 22 Syrian lost their lives, and flood damaged some villages in Syria and cultivated land in Turkey. With good understanding of neighborhood the disaster was handled together with minimum damage to mutual relations (Sabah 2002, Star 2002).

4.3.6 The concept of acquired rights on water

Syria and Iraq claim that they have the acquired rights to use water of the Euphrates and it comes from the history can not be accepted by any means because as the new water projects will be built on the rivers, then the flow will be regulated and sharing conditions should be re-fixed. Another point is that the doctrine of acquired rights has not been shared very widely elsewhere (MFA, 1995, and 1996).

4.3.7 International laws on trans-boundary rivers

There are few international laws defining the rights and obligations of the riparian states of transboundary rivers. In this connection it is possible to mention the names of "The Helsinki Rules on the Uses of the Waters of International Rivers" adopted in 1966, "Complementary Rules Applicable to International resources" adopted in 1986 at Seoul, and "United Nations Convention on the law of Non-Navigational uses of International Watercourses" adopted 1997 but not entered into vigor yet, as examples of the multilateral approaches to deal with water issue. The treaty signed between the United States of America and Mexico in 1944, on "The Utilization of Waters of the Colorado and Tijuana Rivers and of the Rio Grande ", and the treaty signed between the Kingdom of Lesotho and the Republic of South Africa on "The Utilization of the waters of the Senqu / Orange River system "can be given as the examples of bilateral agreements (SAM, 1996, MFA, 1995 and 1996). The Nile Basin Initiative Shared Vision Program (SVP) project was started by the riparian states , so all Technical Advisory committee (TAC) members will provide inputs to lead transparency, proper accountability and smooth implementation of the project along Nile river (NBI Press Release, 2002).

The Euphrates, Tigris and Orontes rivers have been recognized as transboundary rivers by a series of bilateral and multilateral agreements such as Ankara Agreements at 1921, 1939, 1946 and the treaty of Lausanne of 1923 and according to Joint Technical Committee, the rivers of Euphrates and Tigris are "regional waters". The regional water rights are also studied by United Nation (UN) to develop the guidelines to settle the conflicts. In 1970 the general assembly of United Nations asked the International Law Commission (ILC) to develop international law on the non-navigational use of international waters by riparian countries, the text submitted by ILO to UN in 1994 and the Convention was finalized in 1997 but not entered in vigor yet. It will not have a direct bearing on possible disputes between countries using a common transboundary watercourse but it will embody the latest legal norms and guidelines on the issues involved.

4.3.8 Water Management Approach of Turkey

The flow regime of the Euphrates and Tigris rivers is quite variable because the mean annual average flow and the difference between the Q_{max} and Q_{min} show great variations from year to year therefore " Cooperative Water Management " approach considering the existing natural conditions is proposed at present by Turkey to deal with this issue.

In 1984, a three stage plan was proposed by Turkey to solve the dispute with an aim not only to fulfill the expectation of the present generations but also of the future generations. it is expressed that this plan is based on two simple principles:

- The Euphrates and Tigris rivers form a single transboundary river system

- A common inventory of water and land resources to define and answer the real needs of each riparian country is to be prepared scientifically together by using the same principles and methodology in collecting and interpreting the available data.

Stage One: Compiling an Inventory of Water Resources of available data on river discharges, stages, rainfall, evaporation, temperature and other related data and the quality of the data will be checked

Stage Two: Compiling an Inventory of Land Resources of land classification and existing drainage conditions, appropriate crop pattern and corresponding irrigation water requirements will be defined

Stage Three: Analysis of Water and Land Resources. In this connection it will be necessary to define the potential method of irrigation with minimum loss of water according the data of stage one and two and also the modernization and upgrading of the existing irrigation networks will be done according to international norms and standards.

4.3.9 Conclusion

All three riparian countries when agreed on a common agreement they will get equitable and optimal use of the available waters of the region that means Euphrates and Tigris River system will help the solve the water problems of Turkey, Syria and Iraq and Orontes river for Syria and Turkey. In order to reach an agreement on the water issue of Euphrates and Tigris river system, and also of Orontes river, it is very important to use the same scientific and engineering and social criteria adopted internationally, in assessing the water and soil resources and also the needs of all the riparian states.

References

DSI (1998) Southeastern Anatolian Project, GAP, pamphlet, Ankara

DSI (1999) Haritali Istatistik Bulteni, Ankara, Turkey (In Turkish)

DSI (1995) Guneydogu Anadolu Projesi, GAP, Ankara (In Turkish)

DSI (1996) Dams and Hydroelectric power production plants in Turkey, Ankara

Gurer I (1996) "Southeastern Anatolian Project and Transboundry Waters in this Region" Lecture notes in Post Graduate Course IM 576 (Unpublished), Ankara, Turkey

MFA (1995) Water Issue Between Turkey, Syria, and Iraq, Ankara, Turkey

MFA (1996) Water Issue between Turkey, Syria and Iraq, Ministry of Foreign Affairs, Department of Regional and Transboundary Waters, Ankara, Turkey.

NBI Press Release (2002) Arusha, Tanzania,21 June 2002

SAM (1996) Facts About Euphrates and Tigris Basin, Center For Strategic Research, Ankara, Turkey

Sabah (2002) June 4, News in newspaper (in Turkish)

Star (2002) June 4 News in newspaper (in Turkish)

4.4 Control, Allocation and Consumption of Water Resources in the Jordan Basin: The Struggle over a Scarce and Vital Resource in the Context of the Middle East Conflict

Margret Johannsen
Institute for Peace Research and Security Policy at the University of Hamburg (IFSH), Hamburg

4.4.1 The struggle over territory is also a struggle over water

Military clashes at the Israeli-Syrian border, which were caused by disputes over the consumption of water in the Jordan Basin, preceded the 1967 Six-Day or June War whose outcome has significantly shaped the conflictual configuration in the Middle East. Part of the territorial dispute in the Middle East is due to the struggle over the scarce water resources in the Jordan River. Whatever the issue is – whether it is the powers of the Palestinian Authority and its borders, the Jewish settlements in the West Bank or the capacity of a future Palestinian state to give a home to returning refugees – the question of the control over water resources is always part of the equation. Moreover, it transcends the bilateral relationship between Israel and the Palestinians and involves most of Israel's neighbouring states, Jordan, Syria, and Lebanon, as well as Iraq and Turkey, which are also affected by the hydrological dilemmas of the Jordan region albeit indirectly.

4.4.2 Force instead of law

In this semi-arid region, Israel, Jordan and the Palestinian Authority have much less water at their disposal than the amount recognized as the minimum to meet basic needs. The scarce supply of natural fresh water cannot satisfy the ever-growing demand caused by a more than average population growth, industrial and agricultural expansion, high urbanization rates and a rising living standard. All these factors result in continuous overconsumption and pollution of the existing resources of the Jordan Basin. For a number of reasons, this scarcity of water, which is a vital resource, results in conflicts over allocation and consumption. Almost all significant water resources in the Jordan Basin are of a transborder nature, i.e. they cross international borders or ceasefire lines. However, most of the riparian states have not concluded internationally binding treaties regulating the allocation and consumption of water resources. Moreover, the majority of water resources are to be found in localities over which political control is disputed and actual control is based not on law but on force. Furthermore, foreign relations between most of the regional states follow the rules of strategic zero sum games which constitute an additional impediment to the settlement of the water disputes.

4.4.3 Disputed surface water resources

Due to borders drawn by the colonial powers after World War I and subsequent border changes resulting from a number of Israeli-Arab wars which, however, were never formally recognized by the parties involved, today's water system in the Jordan Basin is geopolitically fragmented. Its surface waters are the Jordan; its source rivers the Dan, the Hasbani and the Banyas; and its tributaries are the Yarmouk as well as Lake Tiberias which separates the Upper from the Lower Jordan. The source rivers of the Jordan originate in Mount Hermon; the Dan is on Israeli territory, the Hasbani in Southern Lebanon, and the Banyas on the Syrian Golan Heights occupied since 1967 by Israel. According to Israel, Lake Tiberias is entirely in Israeli territory whereas Syria claims riparian rights to the eastern shore. Since the opening of the Madrid peace negotiations in 1991, Syria therefore has demanded Israel's withdrawal to its de facto control as of June 4, 1967, i.e. the eve of the Six-Day War. An Israeli withdrawal of this kind would give Syria access to Lake Tiberias and the Upper Jordan. A withdrawal to the "international border", i.e. the border between the mandatory territories Syria and Palestine on which France and Great Britain agreed in 1923, would be more advantageous to Israel because it runs a few metres east of the lake shore. South of Lake Tiberias, the Yarmouk feeds into the Jordan; its main sources lie in Syrian territory, it is the border river between Jordan and Syria for a few kilometres, in its lower part it runs along the occupied Golan Heights, and finally, before it flows into the Jordan, it is the border river between Jordan and Israel. The Lower Jordan is the border river between Israel and Jordan, and it separates the occupied West Bank and the Jordan before it flows into the Dead Sea. Before 2001, when Israel withdrew from Southern Lebanon, the Lebanese Litani River could also have been counted as being under dispute, because its lower part flows through the "security zone" which Israel established in 1982. Israeli governments have repeatedly shown an interest in diverting the Litani to Israel.

4.4.4 Disputed ground water resources

In addition to its Coastal Aquifer, the region's largest ground water resources are to be found beneath the West Bank, that is, in the form of the Mountain Aquifer which is sub-divided into the Western, the North Eastern and the Eastern Basins. This aquifer is fed by the precipitation over the mountains of the West Bank. The Western and the North Eastern Basins extend underground across the Green Line to Israel proper. Hence, just as the rivers mentioned above, they constitute a trans-boundary water resource.

4.4.5 A pan-regional security problem

Israel, Syria, Jordan and Lebanon as well as the Palestinians compete over control and utilization of the surface waters in the region; Israel and the Palestinians compete over the ground water resources in the West Bank. For Israel, Jordan and the Palestinians, the Jordan River system and the neighbouring Coastal- and Mountain Aquifers are the only water supply (Jordan has no share in the coastal or mountain aquifers, it has its own aquifers), whereas Syria and Lebanon can use other rivers over which, however, they do not exercise exclusive control. Because the Jordan basin has scarce natural water reserves, demand exceeds the supply of renewable resources and the geopolitical situation is complex – control over water resources constitutes a pan-regional security problem.[1]

4.4.6 Water supply as a national security issue

In this water dispute, there are two legal principles which can be seen as competing with each other: the principle of territorial sovereignty and the principle of territorial integrity. Within international law, these principles constitute the extreme position which in the Middle Eastern reality of conflict are bound to clash. Upper riparians base their claims on the principle of territorial sovereignty as it would allow them to use water without restrictions. Lower riparians operate on the principle of territorial integrity as it or would serve to avert damage resulting from reckless water use by the upper riparian.

4.4.7 Allocation disputes and armed conflict

As a result of a number of bilateral and multilateral treaties which treat the utilization of international waterways according to the limited sovereignty doctrine thereby overcoming the difficulties surrounding the above-mentioned legal principles, the validity of these two extremes has eroded within the international community. However, with one exception, such treaties are absent in the Middle East. Here, water supply is treated as a national security issue and the disputes over the control and consumption of water are still, as in the past, handled in an antagonistic manner. Just as was the case on the eve of the 1967 war, whose outcome enabled Israel to secure more than fifty per cent of her water supply by means of control over the occupied territories, threat perceptions dominate dealing with competition

1. For the various dimensions of the competition over water in the context of the Middle East conflict, see Stephan Libiszewski, *Water Disputes in the Jordan Basin Region and their Role in the Resolution of the Arab-Israeli Conflict.* ENCOP Occasional Paper No. 13, Bern/Zürich: Swiss Peace Foundation/Swiss Federal Institute of Technology, 1995.

over water supplies. Everyone suspects everyone else of stealing ground water, cutting off springs, manipulating river flow through dams and even diverting entire rivers.

4.4.8 Contractual arrangements between Israel and Jordan

In their peace treaty of 1994, Israel and Jordan were able to agree on the allocation and consumption of Jordan River water for basically two reasons: Firstly, both states are in a position to cut each other off from their water sources – Jordan as the upper riparian of the Yarmouk and Israel as the upper riparian of the Jordan. Thus, it was obviously advisable to seek an agreement. And secondly, since 1988, when Jordan renounced its claim to sovereignty over the West Bank in favour of the Palestinians, the Israeli-Jordanian water dispute was no longer linked to a dispute over borders; from that date it was a genuine allocation conflict. Consequently, the parties were all the more amenable to a negotiated settlement as it was now linked to co-operative agreements on more efficient water utilization.

4.4.9 Linking the water dispute with other conflicts

The unresolved territorial dispute between the other parties to the conflict, however, has as yet prevented arrangements of the kind agreed upon in the Israeli-Jordanian peace treaty.[1] Control of the water resources in the Golan Heights is linked with military aspects of Israeli or Syrian control over the mountainous region. An agreement between Israel and Lebanon pertaining to the utilization of the Litani and Hasbani could not be accomplished. As for the Litani, Lebanon considers it a national river, as it is the most important water resource for the country and has always refused to discuss sharing it with Israel. As for the Hasbani, Lebanon's refusal to discuss this matter with Israel may be due to Syria's veto power in strategic matters of Lebanese policy. However, it is above all the competition between Israel and the Palestinians regarding the West Bank groundwater that touches the core of the Middle East conflict and cannot be settled without solving the major elements of the territorial conflict.[2]

1. Alwyn Rudolf Rouyer, *Turning water into politics: The water issue in the Palestinian-Israeli conflict*, Basinstoke: Macmillan, 2000.
2. Royer, however, argues in the opposite direction stating that, unless the Palestinians receive a fairer share of the region's water, an overall sustainable peace agreement between Israelis and Palestinians will not be achieved. See Alwyn R. Royer, "The Water Issue in the Palestinian-Israeli Peace process, *Survival*, vol. 39, no. 2, Summer 1997, pp 57-81, here p 77.

4.4.10 Access to water as an imperative of the Zionist land rush

Even those segments of the Israeli public which are seen as part of the "peace camp" find it difficult to accept compromises on the water issue as they would touch upon a vital Israeli interest. The Zionist documents presented on the occasion of the Paris peace conference of 1919, which within the borders demanded for a national homestead for the Jewish people included the entire Jordan Basin, already put forward the argument that access to water was indispensable for the very existence of the Jewish people. Control over water lies at the core of the developmental and strategic settlement plans of the Zionist movement. On the Arab side, the Israeli water strategy, namely the diversion in significant quantities of water from the Upper Jordan to the urban centres on the coast as well as into the Negev Desert in order to pursue the "agrarian imperative"[1] dictated by Zionist ideals, led to countermeasures which, *inter alia*, included the founding of the Palestinian Liberation Organization (PLO) in 1964. The first Palestinian guerrilla operations were directed against the "National Water Carrier", the Israeli water supply system whose construction began in 1953 and was completed in 1964.

4.4.11 Control over water through occupation

As of today Israel secures over fifty per cent of its water supply through control of occupied territory, namely the Syrian Golan Heights and the Palestinian West Bank. Through its occupation of the former, Israel can prevent Syrian manipulation of the Upper Jordan and hence Lake Tiberias, the region's only fresh water reservoir; through its occupation of the latter, Israel exercises physical control over the West Bank's ground water and political control over its competitor regarding the utilization of these water resources. The Jewish settlement policy, especially the chain of settlements east of the Green Line secures this control. There are auspicious zones for drilling wells on both sides of the Green Line. Israel utilizes 80 per cent of West Bank ground water, either through a system of wells west of the Green Line which tap the waters running westward or through local wells inside the West Bank controlled by Jewish settlers and Mekorot, the Israeli water company.

4.4.12 Asymmetrical allocation of water in the Palestine conflict

The Mountain Aquifer supplies approximately a quarter of the total water for Israeli consumption and an even higher percentage of Israeli drinking water. Through military regulations and discriminatory allocation quotas and water

1. Alwyn R. Rouyer, "Zionism and Water: Influences on Israel's Future Water Policy During the Pre-State Period", *Arab Studies Quarterly*, Winter 1997, vol 19, no 1.

prices, the State of Israel severely restricts the utilization of water by the Palestinians in order to guarantee supply to Israel proper and to the settlements in the occupied territories at the desired level. This practice results in an extremely asymmetrical allocation of water and per capita consumption.[1] In the mid 1990s, the amount of water from the West Bank supplied to Israel (including the Jewish settlers) was four times as high as the quantity supplied to the Palestinians.[2] As a result of inequitable allocation, the annual Israeli per capita consumption of water is three to four times as high as the Palestinian consumption. In the occupied territories Jewish settlers have ten times as many irrigated areas at their disposal as the Palestinians. The limitation of the Palestinian utilization of water – the Interim Agreement of 1995, however, increased it by 28.6 mcm/year (which constitutes 10 per cent of the overall water supply of the Occuppied Territories and 30 per cent of domestic and urban use) – contributes significantly to blocking the development of the Palestinian community. In the Israeli-Palestinian agreements concluded thus far, Israel has, in principle, recognized Palestinian water rights. However, the status of the resources and the allocation of quotas have not been agreed upon and remain to be resolved in the final status negotiations.

4.4.13 Ways to a solution of the water dispute

In addition to more equitable allocation, a sustainable solution of the water conflict requires a whole range of measures: Technical deficiencies (e.g. leakages in urban water pipes) causing loss of water must be remedied. Highly efficient irrigation methods, which especially Israel has developed, ought to be introduced in the agricultural sector on a larger scale. However, the high demand for water, particularly of Israeli agriculture, ought to be reduced – 64 per cent of Israeli fresh water consumption occurs in the agricultural sector but its contribution to the gross national product is less than three per cent. Multilateral co-operation would benefit the exploitation of additional water resources as well as a more efficient utilization of existing water resources. This co-operation is a necessity if the water supply is to be increased. Considering the expected population growth, an increase is imperative in order to cope with the Israeli, Jordanian and Palestinian water deficits in the long term. However, such co-operation requires the resolution of the Middle Eastern territorial conflicts, which, in turn, would significantly improve the climate for investing in large-scale projects such as pipelines and desalination plants.

1. For a survey of discrepancies in consumption see B'Tselem, *Thirsty for a Solution. The Water Crisis in the Occupied Territories and its Resolution in the Final-Status Agreement*, Jerusalem, July 2000, pp 38-40.
2. See *Israeli-Palestinian Interim Agreement on the West Bank and the Gaza Strip*, Washington, D.C., September 28, 1995, Annex III, Appendix 1, Schedule 10, p 199.

References

B'Tselem (2000) Thirsty for a Solution. The Water Crisis in the Occupied Territories and its Resolution in the Final-Status Agreement, Jerusalem, July 2000, pp 38-40

Israeli-Palestinian Interim Agreement on the West Bank and the Gaza Strip, Washington, D.C., September 28, 1995: Annex III, Appendix 1, Schedule 10, p 199

Libiszewski S (1995): Water Disputes in the Jordan Basin Region and their Role in the Resolution of the Arab-Israeli Conflict. ENCOP Occasional Paper No.*13,* Bern/Zürich: Swiss Peace Foundation/Swiss Federal Institute of Technology

Rouyer AR (2000) Turning water into politics: The water issue in the Palestinian-Israeli conflict, Basinstoke: Macmillan

Royer AR (1997) "The Water Issue in the Palestinian-Israeli Peace process, Survival, vol. 39, no. 2, Summer

Rouyer AR (1997) "Zionism and Water: Influences on Israel's Future Water Policy During the Pre-State Period", Arab Studies Quarterly, Winter 1997, vol *.19*, no. 1

4.5 Water Conflicts in the Middle East: The Twin Rivers Orontes and Litani

Ghassan Abu Ju'ub & Rafig Azzam
Department of Engineering Geology and Hydrogeology (LIH), Aachen University of Technology (RWTH), Lochnerstr. 4-20, 52064 Aachen-Germany

Key Words

Orontes, Litani, Syria, Lebanon, Turkey, Israel, Middle East, Water Conflicts.

Abstract

In areas with water scarcity like in the Middle East, conflicts and disputes over water resources are common, since water is vital to the daily life and the economy.

Some Middle Eastern water conflicts are well known, like those over the Jordan River, Euphrates and Tigris, Groundwater in Palestine, Golan Heights and its water resources, Nile, etc. Others are less known, but not less serious, like the Orontes and Litani among others.

In this paper, both cases of Orontes and Litani are described briefly. The two rivers are conjointly considered in this paper, since they originate from the same area but end up too far from each other causing two separate conflicts, different in some aspects and similar in others. The physical properties of the two rivers such as course, discharge and length, as well as the conflict over the usage of their water are described. Furthermore, the main reasons for the water conflicts in the Middle East are discussed and a general overview about how to solve, or at least to reduce them, is given.

4.5.1 Introduction

The two rivers Orontes and Litani, originate from the same area: Al-Biqaa' Valley in Lebanon, near the city of Ba'labak.

Al-Biqaa' Valley is the part of the Syrian African Rift located between the two mountainous series of Lebanon and Anti-Lebanon. It is around 121 km long and 8-14.5 km wide [10]. The southern part of the valley is drained by the Litani and it is the most fertile and productive part of Lebanon. The northern part is drained by the Orontes and is located in the rain shadow of the Lebanon Mountains, it is less important for Lebanon, agriculturally.

The sources of the two rivers are not far from each other, but they flow in opposite directions. The Orontes flows northwards before turning finally to the west to discharge in the Mediterranean Sea. The course of the Litani, on the contrary, is directed southwards then finally westwards to discharge also in the Mediterranean.

The Orontes is almost two times larger than the Litani, in terms of length and flow quantities. But both remain small compared to the large rivers of the Middle East, like Euphrates, Tigris and Nile. Still, the location of both rivers in such an area with very limited water resources gives them special importance and moves them into the centre of focus.

The Orontes flows from Lebanon into Syria and passes Al-Iskandarun (Alexandretta) Province in Turkey before ending in the Mediterranean. There is a conflict over it between Syria and Lebanon on one hand and with Turkey on the other.

The Litani flows entirely inside Lebanon, so it is not a trans-boundary river. But at a certain location, it becomes very close to the Israeli border. The conflict over this river is between Lebanon and Israel.

The conflict over both rivers is part of the wider water conflict in the Middle East, which is the result of several reasons, as described next.

4.5.2 Reasons for the Water Problems in the Middle East

The Middle East belongs generally to the arid and semi arid areas, which are characterised by low rainfall, high temperature and evapotranspiration.

To underline the seriousness of the water problem there, it is sufficient to mention that the Arab World in the Middle East and North Africa (MENA) makes around 10% of the world's total area and 5% of its total population, but it contains less than 1% of its renewable fresh water resources [2].

The water problems in the Middle East are basically because of the following main reasons [1]:

1. Scarcity of the naturally available water resources.
2. Rapid increasing demand on these resources due to the rapid population growth and the ambitious development programs of the different countries in the region, including programs in the irrigation sector.
3. Lack of proper management of these resources.
4. Inefficient usage of these resources in some cases.
5. Old water networks with high losses and unaccounted-for water.
6. Wastefulness of fresh water as used in swimming pools, etc.
7. Using old irrigation techniques in some cases.
8. Cultivating crops with high water demands like banana, citrus trees and flowers.
9. Lack of cooperation between the countries of the region, and lack of agreements between them on the distribution of their shared water resources.
10. Lack of efficiency regarding the use of alternative water resources like treated wastewater, salt and brackish water desalination, cisterns, storm water harvesting, etc.

All these factors result in increasing the tension and conflicts between the Middle Eastern countries over the usage of their limited water quantities. They also result in the lack of perspectives for the future development of the area and how to use its resources in the most efficient way.

The locations of the main rivers under dispute in the Middle East are shown in Figure 1. In addition to these conflicts there are other conflicts over the water resources in other parts of the region, like the resources of the West Bank and the Golan Heights [2].

Fig. 4.5.1. The Middle East with the main rivers under dispute.

4.5.3 Orontes (Al-'Asi) River

Physical Environment

The Orontes is known in Arabic as "Nahr Al-'Asi", Nahr means river and Al-'Asi means "against the rule" or "rebel". The Arabs called it by this name because the river flows in the opposite direction to all other major rivers in "The Fertile Crescent" like the Jordan, Litani, Euphrates and Tigris. All these rivers are directed southwards while the Orontes flows northwards [2].

It is a small river in the international standards and compared to the big rivers in the Middle East. It does not make more than 3.3% of the Euphrates and 2.5% of the Tigris. Still it is one of the main and important rivers in the area.

This biblical river is around 450 km long and has an annual discharge of around 1.2 Mrd.m^3/yr. (Milliard cubic meters per year) [2, 8] at the Syrian-Turkish border. It receives additional quantities from inside Turkey. It drains large parts of the northern Levant. The vast majority of it is in Syria (figure 2).

The source of the Orontes is in the central northern part of Al-Biqaa' Valley in Lebanon, not far from the city of Ba'labak and very close to the source of the Litani. It flows then northwards through the Biqaa', between the two parallel mountainous series of Lebanon and Anti Lebanon.

After that it flows into Syria, where it crosses the western part of the country. Then it forms the border between Syria and Turkey for about 31 km before crossing into Al-Iskandarun Province. There it bends west and flows until discharging finally in the Mediterranean Sea near the city of Antakya (Antioch).

In Syria the river receives quite large number of tributaries, many of them are seasonal. Along its stream, Lake Hims (Qattineh) is located. North west of Hamah, it crosses the fertile Al-Ghab Plain.

The Oronets is mostly unnavigable, but it is very important for agriculture especially in Syria. The agricultural sector as well as the domestic water consumption in western Syria depends to a large extent on its water. Actually, the Orontes together with the Euphrates constitute Syria's main water resources.

The river is very famous with the ancient giant water wheels along its stream, especially in the cities of Hamah and Hims. These wheels vary between 6 to 21 m in diameter [10] and are installed to lift the water of the river to the surrounding plains.

The Conflict

The conflict over the Orontes is between Syria and Lebanon on one side with Turkey on the other. There are disputes between the two neighbours -Syria and Turkey- not only over water, but also over several issues. In addition to the conflict over the Euphrates and Tigris -which is indeed the most hot water issue between the two countries-, there is also a conflict over the Orontes, which is a much smaller river.

Syria and Lebanon signed a treaty in 1994 regarding the usage of the Orontes. According to it, Lebanon is entitled to use 80 Mm^3/yr. (Million cubic meters per year) from a total of 430 Mm^3/yr. of water flowing in the Oronets inside the Lebanese territories. The remaining quantity of 350 Mm^3/yr flows into Syria [2].

The problem with Turkey is quite complicated, because the river ends in the Province of Al-Iskandarun (Alexandretta), which itself is under dispute since a very long time. So here, the water conflict is mixed with territorial dispute.

Syria considers this province to be a Syrian land and does not recognize it to be part of Turkey. Actually the Syrian maps show it always as part of Syria. This province was split from Syria by France and given to Turkey in 1939 when Syria was a French colony. Thus, the annexation of the province by Turkey is illegal, the Syrians argue. At that time the French colonisation power wanted to increase its strength in the area by weakening the Arab Nationalism Movement, which was growing rapidly and strongly. The French did that through supporting and separating the religious and ethnic minorities and dividing (Greater Syria).

Based on the above, Syria does not consider Turkey as a riparian to the Orontes. The Syrians say also that Turkey is not in need for the water of this river, because it has plenty of water which it even does not use and is trying to sell to other countries. Therefore it is unfair that Turkey is asking Syria to give up some of its limited water resources, which the country is in bad need for.

The Syrians complain continuously from the Turkish behaviour in the Euphrates-Tigris basin. According to Syria, Turkey is using the water of the two rivers and does not give Syria and Iraq their fair share and some times it releases bad quality water to Syria. Turkey is implementing several projects and dams on the streams of both rivers without informing the downstream countries (Syria and Iraq). Syria wants to reach an agreement on the Euphrates and Tigirs, but Turkey is still refusing to discuss the issue and consider the water of the two rivers as "Turkish Water".

On the other hand, the Turkish insist that the Province of Al-Iskandarun is part of Turkey and refuse the Syrian argument that it belonged historically to Syria and was split from Syria by force. They want Syria to give up its claims in the province and to recognize that Al-Iskandarun belongs to Turkey.

Based on that, Turkey sees itself as a riparian to the Orontes and demands that it gets its fair share in the river's water.

Turkey makes a link between the issue of the Orontes and that of the Euphrates and Tigris. The Turkish say that they are giving Syria and Iraq their fair share in the Euphrates and Tigris but they are not using that water efficiently. At the same time they accuse Syria of not paying attention to the needs of the downstream country (Turkey) when it comes to the Orontes. The Turkish argue that Turkey is not really rich in water and that the country needs the water of the Orontes to use it for the irrigation of that region. Turkey also complains that Lebanon and Syria did not inform it when both countries signed their treaty over the Orontes (Al-'Asi) in 1994.

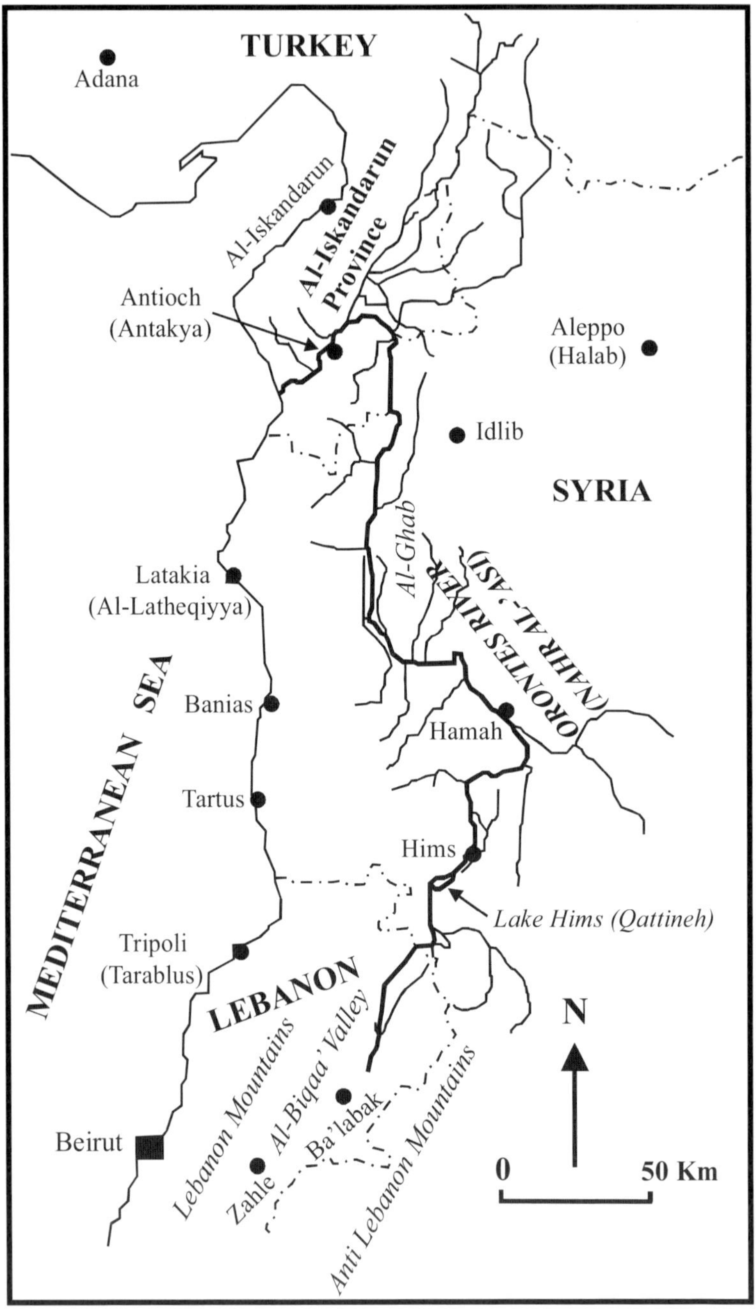

Fig. 4.5.2. Orontes (Al-'Asi) River.

4.5.4 Litani river

Physical Environment

The Litani originates nearby the city of Ba'labak in the central north of Al-Biqaa' Valley in Lebanon. It flows southwards through the valley, then enters a gorge at Qaraun and flows through it for about 30 km. Near the town of Al-Nabateyyeh and the Beaufort Castle the river bends abruptly to the west, makes a cut through the Lebanon Mountains Series and continues flowing westwards until it empties in the Mediterranean Sea north of the city of Tyre [6].

The source of the Litani is very close to that of the Orontes. While the Orontes flows northwards, the Litani flows southwards. The Litani is a small river, even when compared to the Jordan and Orontes. It is about 170 km long and has a total annual flow of around 0.6 Mrd.m^3 in average [2]. It flows entirely inside Lebanon, its total catchment area is there and it forms one of the major water resources for that country. Actually the Litani is the only "major" river in the Middle East, which is not trans-boundary and not shared between more than one country (figure 3).

In some locations the river flows not far from Israel. At its bending point, where it changes direction from southwards to westwards, it is only 4 km far from the Israeli border [2, 6]. The Litani flows also parallel and close to the Hasbani River, which is a tributary to the Jordan River. At a certain point the two rivers are only 5 km apart [6].

There is a dam on the Lower Litani at Qaraun, behind which the Qaraun Lake is formed. Around 236 Mm^3/yr. are diverted from the Litani to the Awwali River (another main water resource for Lebanon) through Markaba Tunnel [6]. The Litani is used to irrigate the southern half of Al-Biqaa' Valley and the southern Lebanese coastal plain. It is also important for the electro-power generation and is used to supply Beirut and other coastal towns with electricity.

The Conflict

The conflict between Israel and Lebanon over the Litani has a long history. Although the river is located entirely inside Lebanon, the Zionist movement planned to use it since the early 20th century, long before Israel was established.

The first Zionists recognised from the very beginning the importance of water resources in the region. Therefore they asked Great Britain in several occasions to extend the borders of Palestine to include the whole catchment of the Jordan River (including the Yarmuk) and the Litani River (or most of it). The first Zionist demand was made in 1905 and was repeated in 1919-1920 through two letters sent by Chaim Weizmann, the head of World Zionist Organisation (later on the president of Israel) to the British Prime Minister and Foreign Secretary. He specified the demand by asking to extend the borders of Palestine several kilometres to the east of the Jordan River and 40 km to the north of the bend of Litani [3]. That means having the Jordan River (including the Yarmuk and Upper Jordan) and the Litani inside the demanded Jewish Home Land. Weizmann argued that the area north of the proposed borders has more than enough water, therefore the Litani and Al-Biqaa' Valley are much more important for Palestine (to be established as state of Israel). According to him, Palestine will be very poor and economically dependant, if it does not include the Jordan, Litani and Lake Tiberias [3].

However, the British (and French) did not satisfy the wish of the Zionists and the borders of Palestine were decided according to colonial interests (these are the borders of the current Israel and Palestine). The same decision was taken by the League of Nations. But the Zionists did not drop their demands and in the 1940's there were two main suggestions for the usage of the Litani:

- An international commission recommended in 1941 that 7/8 of the Litani's water should be leased to Palestine (Israel).
- In 1945 Professor Lowdermilk worked out a water Plan for the whole region. He suggested to change the course of the Litani and to divert its water to the Jordan River, i.e. making it a tributary of the Jordan. He wanted to use its water inside Palestine (Israel) for irrigation along the Jordan Valley and in the centre of the country and to supply Lebanon with the generated electricity. The plan was not implemented because of the Arab refusal.

In 1947 David Ben Gurion considered to make the Litani Israel's northern border. After Israel was established in 1948 and during the 1950's and 1960's the Israeli Leaders, especially Ben Gurion and Moshe Dayan, continued thinking of occupying southern Lebanon and the Litani. After the 1967 war, in which Israel controlled almost all the Jordan River Basin, Dayan said that Israel achieved satisfying borders except with Lebanon [3]. Later on there were several Israeli attempts to control the Litani. Israel's official reason for invading Lebanon in 1978 and 1982 was to move the PLO (Palestine Liberation Organisation) forces far from the Israeli borders. But it is believed that another main reason was controlling the Litani. The official Israeli name given to the 1978 invasion was "Litani Operation" [2]. The Israelis controlled south Lebanon up to the bend of the Litani until year 2000.

Israel believes that Lebanon is rich in water in contrast with Israel and that Lebanon is not using the water of the Litani completely and efficiently. The Israeli thinking was always to make the Litani a tributary of the Jordan River by diverting it southwards. The location of the river very close to the Israeli borders (4 km far from Israel at the bend), the simplicity of diverting its water to Israel (A tunnel at the bend will make the river flow by gravity to Israel) and its relatively good water quality are three main reasons to make the Litani of particular importance to Israel and may be the best option for the Israelis to satisfy their water needs.

Lebanon says that its water resources are hardly enough to satisfy its own needs. The Litani is the most important resource for Lebanon upon which irrigation and electricity generation in large part of the country depends. The Lebanese argue that the political conditions, under which part of Lebanon was occupied by Isreal, were the reason, why Lebanon could not use the Litani completely until now. After all, the Litani is a national Lebanese River, it is only Lebanon who decides about this water resource and Israel has no right to interfere, the Lebanese say.

The conflict over Litani is not the only one between Israel and Lebanon. This year there was a lot of tension because of the Lebanese plan to supply some villages in the south with water from Al-Wazzani (a tributary of Al-Hasbani, which is a tributary to the Jordan). Israel objected that and threatened the usage of force.

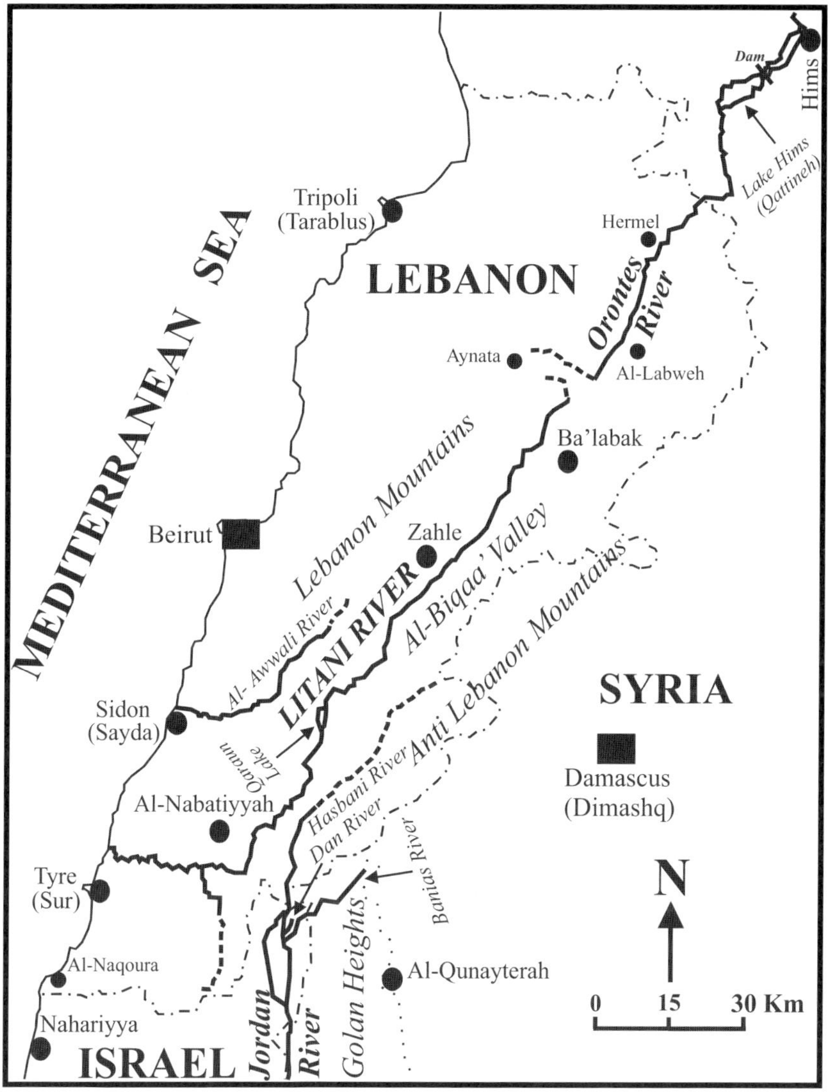

Fig. 4.5.3. Litani River.

4.5.5 The Way Forward

Water resources in the Middle East are for sure scarce and not enough to satisfy the demands of many countries there. This scarcity will become more serious in the near future. Therefore, the Middle Eastern countries have one of two options to follow [1]:

To compete for the existing water resources without developing new ones. This will result in more conflicts and may be wars between them.

To cooperate on the usage of the existing water resources in most efficient way and to develop new resources in close collaboration. This would not only solve the water problem, but would also encourage the cooperation in other fields.

However, the precondition for a successful cooperation is a fair distribution of the existing resources from the very beginning according to the international water law. This gives every country the feeling of satisfaction and willingness to cooperate.

References

1. Abu Ju'ub Gh (2003) Water Conflicts in the Middle East Between the Present and the Future. Published on the internet by Ibn Rushd Forum for Freedom of Thought. Internet site: http://www.ibn-rushd.org/forum/water.pdf
2. Abu Ju'ub Gh (2002) Water Resources in Jenin Governorate / North of the West Bank - Palestine, A Part of the Water Problem in the Middle East. Ph.D. thesis, Heft 83, Lehrstuhl fuer Ingenieurgeologie und Hydrogeologie-Aachen University of Technology (RWTH), Aachen.
3. Amery HA, Kubursi AA The Litani River Basin: The Politics and Economics of Water. Internet site: www.lcps-lebanon.org/pub/breview/br3/amery3.html
4. Daoudy M (1999) Water, Institutions and Development in Syria: a Down-stream Perspective from the Euphrates and Tigris. WCD Regional Consulta-tion, Cairo. Internet site: www.dams.org/sub_p_af4.asp
5. Feitelson E (2000) The ebb and flow of Arab-Israeli water conflicts: are past confrontations likely to resurface? In: Water Policy 2. Elsevier, Amsterdam, pp 343-363.
6. Joy Moss A. Litani River and Israel-Lebanon. ICE Case Studies. Internet site: www.american.edu/TED/ice/litani.htm
7. Libiszewski S (1995) Water Disputes in the Jordan Basin Region and Their Role in the Resolution of the Arab-Israeli Conflict. ENCOP Occasional Paper No.13. ETH, Zurich.
8. Ministry of Foreign Affairs-Turkey. Water: A source of conflict or coopeari-ton in the Middle East? Internet site: http://www.mfa.gov.tr/grupa/ac/aci/default.htm
9. Schetelig K, Abu Ju'ub Gh (2001) Konfliktpotential Wasser in Nahost, insbe-sondere in Israel und Palaestina (in German). In: Nahost Konfliktpotential Wasser (Seminar), Aachen.
10. The Columbia Encyclopedia, sixth edition (2001): Information over Orontes and Litani. Internet site: http://www.bartleby.com/65/

Index

A

AAS 191
ACSAD 142
Agricultural irrigation 221
- production 149
Agriculture 15, 49
Al Arroub 76
Algerian 231
Ammonium 44
Anaerobic conditions 183
- digestion 183
Anoxic hypolimnion 43
Anthropogenic 196
Anticlinal structure 75
Aquifer Systems 22, 102
aquifers 7
Arid 3
- belt 4
- climates 102
Ataturk dam 338
Atmosphere 101
Atmospheric 4
- dust 215
- variability 279
Azraq groundwater 108
- spring 106

B

Bacteria 158
Bacterial contamination 157
Bacteriological contamination 146
Banias 48
- River 31
Banyas 208, 346
Beka'a valley 140
Benin 279
Bethlehem Formation 75
Big water 117
Biofilms 158
Biosphere 101
Birds monitoring 220
Brackish ground water 243
- water 13

C

Cairo Waste Water 192
Calcium Chloride 37
Carbonate contents 195
- rocks 76
Carmel Mountain 240
Chemical fertilizers 267
Chloride 37, 108, 199
Chlorophyll 48, 212
Cisterns 243
Civil movement 122
- society 122
Clay 152, 173
Cold War 113
Conflict 113
Conglomerates 149
Continental groundwater 103
- hydrosphere 279
Cr 192
Cu 192
Cyperus papyrus 207
Cyprus 57

D

Damascus 149
Dan 31, 48, 208, 346
Dead Sea 6, 31, 36, 108
Delta floodplain 191
Desert 35
- climate 88
Dialogue on water 338
Djefet 184
Dolomite 75
- chalk 76
Drâa valley 279
Drinking water 41

E

Earth 101
Eastern-Hula Canal 210
Ecological sanitation 263
Economic fundamentals 117
Ecosan 263

Ecosystem 49, 101
Effluent irrigation 167
Egypt 146
El Lajjun Model 133
Endoreic system 106
Environmental 117
- education 20, 189
- impact assessment 290
- problems 152
- Quality Authority 94
ESCWA 142
Euphrates 15, 319, 333, 342
Eutrophic 210
Evaporation rate 49

F

Fauna 209
Fe 192
Fertilizers 99, 149
Fish 219
- production 46
Floodwater 13
Flora 209
food 41
Fossil groundwater 15, 103
- water 10, 18
Freshwater 6, 87, 109, 112

G

Gabal el Asfar 192
Gaza Strip 87, 243
Geo-Modelling-Software 251
Ghana 279
Ghouta 140
- Plain 149
Global trade 112
- warming 55
Great Anatolian Project 320
- Cairo 191
Greywater 270
Groundwater 7, 13, 87
- contamination 141
- extraction 300
- management 107
- mining 136
- Model 133
- monitoring 145
- protection 99, 129, 130, 295, 300
- resources management 129
- Vulnerability 146
Guinea coast 279
Guneydogu Anadolu Projesi 320
Gypsum 217

H

Hasbani 33, 346
- River 31
Hatzbani 48, 208
Hazardous substances 265
Hazards 145
Health risks 265
Heavy metals 149, 195
Hebron 79, 258
Herodion Beit-Fajjar 75
High Atlas 280
Holocene 193
Hormones 265
Household 17, 157
- storage 160
Hula Project 209
- Valley 41
Hydraulic mission 119
Hydrochemistry 80
Hydrogeology 251
Hydrological balance 213
- cycle 275
- parameters 28
Hydrology 111
Hydro-politics 111
Hygiene 266
Hyper-aridity 15
Hyper-dry areas 103
Hypertrophic 210

I

ICP 191
IMPETUS project 275
Industrial modernity 119
- waste water 98
- wastewater 149
- zones 145
Infiltration 161
Infrastructure 12, 16
Inorganic pollutants 192
- solids 168
Insel Hengsen 235
InWEnt 289
Iraq 319
Irbid area 138
- region 145
Irrigation 149, 167
Israel 5, 31, 167, 243, 353
- agriculture 246
Israeli settlers 93

J

Jenin City 69
Jenin Governorate 63
Jericho area 9
Jerusalem 258
- Formation 75
Jordan 5, 31, 56, 243
Jordan desert 6
- Rift Valley 5, 6
- River 5, 16, 31, 208, 345

K

Kafrien Project Wells 28
Kinneret waters 222
Kurnub Formation 28

L

Lake Agmon 209
- Hula 207
- Kinneret 41, 208
- Tiberias 6, 31, 346
Lebanon 31, 145, 243, 353
- Mountains 6
Limestone 146
Limnological data 210
Litani 16, 353
Lithosphere 101

M

Macro-algae 218
Macrophytes 218
Magnesium 37
Management 85
- training 290
Marj Bin Amer 69
- Sanur 63, 72
Marls 75, 76
Mechanical treatment 168
Mediterranean 56
- Meteoric Water 80
Microbial quality 158
Microbiological quality 159
Middle East 15, 31, 87, 102, 111, 112, 309, 333, 353
- conflict 348
Minerals 37
Miocene 194
Mn 192
Monitoring system 99
Morocco 275
Mount Hermon 32
Mountain Aquifer 7, 75

N

Naftali Mountains 48
Nahr Al-Mufjir 63
- Al-Muqatta 63
- Jalut 63, 64
National guidelines 139
Natural resource 87
Nigeria 279
Nile 16
- Delta 146
- river 342

Nitrate concentrations 289
- contents 146
Nitrates 198
Nitrogen 177
North Africa 15, 87, 111, 112
- Atlantic Oscillation 56, 282
Nutrient recycling 263

O

Oligocene 194
Olive Mills 181
- oil 181
Olives 181
Organic compounds 149
- matter 195, 196
- solids 168
Orontes 342, 353
Oslo II Agreement 247

P

Palestine 16, 31, 63, 87, 181, 243
Palestinian Authority 84
- National Authority 88
- Water Authority 93, 288
- Water Policy 94
- water strategy 243
Palms 192
Palmyra basin 139
Pathogens 265
Peace process 317
Pesticides 99, 149
Petra 16
pH 200
Pharmaceutical residues 265
Phosphate 198
Phosphorus 41, 44, 177
Phragmites australis 207
Phytoplankton 48, 210
Plankton 41
Pleistocene 193
Political economy 111, 112
Politics 112
Pollution 92, 102
- sources 99
Polyphenols 183
Potable water 11
Pre-cambrian mountains 281
Public health 157
-- risks 96

Q

Quality management 295

R

Radioactive 102
Rainfall 5, 49
Rainwater 159
Red Sea 57
Riparian States 333
Risk classification 304
- management 295
River Jordan 88
Rock 107
- type 145
Runoff 65

S

Sahara 4
Sandstone 21
Sanitation systems 263
Saqiye group 252
Sea water 13
Seasonal lake 63
- wadi 63
Security 112
Sedimentation 103
Seeps 243
Semi humid 88
Semi-arid 3
- areas 103
- climates 102, 103
- zones 4
Sewage pollution 157
- water 195
Shatt-Al-Arab 319
Siwaqa-Qatrana-Hasa 137

Sludge treatment 290
Small water 117
Socio-economic 3
Sodium adsorption ratio 168
- Chloride 37
Soil contamination 152
- science 140
- solution 170
- water 112
Solid waste management 290
Southeastern Anatolian Project 338
Spring waters 23
springs 88
Stable isotopes 26
Subsurface water 101
- waters 105
Subtropical 277
- zone 4
Sulfides 44
Sulphate 198
Surface runoff 6, 174
- water 3, 6, 63, 64, 88, 90, 174, 347
Sweimeh Area 21
Syria 5, 31, 56, 243, 319, 353

T

Tanks 159
Tectonic 107
Temperatures 55
Tigris 15, 319, 333, 342
Togo 279
Topography 64
Transboundary rivers 333
Tritium 106
Tropical zone 4
Tubas 73
Tulkarem District 185
Turkey 56, 319, 333, 353
Typha domingensis 207, 218

U

Urine separation systems 270

V

Virtual water 111, 112

W

Wadi al Arab 146
- Al-Bassa 71
- Al-Fara'a 63, 64
- Al-Jamus 71
- Al-Khudairah 63
- Al-Maleh 63, 64
- Bardalah 63, 64
- Kafrien 29
- Mujib 131
- Shubash 63, 64
- Ti'innik 71
--Musa 16
Wadis 243
Waste import 99
- management 99
- water 13, 19, 93
Wastewater Concepts 181
- Management 246
- management 263
- reuse 263
- treatment 142
Water abundance 310
- Authority of Jordan 130
- balance 101, 131, 314
- budget 211
- conflict 31, 309, 353
- consumer 245
- crisis 11
- cycle 267
- deprivation 310
- Diplomacy 316, 319
- dispute 347
- distribution 31
- inflows 210
- Law 96
- level 36, 81
- Management 13, 31, 93, 119, 316
- policy 98

Water pollution 152
- pollution control 97
- price 142
- Quality 17, 46, 49, 50, 80, 92
- quality standards 96
- Resources 12, 17, 52, 63, 87, 90, 333, 345
- resources management 97, 295
- rights 143
- Scarcity 88
- supply 109, 143
- supply management 291
- users 116
- Utilization 207, 220
Water-balance 5
Watersheds 112
Wells 289
West Africa 276
- Bank 63, 181, 243
Western Aquifer Basin 251
World Health Organization 244

X

XRF 191
XRF analysis 194

Y

Yarmouk 346
- River 6, 31, 36

Z

Zibar 183, 184
Zn 192
Zoo plankton 48
Zooplankton 210

Printing: Mercedes-Druck, Berlin
Binding: Stein + Lehmann, Berlin